AF880573

China Seismic Experimental Site

Yong-Gang Li · Yongxian Zhang · Zhongliang Wu
Editors

China Seismic Experimental Site

Theoretical Framework and Ongoing Practice

Higher Education Press

Springer

Editors
Yong-Gang Li
Department of Earth Sciences
University of Southern California
Los Angeles, CA, USA

Yongxian Zhang
Institute of Earthquake Forecasting
China Earthquake Administration
Beijing, China

Zhongliang Wu
Institute of Earthquake Forecasting
China Earthquake Administration
Beijing, China

ISBN 978-981-16-8606-1 ISBN 978-981-16-8607-8 (eBook)
https://doi.org/10.1007/978-981-16-8607-8

Jointly published with Higher Education Press
The print edition is not for sale in China (Mainland). Customers from China (Mainland) please order the print book from: Higher Education Press.

© Higher Education Press 2022, corrected publication 2022
This work is subject to copyright. All rights are solely and exclusively licensed by the Publisher, whether the whole or part of the material is concerned, specifically the rights of reprinting, reuse of illustrations, recitation, broadcasting, reproduction on microfilms or in any other physical way, and transmission or information storage and retrieval, electronic adaptation, computer software, or by similar or dissimilar methodology now known or hereafter developed.
The use of general descriptive names, registered names, trademarks, service marks, etc. in this publication does not imply, even in the absence of a specific statement, that such names are exempt from the relevant protective laws and regulations and therefore free for general use.
The publishers, the authors, and the editors are safe to assume that the advice and information in this book are believed to be true and accurate at the date of publication. Neither the publishers nor the authors or the editors give a warranty, express or implied, with respect to the material contained herein or for any errors or omissions that may have been made. The publishers remain neutral with regard to jurisdictional claims in published maps and institutional affiliations.

This Springer imprint is published by the registered company Springer Nature Singapore Pte Ltd.
The registered company address is: 152 Beach Road, #21-01/04 Gateway East, Singapore 189721, Singapore

Preface

On May 12, 2018, the International Conference for the Decade Memory of the Wenchuan Earthquake with the 4th International Conference on Continental Earthquakes announced the launching of the China Seismic Experimental Site (CSES). The subject of this book is to illustrate a conceptual framework of the CSES for fostering the international and regional collaboration in earthquake research and disaster mitigation. The CSES provides a principal natural laboratory in the Sichuan–Yunnan region to support core researches in earthquake geology, tectonic geodesy, seismology, earthquake engineering, and computational science. The CSES is carried out by geoscientists from Chinese scientific community with colleagues abroad and international organizations.

The CSES, as a natural laboratory in earthquake science and technology, is to facilitate investigator-driven core researches in earthquake geology, tectonic geodesy, earthquake and engineering seismology, earthquake and civil engineering, and computational science for the enhancement of seismic disaster resilience of the society. Scientific focus of the CSES lies in three aspects, namely the geodynamic environment of continental earthquakes, the dynamics of continental earthquakes, and the mechanism of the disaster chain related to continental earthquakes. Quality intensified observations with well-defined scientific objectives and modeling aided by high-performance computation lay the basis for the success of the CSES. Researchers and graduate students in earthquake science will broaden their horizons about observational, computational, and applied seismology and earthquake physics via learning from the CSES theoretical investigation and practice to aid in a global and regional cooperation for preparedness, mitigation, and management of seismic risk.

Los Angeles, USA — Yong-Gang Li
Beijing, China — Yongxian Zhang
Beijing, China — Zhongliang Wu

China Seismic Experimental Site—Theoretical Framework and Ongoing Practice: An Induction

This book introduces an integrated conceptual framework of the China Seismic Experimental Site (CSES), describes its scientific challenges and research priorities, and reports preliminary results coming out of observational infrastructure in seismology, tectonophysics, geodesy, geophysics, and geochemistry. Preliminary community velocity model, community fault model, community strain rate model, and earthquake risk model in the CSES are described in this book. A multi-disciplinary test observation system consisting of GNSS, seismic array, and deep drilling system under construct around middle segment of the Xiansuihe-Xiaojiang fault and other seismogenic faults in the CSES is also mentioned. This book includes multi-disciplinary topics and a wide spectrum of solid earth system to describe various disciplines, methods, and techniques through the CSES. The preliminary results from a conceptual framework of the CSES will benefit domestic and abroad scientists and policy-making authority agents in earthquake science for fostering the international and regional collaboration in earthquake research and disaster mitigation. It also benefits researchers and graduate students of institutes and universities in earthquake sciences for broadening and deepening their knowledge and skills in observational, computational and modeling technology and methods in seismology, geophysics, geology, geodesy, and tectonophysics.

The major components of the CSES are Community X Models (CXM), Community Forecast for Verification (CFV), and Community Accepted Technical Test (CATT), etc. The lower layer consists of components that describe capacity building of observation and data interpretation, act as the backbone of intensified observational activities, and play the central role in the construction of the CSES. The major observation infrastructure includes Coordinated Distributed Capacity-Building Plan (CDCP), Coordinated Distributed Data Centers (CDDC), and Coordinated Distributed Computational Bases (CDCB). In twelve chapters of this book, authors describe the launching of CSES project, including the regional seismicity; existed earthquake monitoring networks, historical seismic disasters and implication of CSES; seismotectonics and geodynamics of the eastern margin of the Tibetan Plateau with implication for the CSES; theoretical framework of the CSES in view of natural science and social science; updated earthquake monitoring network; CSES community models of geology, structure, deformation, and earthquake hazards. The CSES preliminary products are also demonstrated, including massive data processing and distribution; seismic risk assessment; estimated casualty risk for disaster preparation; earthquake forecasting and nowcasting models; field expedition of the June 17, 2019, Changning, Sichuan $M6.0$ earthquake as well a review of characterization of the San Andreas fault by fault-zone trapped waves at the seismic experiment site near Parkfield, California, for reference to facilitate the high-resolution insights of active faults; CSES products and processes as well as international cooperation and management.

This book includes articles full of perceptive investigation of the seismogenic provinces in Southwest China with cross-disciplinary and multiple thematic aspects in modern seismological, geophysical, geological, and stochastic methodology and technology. The new disciplines are illuminated by case study for readers to better understand the occurrence, dynamics, and likelihood of earthquakes. The editors approach this as a broad interdisciplinary effort, with well-balanced reviewing, policy-making, and research reporting aspects. Linked with these topics, the book highlights the importance for characterizing the geological and tectonic backgrounds of earthquake kinemic and dynamic rupture processes, fault-zone structure, crustal deformation and velocity models, earthquake hazards, and forecast in Southwestern China. These contents make this book a must-read for researchers in the field of earthquake physics and prediction. Authors of twelve chapters describe various disciplines, methods and techniques in earthquake science through their investigations in CSEC. This book includes:

- Seismic Experimental Sites: Challenges and Opportunities.
- Seismotectonics and Geodynamics of the Eastern Margin of the Tibetan Plateau with Implication for the CSES.
- From Plate Loading to Regional Fault Network: Toward Mechanical Understanding of the Kinematics.
- CSES Community Velocity Models.
- Three-Dimensional Community Active Faults Models of CSES.
- CSES Community Deformation Models in Southwest China.
- Interseismic Coupling along the Xianshuihe–Xiaojiang Fault System.
- Three-Dimensional Crustal Structure and Seismicity Characteristics of Changning–Xingwen Area in the Southwestern Sichuan Basin, China.
- Estimated Casualty Risk for Disaster Preparation in Five Scenario Great Earthquakes, Sichuan–Yunnan Region, China.
- China Seismic Experimental Site: Seismicity, Ergodicity and Nowcasting Earthquakes.
- Characterization of the San Andreas Fault by Fault-Zone Trapped Waves at Seismic Experiment Site, Parkfield, California: A Review.
- A Brief Introduction to the Major Science and Technology Products and The 2019 highlights scientific and technological progress of China Seismic Experimental Site (CSES).

This book is the sixth monograph of the earth science specializing in observational, computational, and applied seismology and geophysics. The contents of this book mainly come out of scientists participating and involving in the initiating and developing stages of the CSES and partly comes from presentations on The 2nd Annual Conference of China Seismic Experimental Site held in November of 2020 (via Zoom), which highlights the importance for an international cooperation in investigation of the large earthquakes, their predictability and risk, and how to make full use of the present knowledge and techniques to reduce earthquake disasters. Authors from China Earthquake Administration, and universities and institutions in earthquake sciences in China and abroad present and review principles, methods, and techniques for characterization of earthquakes.

This book covers multi-disciplinary topics to allow readers to grasp the various methods and skills used in data processing, analysis and numerical modeling for structural, geological, geophysical, and mechanical interpretation of earthquake phenomena and physics. This book is a self-contained volume and explores each topic with in-depth detail. Extensive reference lists and cross-references with other volumes to facilitate further research. Content suited for both the senior researchers and graduate students in geosciences who will broaden their horizons about observational, computational and applied seismology, and geophysics. For colleagues in China and abroad, this book also provides informative materials to foster collaborations with the CSES.

The book editors wishes to thank Mr. Kun Wang, Mr. Shi Che and Prof. Zhifeng Ding, Prof. Xiadong Zhang and Prof. Li Li of China Earthquake Administration, for their supports to foster this special book as one of outcomes from the CSES. We are grateful to many chapter authors and reviewers for their contributions and collaborations. We also thank many organizations and individuals, including Director Bingxiang Li and Editor Yan Guan at HEP, who helped to make our books possible.

Yong-Gang Li
Yongxian Zhang
Zhongliang Wu

Contents

Chapter 1
Seismic Experimental Sites: Challenges and Opportunities

Zhongliang Wu

Abstract Seismic experimental sites, or natural laboratories for earthquake science, can be traced back to as early as mid-20th century. In recent years, these experimental sites are facing to newopportunities of development; while the solutions to tackle the long-standing challenges to their sustainability have some new developments. This chapter reviews the related challenges and opportunities.

Keywords Coordinated Distributed Experiments (CDEs) · Seismic Experimental Sites

1.1 New Opportunities

Related to earthquake prediction, there had been several famous experimental sites, such as the earthquake prediction experiment sites in the former Soviet Union (e.g., Semenov, 1969), the Parkfield experiment in the USA (Bakun and Lindh, 1985; Roeloffs, 2000), the TDP project in Turkey (Evans et al., 1987), the Tokai experiment in Japan (Davis and Somerville, 1982; Mogi, 2004)[1], the PRENLAB project in Iceland[2], the earthquake prediction experiment sites in China (Sun and Wu, 2007), and the earthquake prediction experiments in Koyna, India (Gupta, 2001), among others. Southern California Earthquake Center (SCEC)[3], which considers southern California as a natural laboratory, is also a good example of seismic experimental site .

[1] Hoshiba, M. 2006. Current strategy for prediction of Tokai earthquake and itsrecent topics. http://cais.gsi.go.jp/UJNR/6th/orally/O04_UJNR_Hoshiba.pdf, last access: July 31, 2018.

[2] Stefansson, R., F. Bergerat, M. Bonafede, R. Boovarsson, S. Crampin, P. Einarsson, K.L. Feigl, A. Guomundsson, F. Roth, F. Sigmundsson, and R. Slunga. 1998. PRENLAB Final Report. Reykjavik; Stefansson, R., F. Bergerat, M. Bonafede, R. Boovarsson, S. Crampin, K. Feigl, F.Roth, F. Sigmundsson, and R. Slunga. 2001. PRENLAB-TWO-final report. Veourstofa Islands Report.

[3] https://www.scec.org/, last access: April 25, 2021

Z. Wu (✉)
Institute of Earthquake Forecasting, China Earthquake Administration, Beijing 100036, China
e-mail: wuzl@cea-igp.ac.cn

© Higher Education Press 2022
Y.-G. Li et al. (eds.), *China Seismic Experimental Site*,
https://doi.org/10.1007/978-981-16-8607-8_1

In recent years, with the paradigm shift from disaster reduction (DR) to disaster risk reduction (DRR), the role of earthquake prediction in social sustainability, as well as the role of earthquake prediction experiment sites, has been facing to a historic transition. Seen from the representation of *seismic disaster risk*,[4] by *seismic hazard*, *exposure*, and *vulnerability*, traditional earthquake prediction tried to identify the increase of *seismic hazard* and provide warnings to the society. The society in turn reduces the *exposure* by organized evacuation (such as the case of the 1975 Haicheng earthquake) so that *seismic disaster risk* may be significantly mitigated. Traditional earthquake prediction experiment site tried to capture potential precursory anomalies so that such warning becomes feasible. In the perspective of DRR, however, when *seismic hazard* with different spatio-temporal scales is assessed, countermeasures can be taken more efficiently to reduce either the *exposure* (such as in the planning of land use, with decade time scale) or the *vulnerability* (such as engineering reinforcement, with annual time scale). In the extreme case of earthquake early warning (EEW), when *seismic hazard* increases dramatically within a very short time (the S-P travel time difference), countermeasures can still be feasible to reduce the *disaster risk* using real-time seismology. In the context of DRR, therefore, a seismic experimental site can, and has to, consider a much wider range of scientific problems, and the contribution of such experimental sites to social disaster resilience can be more significant and more operational.

For quite a long time, '*experiments*' in earthquake science have had limitation because of the lack of accuracy and precision in measurement, and the lack of controlled processes. In recent years, innovative technologies have been reshaping seismological, geodetic, and geological studies. Emerging possibilities of semi-controlled processes (such as fluid/gas injection which might be able to induce/trigger earthquakes), repeatable and continuous measurements (such as non-explosive source imaging, and the application of waveform cross-correlation techniques), and high performance computation for simulating the geodynamic processes of earthquakes have made the field experiments 'more similar to' the *experiments* in fundamental physics. One of the examples is the temporal variation of Earth structure, and the related temporal variation of the velocity of seismic waves. As early as in the 1960s, pre-earthquake variation of the velocity of seismic waves, with clear implication for earthquake prediction, was discovered and widely discussed (e.g., Semenov, 1969). However, accuracy and precision of the field measurements at that time limited the verification and falsification of scientific conclusions to much extent. Since the twenty-first century, innovative technologies such as waveform interferometry and artificial seismic source have significantly enhanced the capability of observing co-seismic and post-seismic variations of Earth structure (e.g., Li et al., 2006, 2007; Wegler et al., 2009; Chao & Peng, 2009; Audet, 2010), and even the pre-earthquake structural variation (e.g., Niu et al., 2008). The list of innovative technologies is long. Making full use of such opportunities, it is time to consider and develop the 'version 2.0 of seismic experimental sites'.

[4] https://www.preventionweb.net/risk/disaster-risk, last access: April 25, 2021.

1.2 Main Challenges

Main challenge of the natural laboratories for earthquake science has been, and will be, the chance to 'meet' the 'target earthquake'. In the history of earthquake prediction experiments, for most if not all of the test areas, the 'target earthquake' occurred either out of the range of the test area (such as the 1996 Lijiang earthquake which is near to, but just out of, the West-Yunnan Earthquake Prediction Experiment Site), or within the test area but after the test has formally closed (such as the 2004 Parkfield earthquake which occurred long after the predicted time of the target earthquake, 1988 $\pm$ 4 years, and the planned duration of the experiment, 1984–1992, although observational facilities were still functioning). Ironically, the failure of earthquake prediction experiment sites, which were originally designed and implemented for the test of short-term earthquake prediction, was in fact due to the uncertainties of intermediate-term earthquake forecast, and the underestimate of such uncertainties in the planning. Taking this lesson, the newly proposed seismic experimental site tends to be larger. Some of the experimental sites, in their planning stage, consider the countermeasures to 'play games with' earthquakes in the sense of game theory [e.g., Working Group for the Long-term Plan of Earthquake Forecast and Prediction (2010–2020), 2010]. However, dearth of 'earthquake samples', especially 'samples' of large earthquakes, keeps to be an intrinsic problem associated with seismic experimental sites.

Another challenge is the complicated factors controlling the process of the preparation and occurrence of an earthquake, which necessitates and challenges the field experiments. 'Necessitate' means that field experiments are as important as laboratory experiments—just recall the 'heat flow paradox' (e.g. Mora & Place, 1998) in the history of seismology. 'Challenge' means that field experiments, even if with good design, always have to face to the fact that more factors seem to be determining the preparation and occurrence process of an earthquake, and actually most if not all of the factors cannot be fully controlled, or even cannot be controlled at all.

These two difficulties contribute significantly to the threaten to the sustainability of the seismic experimental sites, especially those focusing on earthquake prediction. Indeed the temporal length of an earthquake cycle (or even the uncertainty of such temporal length) is much longer than the length of a govenrmental administrative term. If a seismic experimental site fails to prove its successfulness and effectiveness to the society, then the supporting to the related scientific activities is hardly sustainable.'Earthquake prediction paradox', that to prove a method to be successful generally needs intensified observation, to organize intensified observation generally needs the support from the society, and to get the support of the society generally needs to prove to the society that the method is successful, reflects the difficulty of earthquake prediction study which has to be taken into serious consideration. For specific observations, this paradox leads to the result that the observational facilities which were not purposely designed and deployed for earthquake prediction generally have much better performance in earthquake prediction study. For earthquake prediction experimental sites, this paradox leads to the sustained threaten to the sustainability of the experimental sites themselves.

1.3 CDEs: One of the New Solutions to Meet the Challenges

The challenges dicsussed in the above section are not only in the field of earthquake studies. In ecology and environmental science, for example, 'few samples, many factors' is also one of the difficulties blocking the test of scientific hypotheses. Comparing to ecology and environment science which have the main difficulty in obtaining the laws at the global scale from individual local studies, earthquake science faces to the difficulty in obtaining the laws of the whole process of earthquake preparation and occurrence from the studies with duration much shorter than an earthquake cycle, and in obtaining the general laws of earthquake preparation and occurrence from the local case studies of earthquakes with different geodynamic/stress environments and different focal mechanisms. Nevertheless, it is common for ecology/environment science and earthquake science that the 'samples' are not sufficient, and there are many controlling parameters to be considered in the test of the hypotheses.

To tackle this problem, in the fields of ecology and environment science, the concept 'Coordinated Distributed Experiments (CDEs)' was proposed as an operational tool for hypothesis testing (Fraser et al., 2012). CDEs in ecology and environmental science have the following features (Fraser et al., 2012): *(1) hypothesis-driven experimental study; (2) multiple geographic locations; (3) standardized research design; (4) standardized data and coordinated data management;* and *(5) intellectual property sharing*, which are called 'defining attributes'; and *(6) synchronized data collection; (7) multiple investigative teams;* and *(8) low cost and low maintenance*, which are called 'probable characteristics'. It can be seen that almost all of these 5 + 3 features are familiar to earthquake science. For earthquake science, the 'multiple investigative teams' forms the 'community' and further the 'community models'. Seismology is even more advanced regarding to 'standardized data and coordinated data management'. In 1991 and 1997, respectively, the International Association of Seismology and Physics of the Earth's Interior (IASPEI) passed the resolutions[5] related to 'the test areas for earthquake prediction', proposing several test areas including north Anatolian strike-slip fault, Aleutian-Alaskan subduction zone, deep mine-induced seismicity in South Africa, and intermediate intraplate earthquakes in Vrancea, Romania, in the framework of the circum-Pannonian Carpathian seismic belts (proposed in 1991); and Kamchatka the zone of plate-subduction, Iceland the zone of plate spreading, Yunnan with intercontinental strike-slip stress environment, Gulf of Corinth the zone of continental rifting, and Beijing characterized by intracontinental earthquakes (proposed in 1997). Therefore, to much extent, in earthquake

[5] http://iaspei.org/documents/resolutions-statements, last access: April 25, 2021.

science the concept similar to CDEs has a long history. On the other hand, however, although the resolutions stated '*the IASPEI-approved sites for international collaborative earthquake prediction research*', not specifying what should be the standards for the 'approval' of the test sites, the seismic experiments seem to be a kind of distributed experiments without a clearly specified coordination mechanism. Or in another word, field experimental sites in earthquake science had already had the concept 'Distributed', but not yet the concept 'Coordinated' at that time.

1.4 CSES: Towards a Version 2.0 Seismic Experimental Site

During May 12–14, 2018, the International Conference for the Decade Memory of the Wenchuan Earthquake, in connection to the 4th International Conference on Continental Earthquakes (4th ICCE) and the 12th General Assembly of the Asian Seismological Commission (ASC), was held in Chengdu, China. In the meeting, the *Wenchuan Revelation* was discussed and passed, summarizing the lessons and experiences obtained from the Wenchuan earthquake as well as other devastating earthquakes. In the meeting, it was officially announced that the China Seismic Experimental Site (CSES) be established in the Sichuan-Yunnan region (Fig. 1.1). This natural laboratory, as a continuation and expansion of the West Yunnan Earthquake Prediction Experiment Site since 1980 and the National Experiment Site for Earthquake Monitoring and Forecast since 2014, is to facilitate investigator-driven core researches on the geodynamic environment, the dynamics, and the mechanism of the disaster chain, of continental earthquakes.

The experimental site is located near the 'eastern Himalayan syntaxis', with a total area of 780,000 km^2. Since 1965, for every decade, there had been on average 14 earthquakes with $M_S \geq 6.0$, among which 3 with $M_S \geq 7.0$, occurred in that region. These earthquakes are mainly continental, among which there were the May 12, 2008, Wenchuan M_S8.0; the April 20, 2013, Lushan M_S7.0; the August 3, 2014, Ludian M_S6.5; and the June 17, 2019, Changning M_S6.0, all causing significant losses of both human lives and social property.

From May 2018 up to the time of writing, CSES has published its scientific design (Working Group on the Scientific Challenges Report, China Seismic Experimental Site, 2019), released its version 1.0 scientific products (China Seismic Experimental Site, 2021), established its data sharing system (with distributed data centers in different institutions), started its first-phase project of capacity building (including active fault mapping, borehole observation experiments, and GNSS observation), initiated its internationally cooperative projects, and started the testing mode CSEScholarship. Although affected by the COVID-19, the stage of 'taking off' of CSES has been finalized, and the planned works are in progress (China Seismic Experimental Site, 2020).

Vision of CSES is to establish an international field laboratory of earthquake science, in which international and interdisciplinary cooperation could be fostered

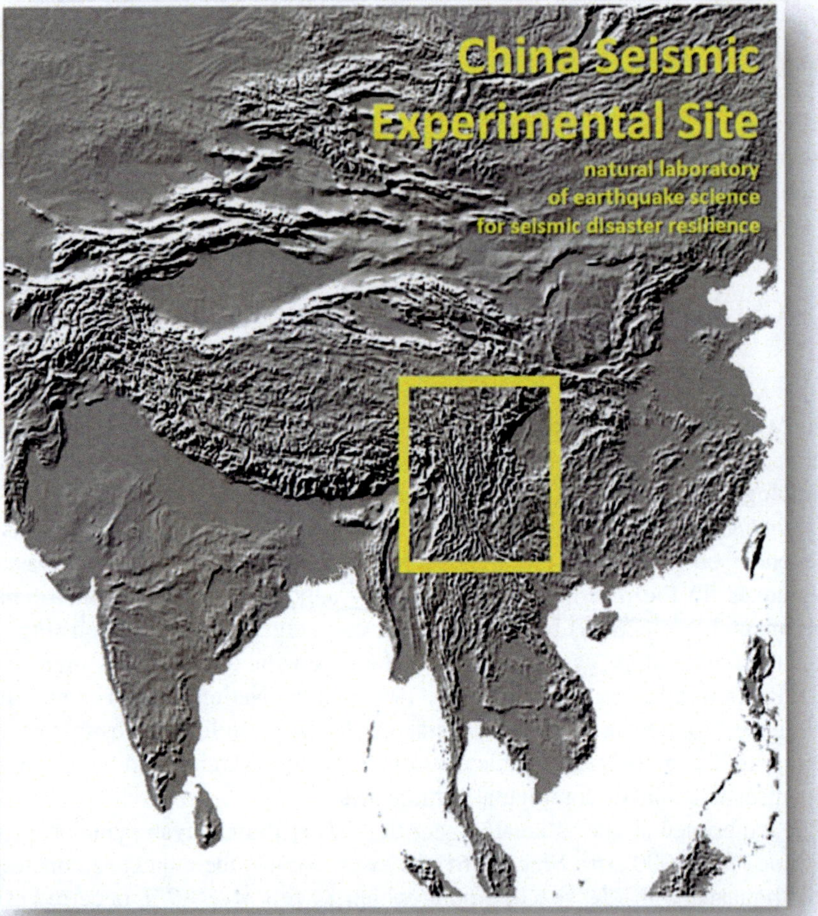

Fig. 1.1 Flyer of the China Seismic Experimental Site (CSES), indicating its location (yellow box) and geodynamic background—from the topography the Himalayas and the Tibetan Plateau can be seen

and supported, so that CDEs jointly with other seismic experimental sites are facilitated. As the 'first order' question in the coordination, the conditions for an area to be a natural laboratory are discussed. In the discussion, the following key factors emerge, sometimes called as a whole as the '$C5^2$ model' of a seismic experimental site.

Lessons and experiences in China (Sun & Wu, 2007), including those of the earthquake prediction experiment sites in west Xinjiang (1971–1975), Shanxi (1971–1974), west Yunnan (1980–), the Beijing-Tianjin-Tangshan-Zhangjiakou area (1980–1984), and North China (1982–1986), all indicate that consideration of sustainability in the perspective of 'big science' engineering plays an essential role in the construction and functioning of seismic experimental sites. Aiming at the goal of sustainability, CSES proposed the *coordinated distributed* capacity-building plan (including national, regional and local levels), and *coordinated distributed* project 'ecology' (indicating projects with different types from fundamental research to geological survey further to engineering application, supported by a variety of stake holders from national funding agencies to industry including private sectors).

The 'C5^2 Model' of a Seismic Experimental Site

- Community Scientific Problems List (CSPL)
- Community X Models (CXM)
- X = structure, stress, deformation, rheology, thermal, etc.
- Community Earthquake Models (CEM)
- Community Forecast for Falsification (CF^3)
- Community Accepted Technical Test (CATT)

- Coordinated Distributed Capacity-Building Plan (CDCP)
- Coordinated Distributed Data Centers (CDDC)
- Coordinated Distributed Computational Bases (CDCB)
- Coordinated Distributed Project 'Ecology' (CDPE)
- Coordinated Distributed Experiments Proposal (CDEsP)

Coordinated distributed data centers might be one of the solutions to the data sharing problem which deals with observational and simulation data in different disciplines and managed by different agencies. Traditional centralized data sharing system was tested to be not as successful as expected, that the construction of the data center itself would use huge resources, and in many cases, after (even during) the construction of the data center, the experimental site itself was over. As a testing mode of the new mechanism of data sharing, now there have been the GNSS data center at the Institute of Earthquake Forecasting of China Earthquake Administration (CEA), the seismological data center at the Institute of Geophysics of CEA, the active fault data center at the Institute of Geology of CEA, and the data center of geophysical exploration at the Chinese Academy of Geological Sciences, with new members of the *coordinated distributed* data centers coming in. CSES in turn organizes the on-line sharing of data products (produced, selected, used, and maintained by the community). The 2019 version of CSES data products includes background data (calibrated historical and instrumental earthquake catalogues, de-clustered earthquake catalogues, and focal mechanism catalogues) and community models (active fault model, deformation model, seismic velocity model, and rheology model), each with digital form for use and DOI for citation (China Seismic Experimental Site, 2020).

The concept CDEs makes it possible to conduct comparative studies among different natural laboratories, not only in China but also in the world, which prevents CSES as an experimental site from being exclusive. In China, what are in mind are west Xinjiang, the Beijing-Tianjin-Tangshan-Zhangjiakou area, the 'Hexi Corridor', the Tancheng-Lujiang fault, and the cross-Taiwan strait area. In the history there have been several seismic experimental sites all over the world. Some of them being still functioning, many of them ceased to work. With the new opportunities provided by innovative technologies and new ideas of earthquake science, as well as new understandings on the seismic experimental sites themselves, it is hopeful to 'reactivate' at least some of the seismic experimental sites so that CDEs could be conducted.

However, the challenge to the sustainability has existed, and will exist for a long time. It has to be noticed that, basically, a seismic experimental site, dealing with multi-disciplinary endeavors and complicated earthquake activity, and needing a long time to reach its scientific goals, has the features of systems engineering. Comparing to other 'big science' engineerings such as space technology, seismic experimental site seemed not done a very good job in the communication with policy makers and the public. For seismic experimental site, opportunities and challenges, especially the problem of sustainability, should be the principal consideration in its planning, construction, maintenance, and upgrading. For CSES, only time can tell whether the presently proposed countermeasures, based on the lessons and experiences in the past, will work or not. The words of Winston Churchill hold: *...this is not the end. This is not even the beginning of the end. But, perhaps, the end of the beginning.*

1.5 Appendix 1: Wenchuan Revelation-*Continental Earthquakes: Living with Seismic Risk*

(passed at the closing plenary of the International Conference for the Decade Memory of the Wenchuan Earthquake, May 14, 2018).

1.5.1 Wenchuan Memory

- The May 12, 2008, Wenchuan ($M_S8.0/M_W7.9$) earthquake of southwest China is the deadliest earthquake in China in the twenty-first century.
- The earthquake initiated at 14:28:04 local time, with a rupture duration of about 100 s along the Longmenshan fault zone. The earthquake ruptured two parallel faults with rupture lengths of 240 km and 70 km, respectively.
- According to official data released on August 25, 2008, the earthquake caused 69,226 deaths, 374,643 injuries, and 17,923 persons missing. Direct economic loss was estimated to be 845,100 million RMB Yuan.

- The studies of Wenchuan earthquake sequence resulted in significant advances in earthquake science.
- Ten years after the earthquake, the International Conference for the Decade Memory of the Wenchuan Earthquake was held on 12–14th May, 2018 in Chengdu, during the 4th International Conference on Continental Earthquakes (ICCE), jointly with the 12th General Assembly of the Asian Seismological Commission (ASC).
- The Conference was especially highlighted by a complimentary message from Chinese President Xi Jinping, calling for developing an earthquake resilient society. His message underlines a historic paradigm shift, not only in China but also in other earthquake-prone countries/regions in Asia and Pacific, even all of the world. He mentioned "China will adhere to the people-centered development philosophy, and pay attention to the disaster prevention and relief work while treating prevention as a major task."
- Based on the exchanges and discussions during the Conference, and while paying homage to people who lost their lives during the Wenchuan earthquake, the following suggestions are made towards mitigating seismic disaster and reducing the risks.

1.5.2 Wenchuan Revelation

- Driven by plate tectonics, most of the great earthquakes occur along the circum-Pacific and Alpine seismic belts. However, as per current state of scientific knowledge, forecasting the place, time and magnitude of an earthquake remains to be a challenge.
- Capability to forecast of major to great earthquakes at a decadal scale deserves our attention and would be a major achievement for reducing earthquake disaster risk. Geology, geodesy, and seismology play major roles in attaining this capability.
- Active faults are major cause of earthquake damages. An important goal is to identify them and avoid civil constructions in the immediate vicinity of these faults.
- Schools should be the safest places during an earthquake. The nationwide School Safety Engineering program was initiated after the 2008 Wenchuan earthquake, and was based on the lessons learnt from the Wenchuan earthquake.
- Secondary disasters, such as tsunamis, floods, landslides and rock falls, are also damaging. Secondary effects of major earthquakes also include nuclear reactor leaks, fires, etc., as observed during earthquakes in the World.
- Implementation of Earthquake Early Warning System (EEWS) plays an important role in reducing seismic risks. With the increase of magnitude of the earthquake, such a system becomes more effective.
- Governmental endeavors for the seismic safety of the major engineering and lifeline systems such as reservoirs and major plants are appreciated and recommended to be strengthened further.

- Strengthening of life-line buildings need to be taken up on high priority basis. At the same time, strengthening of large number of small buildings where most of the people live, needs to be undertaken.
- It is important to focus on basic research to better understand earthquake phenomena. Better observational infrastructure including seismic and geodetic monitoring networks need to be implemented in a regional scale across national boundaries. Studying active faults with aids of modern technologies including LiDAR, SAR etc. needs to be further strengthened.
- Earthquakes do not recognize state boundaries and academic disciplines. Reduction of earthquake disaster risk needs international and interdisciplinary research and collaboration.
- As pointed out in the message from President Xi Jinping, "Scientifically understanding the laws of disasters, effectively reducing disaster risks, and achieving a harmonious coexistence between human beings and nature requires the joint efforts of the international community."

1.5.3 Wenchuan Action

- As an outcome of the International Conference for the Decade Memory of the Wenchuan Earthquake, the inter-governmental-agency collaboration for seismic safety along 'the Belt-and-Road (the Silk Road Economic Belt and the 21st-Century Maritime Silk Road)' was advocated.
- The 4th International Conference on Continental Earthquakes (4th ICCE) announced the launching of the China Seismic Experiment Sites in the Sichuan-Yunnan region and the Tienshan-Pamir region, and invites international cooperation.
- The 12th General Assembly of the Asian Seismological Commission (ASC) proposes to take up a major initiative for international cooperation to establish an integrated Asia–Pacific seismic and geodetic observation and analyses facility.

Acknowledgements This chapter is adapted from the editorial of Jour. Geol. Soc. India, 95(2), 113–116, https://doi.org/10.1007/s12594-020-1400-9, and based on the plenary talk at the ASC Assembly (2018), with new information added up to the time of writing. Thanks are due to Profs. Paramesh Banerjee (ASC), Li Li (ASC), Harsh K. Gupta (GSI), Yongxian (Angie) Zhang (CEA), and epscially Prof. Yong-Gang Li (USC), for his kind invitation and helps in improving the manuscript. Related research works are supported by NSFC U2039207.

References

Audet, P. (2010). Temporal variations in crustal scattering structure near Parkfield, California, using receiver functions. *Bulletin of the Seismological Society of America, 100*, 1356–1362.

Bakun, W. H., & Lindh, A. G. (1985). The Parkfield, California, earthquake prediction experiment. *Science, 229*, 619–624.

Chao, K., & Peng, Z. (2009). Temporal changes of seismic velocity and anisotropy in the shallow crust induced by the 1999 October 22 *M*6.4 Chia-Yi Taiwan Earthquake. *Geophysical Journal International, 179*, 1800–1816.

Davis, J. F., & Somerville, P. (1982). Comparison of earthquake prediction approaches in the Tokai area of Japan and in California. *Bulletin of the Seismological Society of America, 72*, S367–S392.

Evans, R., Beamish, D., Crampin, S., & Ucer, S. B. (1987). The Turkish dilatancy project (TDP3): Multidisciplinary studies of a potential earthquake source region. *Geophysical Journal of Royal Astronomical Society, 91*, 265–286.

Fraser, L. H., Henry, H. A. L., Carlyle, C. N., White, S. R., Beierkuhnlein, C., Cahill, J. F., Jr., Casper, B. B., Cleland, E., Collins, S. L., Dukes, J. S., Knapp, A. K., Lind, E., Long, R., Luo, Y., Reich, P. B., Smith, M. D., Sternberg, M., & Turkington, R. (2012). Coordinated distributed experiments: An emerging tool for testing global hypotheses in ecology and environmental science. *Frontier in Ecology and Environment Science, 11*, 147–155.

Gupta, H. K. (2001). Short-term earthquake forecasting maybe feasible at Koyna, India. *Tectonophysics, 338*, 353–357.

Li, Y. G., Chen, P., Cochran, E. S., Vidale, J. E., Burdette, T. (2006). Seismic evidence for rock damage and healing on the San Andreas fault associated with the 2004 *M*6 Parkfield earthquake, Special issue for Parkfield *M*6 earthquake. *Bulletin of the Seismological Society of America, 96*(4), S1–15. https://doi.org/10.1785/0120050803

Li, Y. G., Chen, P., Cochran, E., & Vidale, J. (2007). Seismic velocity variations on the San Andreas Fault caused by the 2004 M6 Parkfield earthquake and their implications. *Earth, Planets and Space, 59*, 21–31.

Mogi, K. (2004). Two grave issues concerning the expected Tokai Earthquake. *Earth, Plants and Space, 56*, li–lxvi. https://doi.org/10.1186/BF03353074

Mora, P., & Place, D. (1998). Numerical simulation of earthquake faults with gouge: Toward a comprehensive explanation for the heat flow paradox. *Journal of Geophysical Research, 103*, 21067–21089. https://doi.org/10.1029/98JB01490

Niu, F., Silver, P. G., Daley, T. M., Cheng, X., & Majer, E. L. (2008). Preseismic velocity changes observed from active source monitoring at the Parkfield SAFOD drill site. *Nature, 454*, 204–208.

Roeloffs, E. (2000). The Parkfield California earthquake experiment: An update in 2000. *Current Science, 79*, 1226–1236.

Semenov, A. M. (1969). Variations in the travel-time of transverse and longitudal waves before violent earthquakes. *Izvestia Academy Science USSR Physics of the Solid Earth (English Translation), 4*, 245–248.

Site, C. S. E. (2020). *Progress Report of China Seismic Experimental Site (2019)*. Seismological Press. (in Chinese).

Site, C. S. E. (2021). *Annual Data Report of China Seismic Experimental Site (2019)*. Seismological Press. (in Chinese).

Sun, Q. Z., & Wu, S. G. (2007). *Development of the earthquake monitoring and prediction in China during 1966–2006* (pp. 490–509). Seismological Press. (in Chinese).

Wegler, U., Nakahara, H., Sens-Schönfelder, C., Korn, M., Shiomi, K. (2009). Sudden drop of seismic velocity after the 2004 Mw6.6 mid-Niigata earthquake, Japan, observed with passive image interferometry. *Journal of Geophysical Research, 114*, B06305. https://doi.org/10.1029/2008JB005869

Working Group for the Long-term Plan of Earthquake Forecast and Prediction (2010–2020). (2010). Earthquake prediction experiment site: Design, performance evaluation, and implementation. *Earthquake Research in China, 26*, 1–13. (in Chinese with English Abstract).

Working Group on the Scientific Challenges Report, & Site, C. S. E. (2019). *China seismic experimental site: Scientific challenges*. China Standard Press. (in Chinese).

Chapter 2
Seismotectonics and Geodynamics of the Eastern Margin of the Tibetan Plateau with Implication for the CSES

Junmeng Zhao

Abstract In recent decades, global geoscientists have conducted a great deal of research on the formation and evolution mechanism of the Tibetan Plateau in terms of geology, geophysics, geodesy and geochemistry. In 2003, the Institute of Tibetan Plateau Research, Chinese Academy of Sciences was established. We proposed and led the implementation of the Array Network of Tibetan International Lithospheric Observation and Probe Experiment (ANTILOPE) research Project. Under the general framework of this project, we have carried out systematic and comprehensive exploration and research in the basin-mountain-plateau of Western China and obtained massive geophysical observation data and significant research results. The ANTILOPE has three contributions to the geoscience: (1) The deep subduction of the Indian continent under the Tibetan Plateau has been proved by means of natural seismology, and the subduction depth. Angle and the east–west variation law have been determined. The collision boundary between India and Eurasia below the plateau was determined. This research further developed the theory of plate tectonics. (2) The deep fine structure of the basin and range system in Western China has been obtained by means of integrated geophysical methods. The contact relationship and activity law of the micro-landmass group in Northwest China has been revealed. (3) The synergistic evolution of the basin-mountain system and the Tibetan Plateau in western China is revealed by the earth system science method. The circles-layer coupling model of the deep earth processes restricting the surface topographic evolution and environmental change is established. The China Seismic Experimental Site (CSES) is located in the southeastern margin of the Tibetan Plateau. The ANTILOPE provides the basis for the site selection of the CSES and the geodynamic background for the construction of the CSES.

J. Zhao (✉)
State Key Laboratory of Tibetan Plateau Earth System Science (LATPES), Institute of Tibetan Plateau Research, Chinese Academy of Sciences, Beijing 100101, China
e-mail: zhaojm@itpcas.ac.cn

CAS Center for Excellence in Tibetan Plateau Earth Sciences, Beijing 100101, China

Key Laboratory of Continental Collision and Plateau Uplift, Institute of Tibetan Plateau Research, Chinese Academy of Sciences, Beijing 100101, China

University of Chinese Academy of Sciences, Beijing 100101, China

© Higher Education Press 2022
Y.-G. Li et al. (eds.), *China Seismic Experimental Site*,
https://doi.org/10.1007/978-981-16-8607-8_2

Keywords Tibetan Plateau · Crustal structure · Lithospheric structure · Deep geodynamic processes · Eastern margin of Tibetan Plateau · CSES

2.1 Introduction

Since ~ 65 Ma, the constant collision between Indian plate and Eurasian plate has shaped not only high altitude Tibetan Plateau, but also driven plateau material moving eastward at the same time (Ding et al., 2005). Blocked by the rigid Sichuan basin, a large-scale thrust nappe structure with northeast-southwest direction was formed in the eastern margin of the Tibetan Plateau. The Longmen Shan fault zone with most steep terrain between the Sichuan basin and Songpan-Ganze block is a typical tectonic active region on the eastern margin of the Tibetan Plateau where the 2008 Wenchuan earthquake of $M_S 8.0$ and the 2013 Lushan earthquake of $M_S 7.0$ occurred here recently. The eastern margin of the Tibetan Plateau serves as a transition zone for the eastward expansion of plateau materials. Since the late Cenozoic, the Longmen Shan uplifted rapidly, forming an abrupt relief of terrain, mainly showing that the Songpan-Ganze block uplifts ~ 4 km within only ~ 30 km range from Sichuan basin to Songpan-Ganzi block, while the amount of shortened crust of the Longmen Shan Mountain since the late Cenozoic is not enough to produce the plateau edge uplifting and crustal thickening so much. It is an interesting question: what is related lithospheric structure and deep tectonic processes for that deficit?

The Studies of **A**rray **N**etwork of **T**ibetan **I**nternational **L**ithospheric **O**bservation and **P**robe **E**xperiment (**ANTILOPE**) (Fig. 2.1) project show that the boundary between the Indian and Asian lithospheres below Tibet runs roughly from the western Tarim Basin to the eastern Himalayan Syntaxis. A particular lithospheric region (so called "Tibetan Plate") was formed in northern and eastern Tibet as a crush zone between the two colliding plates, the existence of which is marked by high temperature, low mantle seismic velocity, poor Sn propagation, larger seismic anisotropy, and east and southeast oriented displacements. The crustal shortening in the southern Tibet is accommodated by under-thrusting of the Indian crust below the Asian crust that may reach further north than the Yarlung-Zangpo suture (YZS). In northern Tibet, crustal shortening is accommodated by homogeneous crustal thickening. The more rugged and higher topography in west Tibet can be supported by the rigid mantle lithosphere there, whereas to the east the lithosphere is weaker due to the existence of the crush zone, causing the topography there lower and flat (Zhao, 2010). Under north–south compress, the soft material of the crush zone moves eastward and divided into four directions when meeting the rigid Sichuan Basin: to the southeast forming the Yunguichuan plateau; to the northeast causing Erdos basin partial craton destruction; to the upward forming Longmen Shan Mountain; and to the deep mantle which still needs more evidences (Fig. 2.2).

The study of the structure and deformation of the two plate collision zones plays a key role in revealing the tectonic evolution process including plateau uplift and expansion. These research results are undoubtedly the growth points of the new

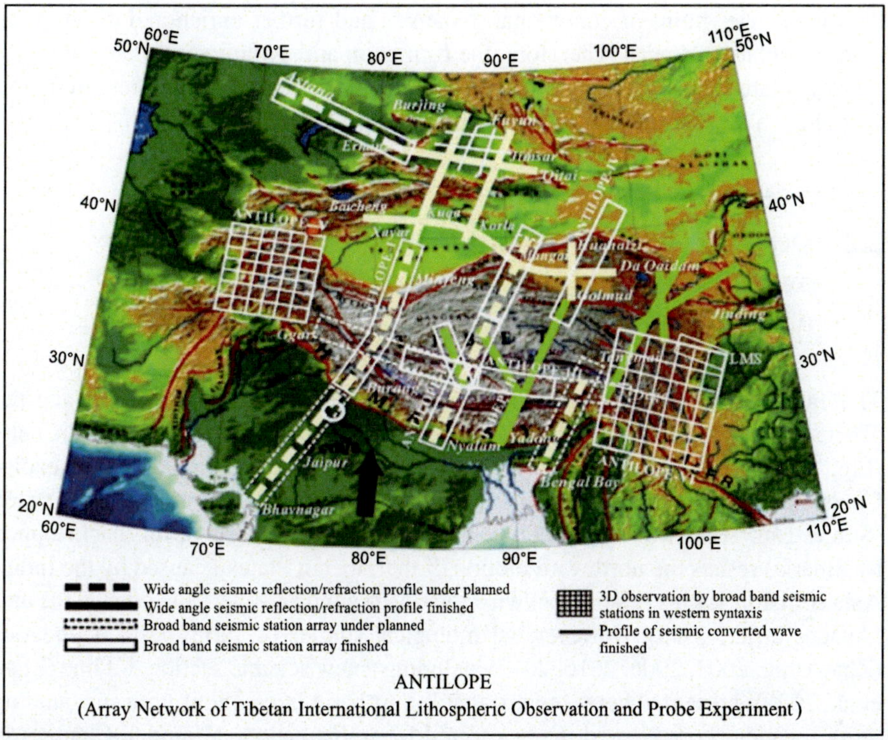

Fig. 2.1 Map shows locations of seismic lines and arrays in the ANTILOPE. Red lines denote active faults in the region. XB (Xayar-Burjing), KJ (Korla-Jimsar) and BD (Baicheng-Da Qaidam)

Fig. 2.2 A conceptual model of Western China

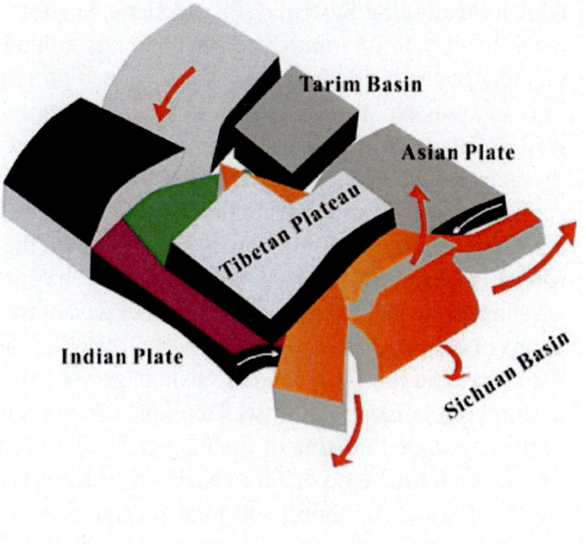

theory and viewpoint of continental dynamics and further enrich and develop the theory of plate tectonics. Therefore, the formation and evolution mechanism of the Tibetan Plateau has always been the focus of global earth science research, and absolutely a key for the Theoretical Framework of CSES.

2.2 Deep Structure, Dynamic Processes and Constraints on Resources and Environment of the Basin-Mountain-Plateau in Western China

The Tarim Basin is a cratonic block located between the Siberian craton and the Tibetan/Pamir plateau, bounded on all sides by major faults and mountain belts (Fig. 2.1). Its Precambrian basement is covered by thick Paleozoic–Mesozoic-Cenozoic marine sediments and post-Miocene Taklimakan Desert eolian dune sands (Sun & Liu, 2006, Sun et al., 2008). In the classical view, the basin (backstopped by Siberia) resists the northward motion of the Tibetan Plateau caused by the India-Asia collision, with a minor clockwise rotation around an Euler pole outside the unit (Avouac et al., 1993). However, seismological studies of Tarim basin boundaries (Zhao et al., 2003, 2006, 2010, 2019) including a new seismic section, KJ line (Zhao et al., 2020), presented here, suggest that lithospheric interactions with surrounding units are more complicated than expected from this canonical model. Changes in environment, such as the Paleogene aridification and desertification, attributed to a combination of Tibetan Plateau uplift and seawater retreat from the area, also suggest higher-order complexity. Here we report evidences from four seismic profiles and Global Positioning System (GPS) geodetic data for clockwise rotation of the Tarim block relative to its immediate surroundings around a virtual axis within the unit, yielding larger strain gradients than predicted previously from paleomagnetic data. This rotation explains the observed structural complexity along the boundaries, as well as changes in paleoenvironment, and is itself a logical consequence of the asymmetry and diachronous nature of the India-Asia collision zone.

Over the past two decades a number of seismic experiments have been conducted across the borders of Tarim basin (Fig. 2.1) with the purpose of studying the lithospheric structure of the basin with respect to the surrounding mountain belts. Here we summarize the results insofar they are relevant for the discussion of regional variations in deformation style. In the north, the observation of a double Moho beneath Tien Shan and joint inversion of seismic, gravity, and geomagnetic data suggest that Tarim basin crust underthrusts Tien Shan. In the northeast, the seismic data reveal southward underthrusting of the Junggar basin lower crust beneath Tien Shan and a gap between the Moho of Tarim basin and eastern Tien Shan. In the southeast, wide-angle reflection/refraction and joint inversion with gravity data reveal southward underthrusting of the Tarim lithosphere beneath the Altyn Shan. In the southwest, pure shear deformation is accommodated by deep faults and receiver function profiles reveal a Moho step of ~ 20 km, from ~ 80 km beneath Tibet to ~ 60 km beneath Tarim basin.

From the northern part of the Tarim Basin to the northern margin of the Junggar Basin, the XB Line crosses northern Tarim Basin, the central Tian Shan, and the Junggar Basin, with a length of 900 km. Along the profile, wide-angle seismic reflection/refraction and a joint inversion of gravity and geomagnetism have been carried out. The results reveal a double Moho beneath the Tian Shan, which indicates that the Tarim Basin underthrusts the Tian Shan. We present here results from a recent, previously unpublished, 600 km long seismic profile (KJ line), which crosses the northeastern margin of the Tarim Basin, the east Tien Shan, and the southern part of the Junggar Basin. This comprehensive geophysical profile reveals a double-Moho structure underneath the northern margin of the eastern Tien Shan with underthrusting of the Junggar Basin lower crust beneath the Tien Shan. Beneath the southern Tien Shan margin a lateral gap appears between the Moho of the Tarim Basin and that of the east Tien Shan. Along with previous seismic profiles this reveals a complex configuration along the basin boundaries. The NW–SE striking, 1400-km long seismic profile (BD line) traverses the eastern part of the Tarim Basin, the Altyn Shan, and the Qaidam Basin. Wide-angle reflection/refraction and joint inversion of gravity and geomagnetism revealed underthrusting of the Tarim lithosphere beneath the Altyn Shan to a depth of 90 km. ANTILOPE-I is a passive-source seismic profile in western Tibet, extending from Himalaya to the southern Tarim Basin, with a total length of ~ 800 km. The Moho is seen at a depth of ~ 80 km underneath Tibet and at ~ 60 km underneath the Tarim Basin, suggesting a vertical Moho offset of ~ 20 km across the Altyn Shan. In the same region an earlier Sino-French experiment revealed a similar Moho step. The crustal shortening here is accommodated by pure shear in contrast to simple shear at the southern Tarim margin (along BD line).

The lithospheric interactions between bounding units implied by these observations are inconsistent with the canonical framework of predominantly NS compression but are readily explained if the strain gradients due to clockwise rotation of the Tarim Block are larger than previously suggested. Paleomagnetic data suggest negligible post-Cretaceous rotation of Tarim basin, whereas a kinematic model derived from modeling slip rates along major thrust and strike-slip faults revealed a Cenozoic clockwise rotation of the Tarim basin with respect to Siberia at a rate of $0.65 \pm 0.3°$/Myr and around a pole at (43.5° N, 95.7° E) (Zhao et al., 2019). Present-day GPS observations also reveal a quite similar clockwise rotation of $0.679 \pm 0.059°$/Myr around a pole at ($37.3 \pm 0.3°$ N, $93.3 \pm 0.9°$ E), near the middle of the Qaidam basin, relative to stable Siberia. However, these rotations relative to Siberia are not efficient or sensitive to explain the different deformation styles between Tarim basin and its surroundings inferred from the seismological studies. Euler poles outside Tarim produce rotations that are too small, however, to explain the different deformation styles inferred from the seismological studies.

Here, for the purpose to check if the Tarim Basin itself is rotating relative to its immediate surroundings or not, we use a special strategy to "fix" the immediate surroundings by removing the "net translation and rotation of the surroundings" from the whole GPS velocity field, so as to highlight the "subtle" rigid motion of the Tarim

Basin relative to its fixed immediate surroundings. With the new GPS velocity field relative to the "fix immediate surroundings of Tarim basin", we solve that the Tarim basin is rotating clockwise with an angular rate of $0.461 \pm 0.036°$/Myr around a pole at ($39.2° \pm 0.27°$ N, $82.0° \pm 0.34°$ E) near the geometric center of the unit, which produces strain gradients that can explain morphological changes along the Tien Shan (Zhao et al., 2019) and the otherwise puzzling styles of deformation in the boundary zones, that is, compression in NW with central Tien Shan, transtension in NE with eastern Tien Shan, compression with Altyn Tagh Shan in the SE, and left lateral shear motions in the SW.

The clockwise rotation of the Tarim block can explain pertinent topographic features of the Tien Shan: that is, the north–south symmetry of the central Tien Shan (CTS) and the asymmetry of the eastern Tien Shan (ETS). Clock-wise rotation of Tarim caused extension underneath the southern margin of ETS, and differential uplift and subsequent weathering and erosion could have caused the asymmetry observed along the KJ line. The geological difference between the southern and northern margins of the ETS influenced deformation of the basins at its two sides. At the southern margin of the Junggar Basin, the deformation is strong, causing steep structures and intense folding. The relatively flat northeastern margin of the Tarim Basin may be caused by extension due to the clockwise rotation of the basin.

Block rotations are not expected in an overall regime of (pure shear) north–south convergence, but we propose that the rotation of the Tarim basin is driven by the asymmetric indentation of the northward moving Indian plate and facilitated by lateral variations in geomechanical properties of plate boundary zone. First, the lithospheric boundary between the Indian Plate and the Eurasian Plate beneath Tibet roughly follows a line from the western Tarim basin to the eastern Himalayan syntaxis. Second, increased seismic anisotropy and high attenuation in the uppermost mantle suggest that in central and east Tibet hot and more easily deformable mantle material fills the gap between Indian and Asian plates. In western Tibet, toward the Pamir, indentation of the Indian plate across the entire width of the Plateau exerts a northward push on the western Tarim basin and the weak zone further east allows the Tarim basin to rotate clockwise in response to the northward push further west.

The rotation of the Tarim block had significant consequences for the paleoenvironment of the Asian interior. Previous studies have indicated that the Northern Pamir has been displaced northward 400–300 km with respect to the rest of Eurasia, leading to collision with the southeastern Tien Shan. Traditional views attribute all crustal shortening in this region simply to the northward motion of the Pamir in response to the India-Eurasia collision. Our new geophysical evidence refines this view and suggests that the collision of the Northern Pamir with the southern Tien Shan is so localized because the asymmetric indentation of India and the concomitant clockwise rotation of the Tarim Block enhanced the northward displacement of the Pamir compared to other parts of the collision zone. The localized convergence influenced the retreat of sea water from the Tarim Basin in the late Eocene and accelerated the final collision in the latest Miocene, which blocked the transport of water–vapor by the prevailing westerlies from the Atlantic and the Mediterranean Sea to the downwind Tarim Basin, leading to the extreme arid climate and the formation of the world second-largest shifting-sand desert (Taklimakan Desert) in the Tarim Basin in the latest Miocene.

The paleoenvironmental changes have led to distinct differences in biological evolution of the Junggar Basin nearby the Tarim basin, and the Mongolian plateau, more than 1000 km farther east, but with the same latitude. In the Junggar basin, mammalian fossils of the Late Eocene are dominated by perissodactyls with low dental crowns, including giant rhinos, but the fossil assemblages of the Early Oligocene are dominated by small rodents and lagomorphs. Therefore, the mammalian fossils of this area reveal a distinct faunal turnover pattern characterized by a shift from Late Eocene, large perissodactyl-dominant fauna to Early Oligocene, small rodent/lagomorph-dominant fauna. The vegetation type during the Late Eocene is a mixed coniferous and broad-leaved forest, but the Early Oligocene flora is a forest-steppe environment with open grasslands in the Junggar basin. Therefore, the pollen results also reveal an abrupt shift in the ecosystem from warm-humid forest in the Late Eocene to dry-temperate forest-steppe in the Early Oligocene. Similar to the Junggar basin, the Late Eocene mammalian fauna of the Mongolian plateau in East Asia is dominated by perissodactyls. However, ungulate mammals, which are represented by Paraceratherium, the largest giant rhino and the largest terrestrial mammal, still lived during the Early Oligocene in the Mongolian plateau, such as Ordos and Erenhot areas of Inner Mongolia as well as central and eastern Mongolia. For example, the Early Oligocene fauna from Ordos contained rhinos Paraceratherium, Aprotodon and Cadurcodon, chalicothere Schizotherium, tapir, deer Lophiomeryx and Eumeryx, and bovid, while the rodent/lagomorph-dominant fauna have not occupied this area until the Late Oligocene.

Block rotation is common in plate boundary zones controlled by strike slip or transpression, but our study indicates that such rotations (and concomitant lateral variations in paleoenvironment) can also result from asymmetry in the structure and mechanical properties of colliding tectonic units in predominantly pure shear regimes. The Tarim Basin rotation resulted in a complex boundary zone at the northern margin of the Tibetan Plateau. Involving rotation of the Tarim Basin block in tectonic models can help understand the along-strike lithospheric variations beneath the Indian-Asian collision zone and debated issues such as whether the northward growth of the Tibetan Plateau is controlled by underthrusting of the Asian plate beneath Tibet or by a mode of pure shear shortening.

2.3 Crust-Mantle Structure and Dynamics of the Eastern Margin of the Tibetan Plateau

The study of the lithosphere and asthenosphere structure in the eastern margin of the Tibetan Plateau is an important means to understand the tectonic deformation of the crust and upper mantle and the eastward migration of materials in the plateau. The deep structure in the eastern margin of the Tibetan Plateau was studied by CCP migration and superposition imaging based on the data collected from the digital and temporary seismic stations in Sichuan region. The results show that the crustal thickness in the eastern margin of the Tibetan Plateau is 10–20 km larger than that in the Sichuan Basin. In the transition zone between the eastern margin of the Qinghai-Tibet Plateau and the Sichuan Basin, there is a large vertical fault and deformation at the Moho surface. From the eastern margin of the Plateau to the Sichuan Basin, the boundary between the lithosphere and asthenosphere (LAB) shows a gradual increase in depth, while the 410 km discontinuity depth tends to be shallower. There are also several discontinuous stratifications between the lower crust interface and LAB interface and between LAB interface and 410 km discontinuity. There are obvious low-velocity layer distribution under the LAB interface of the eastern Tibetan Plateau and Sichuan Basin, but there are differences between them. The low-velocity layer distribution under the LAB interface of the Sichuan Basin is more complete, while the low-velocity layer distribution under the LAB interface of the eastern Tibetan Plateau shows discrete high-velocity block distribution. The obvious difference between the deep structure of the eastern margin of the Qinghai-Tibet Plateau and the Sichuan Basin shows the complexity of the deep geodynamic background in this area.

The teleseismic seismic wave data recorded by passive source profile are inversely imaged by receiver function method, and the velocity interface of the crust and upper mantle beneath each station is obtained. Combined with the S-wave velocity of the station and the crustal stratification and velocity distribution of the artificial seismic profile in the study area, the study of the crustal structure, active fault plane and detachment surface is of great significance for the discussion of the collision contact relationship and dynamics between the Longmenshan and the Yangtze Craton in the eastern margin of the Tibetan Plateau. The results show that: (1) The crustal thickness at the western margin of the Yangtze platform varies laterally from 35 km to 48 km; (2) In the Longmenshan nappe, the Moho surface fluctuates little, and the crustal thickness is about 45 km. The Moho surface of Songpan-Ganze block formed a steep drop zone of 16–20 km after crossing the central fault zone of Longmenshan, and the crustal thickness varied between 45 km and 68 km. The Moho surface formed a 12 km gently descending zone about 30 km across the west side of the Houshan fault zone of Longmenshan, and the crustal thickness varied between 45 km and 58 km. The study of the lithosphere and asthenosphere structure in the eastern margin of the Tibetan Plateau is an important means to understand the tectonic deformation of the crust and upper mantle and the eastward migration of materials in the plateau. The deep structure in the eastern margin of the Tibetan Plateau was studied by CCP migration and superposition imaging based on the data collected from the digital and temporary seismic stations in Sichuan region. The results show that the crustal thickness in the eastern margin of the Tibetan Plateau is 10–20 km larger than that in the Sichuan Basin. In the transition zone between the eastern margin of the Tibetan Plateau and the Sichuan Basin, there is a large vertical fault and deformation at the Moho surface.

From the eastern margin of the Tibetan Plateau to the Sichuan Basin, the depth of the lithosphere-asthenosphere boundary (LAB) increases gradually, while the depth of the 410 km discontinuity tends to be shallower. There are also several discontinuous stratifications between the lower crust interface and LAB interface and between LAB interface and 410 km discontinuity. There is obvious low-velocity layer distribution under the LAB interface of the eastern Tibetan Plateau and Sichuan Basin, but there are differences between them. The low-velocity layer distribution under the LAB interface of the Sichuan Basin is more complete, while the low-velocity layer distribution under the LAB interface of the eastern Tibetan Plateau shows discrete high-velocity block distribution. The obvious difference between the deep structure of the eastern margin of the Tibetan Plateau and the Sichuan Basin shows the complexity of the deep geodynamic background in this area.

2.4 Implication for the CSES

The CSES is located in the national territory between 97.5°–105.5° E, 21°–32° N with a total area of about 780,000 km^2. This area includes the Sichuan-Yunnan rhomboid block near the eastern Himalayan syntaxis and its surrounding area. The tectonic environment is complex, with various fault systems such as compression, shear and tension, and there are both inter-plate earthquakes and intra-plate earthquakes. In the past 50 years, there have been an average of 14 earthquakes with a magnitude of 6.0 or above and three earthquakes with a magnitude of 7.0 or above in every 10 years, among which the Wenchuan *M*8.0 earthquake on May 12, 2008, the Lushan *M*7.0 earthquake on April 20, 2013 and the Ludian *M*6.5 earthquake on August 3, 2014 have all caused serious casualties and property losses.

In recent decades, seismology has gained important understanding in the mechanism of earthquake initiation, occurrence and disaster-causing. However, due to the lack of accurate measurement and control of the experimental process, the field "experiment" of seismological science still has great limitations. In recent years, the application of dense and ultra-dense arrays and new observation techniques (such as LiDAR) has obtained high-precision structure and motion of subsurface media. The study of seismology, geodesy, and geology is being reshaped by the use of new technologies such as partial control of earthquake genesis processes (such as gas or fluid injection), repeatable and continuous measurements based on non-explosive source imaging and waveform cross-correlation techniques, and the simulation of seismic dynamic processes based on supercomputers. Field experiments are also beginning to have elements of what we are familiar with as "physics experiments".

In 2017, the China Earthquake Administration, in collaboration with the Ministry of Science and Technology, the Chinese Academy of Sciences, the Chinese Academy of Engineering and the National Natural Science Foundation of China, launched the National Earthquake Science and Technology Innovation Project, implementing four scientific programs, including "Transparent Crustal", "Anatomy of Earthquake", "Resilient Urban and Rural Areas", and "Intelligent Service". As an important platform for the four scientific programs, the experimental site will attract professionals at home and abroad to jointly promote the innovation and development of seismic science and technology. With the help of high-resolution, comprehensive and multi-dimensional observation and supercomputer simulation technology, the experimental site will strive to solve major scientific problems such as the breeding environment, earthquake occurrence process and earthquake disaster mechanism of mainland earthquakes.

The study of "anatomic earthquake" in the experimental field will deepen the understanding of the physical mechanism of earthquake initiation and rupture, and establish the prediction model of earthquake initiation and occurrence on a more solid physical basis. Through the unified prediction and comparison of strain rate field, seismic activity and strong ground motion, it can be applied to minimize the risk of earthquake disaster.

As a "big science" project, the experimental site pays more attention to the construction of basic observation capabilities such as seismic geology, geophysics, geodesy and geochemistry, and pays more attention to the establishment of data sharing mechanism based on distributed data center, and pays more attention to communication and cooperation among research teams. At the same time, it also highlights the characteristics of China's continental seismic research, and carries out research projects, experiments and the research on risk prevention of the whole chain "from earthquake rupture process to engineering structure response".

Through the experiment site area to carry out scientific research, the CSES will launch the international first-class crust-mantle fine structure, high resolution seismic fault distribution and gravity, magnetic, crustal movement geophysical field changes the image (please rewording), such as production of a batch of earthquake gestation and earthquake risk assessment related science model. We will promote the transformation of earthquake preparedness and disaster reduction from technology-based work to service-oriented work.

References

Avouac, J. P., Tapponnier, P., Bai, M., You, H., & Wang, G. (1993). Active thrusting and folding along the northern Tien-shan and late cenozoic rotation of the tarim relative to Dzungaria and Kazakhstan. *Journal of Geophysics Research—Solid Earth, 98*(B4), 6755–6804.

Ding, L., Kapp, P., Wan, X. Q. (2005). Paleocene-Eocene record of ophiolite obduction and initial India-Asia collision, south central Tibet. *Tectonics, 24*(3), TC3001. https://doi.org/10.1029/2004TC001729

Sun, J. M., & Liu, T. S. (2006). The age of the Taklimakan Desert. *Science, 312*(5780), 1621–1621.

Sun, J. M., Zhang, L. Y., Deng, C. L., Zhu, R. X. (2008). Evidence for enhanced aridity in the Tarim Basin of China since 5.3 Ma. *Quaternary Science Review, 27*(9–10), 1012–1023

Zhao, J. M. (2020). Clockwise rotation of the Tarim basin driven by the Indian plate impact. Part II*. *Earth Sciences and Subsoil Use, 43*(4), 486–498. https://doi.org/10.21285/2686-9993-2020-43-4-486-498

Zhao, J. M., Liu, G. D., Lu, Z. X., Zhang, X. K., & Zhao, G. Z. (2003). Lithospheric structure and dynamic processes of the Tianshan orogenic belt and the Junggar basin. *Tectonophysics, 376*(3–4), 199–239.

Zhao, J. M., Mooney, W. D., Zhang, X. K., Li, Z. C., Jin, Z. J., & Okaya, N. (2006). Crustal structure across the Altyn Tagh Range at the northern margin of the Tibetan Plateau and tectonic implications. *Earth and Planetary Science Letters, 241*(3–4), 804–814.

Zhao, J. M., Yuan, X. H., Liu, H. B., Kumar, P., Pei, S. P., Kind, R., Zhang, Z. J., Teng, J. W., Ding, L., Gao, X., Xu, Q., & Wang, W. (2010). The boundary between the Indian and Asian tectonic plates below Tibet. *PNAS Proceedings of the National Academy of Sciences of the United States of America, 107*(25), 11229–11233.

Zhao, J. M., Zhang, P. Z., Yuan, X. H., Gan, W. J., Sun, J. M., Deng, T., van der Hilst, R. D., Zhang, X. K., Liu, H. B., Pei, S. P., Yang, X. P., Xu, Q., Zhang, H., Deng, G., Ju, C. H., Teng, J. W. (2019). Clockwise rotation of the Tarim basin driven by the Indian plate impact. *Earth Sciences and Subsoil Use, 42*(4), 425–436. https://doi.org/10.21285/2686-9993-2019-42-4-425-436

Chapter 3
From Plate Loading to Regional Fault Network: Toward Mechanical Understanding of the Kinematics

Jing Liu-Zeng, Heng Wang, Jing Xu, Shihu Li, Peng Wang, Longfei Han, Yukun Xing, and Jinyang Li

Abstract The interaction between the Indian Plate, Pacific Plate, Philippine Sea Plate and the Eurasian Plate, combined with the deep geodynamics in the Eurasian Plate, shapes the pattern of active deformation that controls the spatial distribution of strong earthquakes in the mainland of China The southeastern Tibetan Plateau is the fan-shaped area, which is bounded by the Xianshuihe-Xiaojiang fault system in the east and the Sagaing fault in the west. In terms of tectonic boundary condition, Sichuan-Yunnan (also called Chuan-Dian) region i.e. the test ground of China Seismic Experimental Site, is marked with a high degree of complexity, being influenced by the eastward subduction of Myanmar arc in the west, northward subduction of the Indian Plate and clockwise rotation of east and southeast Tibetan Plateau, gravitational effect of over-thickened crust of central Tibetan Plateau, and slab rollback of Sumatra (Simons et al.,2007) (Fig. 3.1).

J. Liu-Zeng (✉) · L. Han · J. Li
School of Earth System Science, Tianjin University, Tianjin 300072, China
e-mail: Liu_zeng@tju.edu.cn

L. Han
e-mail: hanlongfei_2019@tju.edu.cn

J. Li
e-mail: lijinyang101@tju.edu.cn

H. Wang · P. Wang · Y. Xing
Institute of Geology, China Earthquake Administration, Beijing 100029, China
e-mail: wangpeng@ies.ac.cn

Y. Xing
e-mail: yukunxing2018@163.com

J. Xu
The Second Monitoring and Application Center, China Earthquake Administration, Xi'an 710054, China
e-mail: xujing@smac.ac.cn

S. Li
Institute of Geology and Geophysics, Chinese Academy of Science, Beijing 100029, China

© Higher Education Press 2022
Y.-G. Li et al. (eds.), *China Seismic Experimental Site*,
https://doi.org/10.1007/978-981-16-8607-8_3

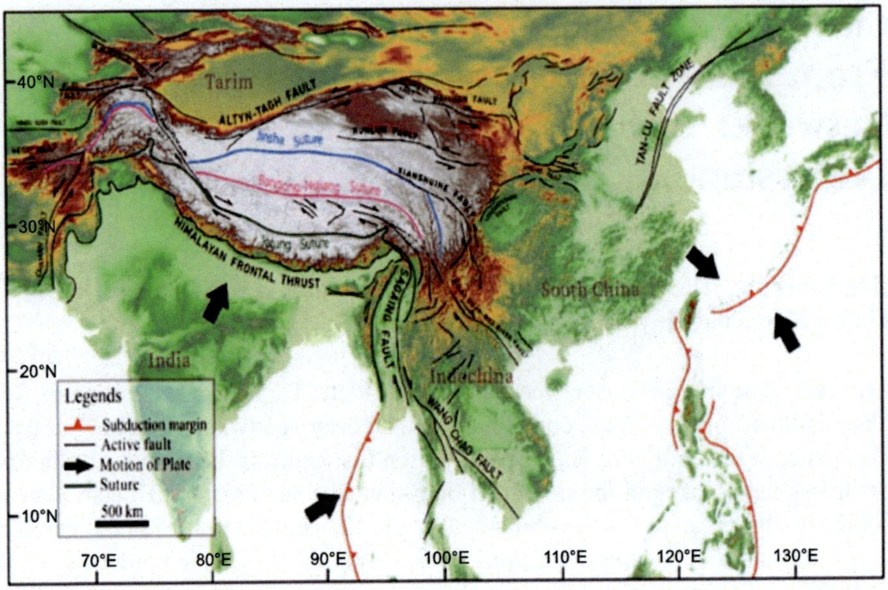

Fig. 3.1 Plate boundary configuration and tectonic setting of SE Asia and China

Keywords Tectonic evolution · Cenozoic · Block rotation · Active tectonic block · Chuan-Dian block · Fault slip rate · Block models · Geologic and geodetic constraints

One striking feature of active tectonics in China is that active faults are well developed and widely distributed, which cuts China's mainland into tectonic blocks and sub-blocks at different levels. Zhang et al. (2003) divided the China's mainland into 6 Level-I tectonic-block regions, including the Qinghai-Tibet, West China, South China, Yunnan-Myanmar, North China and Northeast Asia regions, and 22 Level-II "active tectonic blocks", including Lhasa, Qiangtang, Bayan Har, Qaidam, Qilian, Sichuan-Yunnan, western Yunnan, southern Yunnan, Tarim, Tianshan, Junggar, Sayan, Altai, Alashan, Xing'an-Dongmong, Notheast China, Erdos, Yanshan, North China, East Shandong-Yellow Sea, South China, and South China Sea (Fig. 3.2). Active tectonic block is a geologic entity which is bounded by active fault zones formed in the late Cenozoic and has been active since the late Quaternary period. A block is usually kinematically coherent with a uniform velocity. The active faults bounding a block are characterized of strong tectonic activity, usually being the seismogenic structures of most strong earthquakes. Almost all magnitude 8 earthquakes and 80–90% of magnitude 7 earthquakes in China occurred on the block-bounding faults (Zhang et al., 2003).

The Sichuan-Yunnan region is mainly shaped by the oblique collision between India and Eurasia, but has undergone prolonged multi-phase deformation over

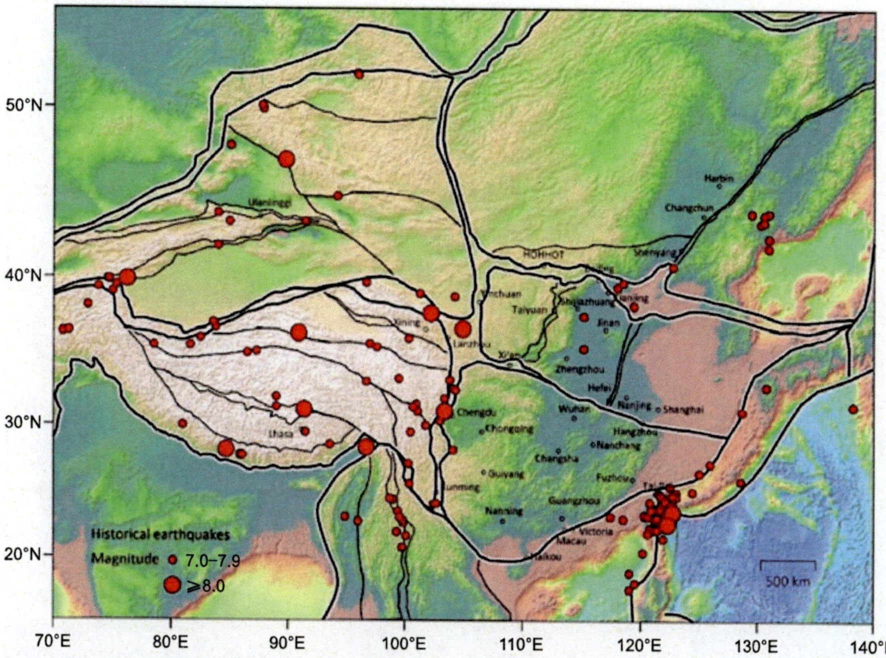

Fig. 3.2 Major active blocks for the mainland of China and surrounding region. Large historical earthquakes ($M > 7$) show close spatial correlation with the boundaries of first-order blocks. Block names and fault names are shown in Fig. 3.3

geologic time scale. The deformation in later phase is often influenced by both inheritance and superimposition of strutures which are formed under a series of changing tectonic regimes starting from oceanic subduction during the Protozoic, through oblique collision and subduction during the Mesozoic, and then to intraplate deformation of mountain uplift and lateral extrusion during the Cenozoic (e.g., Tapponnier et al., 1986, 2001; Pan et al., 1990; Bertrand & Rangin, 2003; Burchfiel & Chen, 2012; Metcalfe, 2013). The constant change in tectonic regime and plate configuration results in a complex pattern of fault system with changing kinematics and tectonic inversion (e.g., Lacassin et al., 1998; Leloup et al., 1995).

3.1 Geologic Records of Block Rotations in the Region

Geologically, the southeastern Tibetan Plateau consists of three major blocks: the Chuan-Dian Block, the Indochina Block, and the Sibumasu Terrane. The Chuan-Dian Block comprises a complex collage of continental fragments, including the southwestern part of the South China Block (Chuxiong Basin), the Songpan-Ganze fold belt, the Yidun Volcanic Arc and the Yanyuan-Lijiang fold belt (Bureau of Geology Mineral Resources of Yunnan Province (BGMRY), 1990; Wang & Burchfiel, 1997)

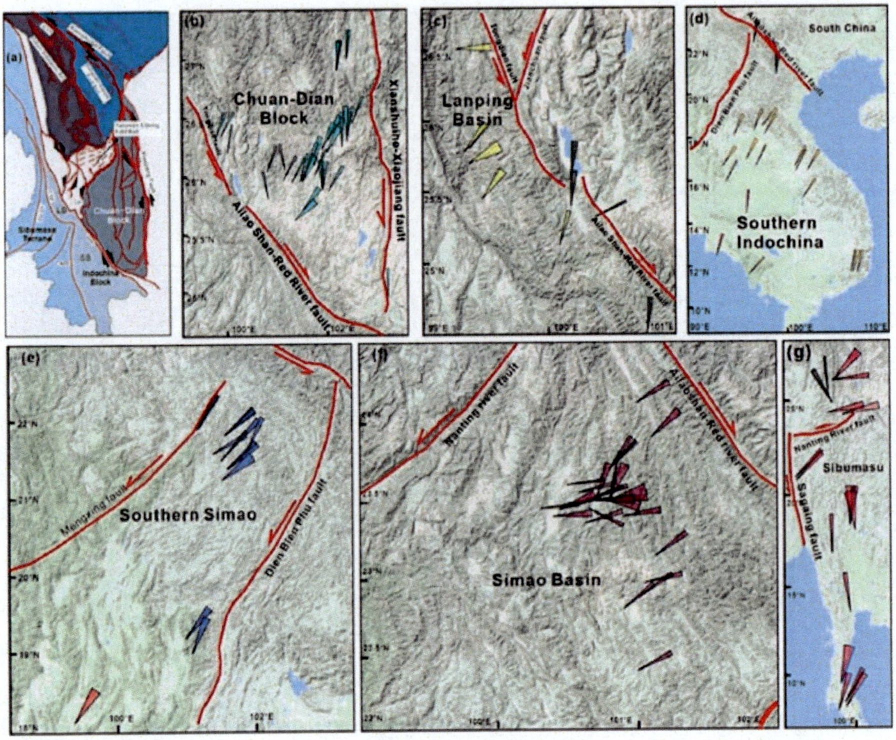

Fig. 3.3 **a** Regional tectonic map showing the major units within the southeastern part of the Tibetan Plateau within China (BGMRY, 1990; Wang et al., 1998). **b** Enlarged map showing the boundaries and variations of paleomagnetic declinations of Chuan-Dian Block, **c** the paleomagnetic declinations of Lanping Basin (LB). **d** The paleomagnetic declinations of Southern Indochina, **e** the paleomagnetic declinations of Southern Simao, **f** the paleomagnetic declinations of Simao Basin (SB), **g** the paleomagnetic declinations of Sibumasu Terrane

(Fig. 3.3a). The South China Block, the Indochina Block and the Sibumasu Terrane are interpreted as derived from Gondwana since the Early Paleozoic, and drifted to the north and eventually accreted to North China and Siberia from Triassic to Late Cretaceous, associated with the successive opening and closure of three major intervening Tethyan oceans, the Paleo-Tethys (Devonian-Triassic), Meso-Tethys (late Early Permian-Late Cretaceous), and Neo-Tethys (Late Triassic-Late Cretaceous) (Metcalfe, 2013). After the collision between the Indian and Eurasian plates in the Early Paleocene, these block in the eastern Tibetan Plateau accumulated differential rotation and latitudinal movement.

The closure of the Neo-Tethys Ocean, the collision between the Indian and Eurasian plates in the Early Paleocene, and the subsequent latitudinal crustal convergence in southern Eurasia, led to significant intracontinental crustal shortening within the Tibetan Plateau, together with crustal rotations of the adjacent areas (Metcalfe, 2013; Tapponnier et al., 2001; Yin & Harrison, 2000). Both geodetic and paleomagnetic data obtained from crustal material in the southeastern margin of the

Tibetan Plateau and its adjacent area indicate the eastern Tibetan Plateau experienced complex crustal movements around the Eastern Himalayan Syntaxis (Li et al., 2017; Maurin et al., 2010; Shen et al., 2005; Zhang et al., 2004) (Fig. 3.3).

3.1.1 *Tectonic Rotation of the Chuan-Dian Block During the Cenozoic*

The clockwise rotation of the Chuan-Dian Block resulted from the large left-lateral displacement of the Xianshuihe-Xiaojiang fault system around the Eastern Himalayan Syntaxis since the middle Miocene. The Paleogene palaeomagnetic data in the central part of the Chuxiong Basin record an observed clockwise rotation of ~ 20° since the Paleogene with respect to the South China Block (Fig. 3.4), and the center of Chuxiong Basin behaved as a relatively rigid block since Miocene (Gao et al., 2017; Otofuji et al., 1998; Tong et al., 2015, 2019, 2020b; Wang & Yang, 2019; Wang et al., 2015, 2016; Yang et al., 2001). The magnetostratigraphic results from a sedimentary sequence of late Miocene-Pleistocene age in the Eryuan area and the Yuanmou area suggest that the Yanyuan-Lijiang fold belt and the central part of the Chuxiong Basin experienced minor rotational movement after the late Miocene (Li et al., 2013; Zhu et al., 2008). Therefore, the Chuxiong Basin and the Yanyuan-Lijiang fold belt accumulated most of the clockwise rotational movement mainly during the Miocene (Fig. 3.5).

The rotational deformation along the eastern and western boundary faults of the Chuan-Dian Block is different from that in the central parts, exhibiting intense and diverse rotations, as inferred by Cretaceous and Paleogene data. In the vicinity of the eastern boundary faults of the Chuan-Dian Block, the magnitude of the rotation gradually increases from the northern part at Xichang to the southern part at Chuxiong. Xichang and Huili are sandwiched in between the Chuxiong Basin and the Sichuan Basin and bounded by left strike-slip faults. These NE-trending boundary faults form a bookshelf structure and caused ~ 10° of counterclockwise rotation to be superposed in the area (Tamai et al., 2004; Wang et al., 2016). In addition, the eastern segment of the Chuxiong section is located at the southern end of the Xiaojiang fault and has accumulated ~ 40° clockwise rotation with respect to East Asia (Funahara et al., 1992). These sections are closer to the eastern boundary fault, and the movement of the shallow crustal material along the Xianshuihe-Xiaojiang fault system is characterized by large and differential rotations.

The magnitude of crustal rotational deformation along the Jinshajiang-Ailaoshan-Red River fault system in the western margin of the Chuan-Dian Block appears to be more complicated than that in the area along the eastern boundary. In the western margin of the Chuan-Dian Block, the Yanyuan-Lijiang fold belt is sandwiched in between the Chenghai-Binchuan fault to the east and the Tongdian fault to the west (BGMRY, 1990; Wang & Burchfiel, 1997). Although the Jianchuan and Yongsheng sections are located in the southwestern segment of the Yanyuan-Lijiang fold belt, the

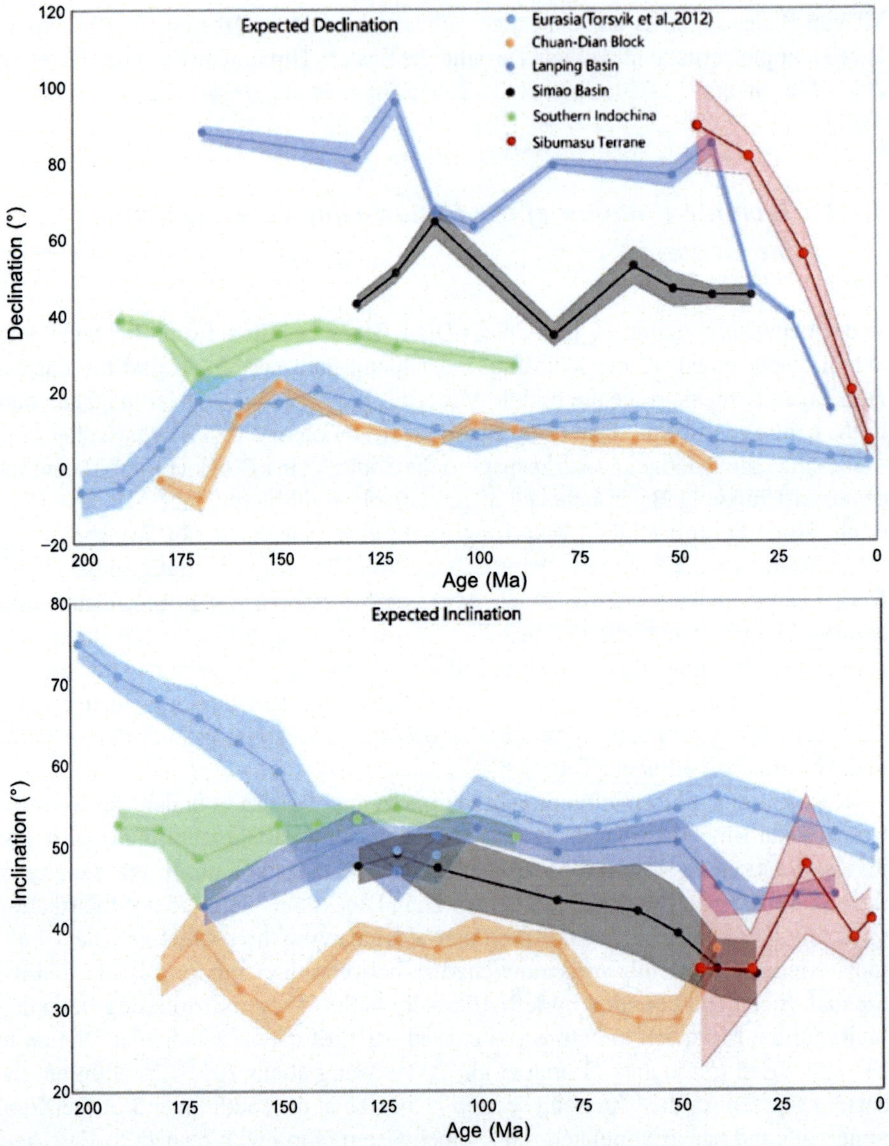

Fig. 3.4 A comparison of the computed Apparent Polar Wander Paths (APWPs) of each sub-terrane using individual direction averages and the APWP of Eurasia (Torsvik et al., 2012). The declinations clearly show differential block rotations for each sub-terrane with respect to Eurasia, while the inclinations are consistently lower than those of Eurasia

rotation of these two segments is consistent with that of the center of the Chuxiong Basin since the Eocene (Gao et al., 2017; Tong et al., 2015).

The paleomagnetic research to the east of the Chenghai-Binchuan fault indicated that the western margin of the Chuan-Dian Block has accumulated large and diverse rotations. In the east of Chenghai fault, Yongsheng, Binchuan, Sanchahe and

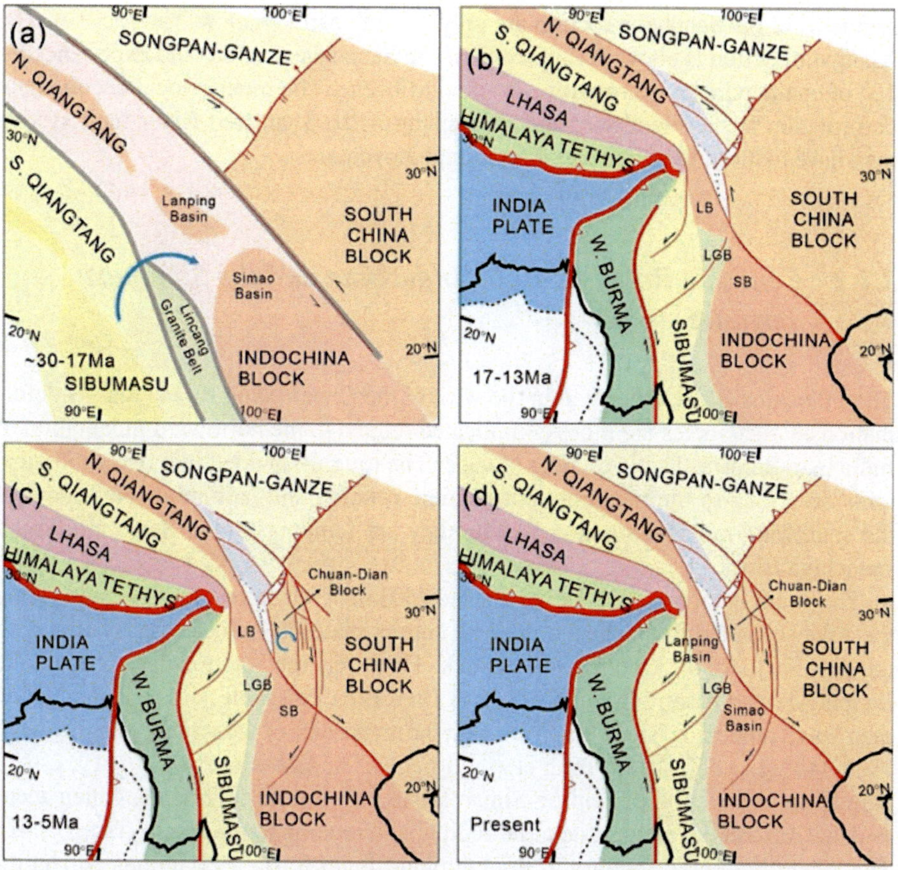

Fig. 3.5 Schematic maps showing the deformation in the Southeastern Tibetan Plateau since early cretaceous. **a** ~30 –17 Ma, the inset indicates the crustal clockwise rotation of the Indochina Block and the major activity phases of block boundary strike-slip faults since the Late Oligocene, **b** 17–13 Ma, the inset indicates the initial rotational time of the Chuan-Dian Block was probably the Miocene together with the action of the Xianshuihe Fault system, **c** 13–5 Ma, the inset indicates the Chuan-Dian Block had experienced clockwise rotation together with the action of the Xiaojiang fault system and reaction of the Red River fault, **d** the Tibetan Plateau and its adjacent area at present and the Chuan-Dian Block experienced minor angle rotation since Late Miocene. LB: Lanping Basin; LGB: Lincang granite belt; SB: Simao Basin. The solid black line marks the sea-land boundary. The black dotted line marks the continental shelf

Xiangyun sections were distributed from north to south (Fig. 3.3b). The magnitude of rotation for the Yongsheng and Binchuan sections is consistent with previous paleomagnetic results from the center sections in the Chuan-Dian Block and accumulated ~ 20° clockwise rotation (Gao et al., 2017; Tong et al., 2015) (Fig. 3.4). The Sanchahe section, about 25 km southeast of Binchuan, experienced a more complicated rotation. With Sanchahe town as the boundary, the northern segment of the Sanchahe section experienced ~ 20° of anticlockwise rotation, whereas the southern segment of the Sanchahe section experienced ~ 20° of clockwise rotation, although the two segments of the Sanchahe section are only 10 km apart (Wang & Yang, 2019). The Xiangyun section is about 30 km south of the Sanchahe section, and experienced ~ 10° of counterclockwise rotation since the Miocene. The intense local decollement deformation in the vicinity of the Jinshajiang-Ailao Shan-Red River fault system may have induced the differential rotational movement.

3.1.2 *Cenozoic Rotation of the Indochina and the Sibumasu Terranes*

The Indochina Block is located in the west of the Chuan-Dian Block, and its deformation characteristics are more complicated related to the northward indentation of India in to Eurasia. The Indochina block can be roughly divided into two geological units according to the magnitude of tectonic rotation and latitude displacement of the southward movement since 50 Ma; they are Lanping-Simao Basin and South Indochina Block.

The South Indochina Block experienced 250 km of left-lateral displacement along the Ailaoshan-Red River shear zone with approximated 15° clockwise rotation with the respect to stable South China Block. However, the rotational pattern of the Lanping-Simao Basin in the north of the Dien Bien Phu fault is more complicated than South Indochina Block, involved with the magnitude of the rotation ranging from 7° to 144° (Li et al., 2017; Tong et al., 2017, 2020a; Yan et al., 2017, 2019, 2020). Numerous faults cut the Lanping-Simao Basin into several fragments, which accumulated differential magnitude of latitudinal and rotational movement (Fig. 3.3c–f). The local differential rotation in the Lanping-Simao Basin accommodated significant latitudinal movement of the shallow crust related to northward indentation of the India into Eurasia. The north end of the Lanping-Simao Basin accumulated 360 km southward offset displacement along the Ailaoshan-Red River fault system more than the south Indochina Block (Tong et al., 2020b; Yan et al., 2019, 2020). The Sibumasu Terrane and Indochina Block are separated by the Changning-Menglian suture zones. Paleomagnetic data reveal ~ 70–80° clockwise rotation of northern Sibumasu with respect to Eurasia (Fig. 3.3g), indicating that the northern Sibumasu region experienced a much larger clockwise rotation than the Chuan-Dian Block and the Indochina Block (Huang et al., 2018; Li et al., 2018; Tong et al., 2016, 2017; Zhao et al., 2015).

The available paleomagnetic data from Chuan-Dian Block, the Indochina Block and the Sibumasu Terrane were reviewed in this section. Based on the paleomagnetic data and geological observations, the terranes movement after the closure of the Neo-Tethys Ocean and the subsequent collision between the Indian and Eurasian plates was reconstruct (Fig. 3.5). The center of the Chuan-Dian Block, behaves as a rigid block, but the margin of the Block experienced more complex rotational movement (Fig. 3.3b). The Indochina Block and the Sibumasu Terrane are closer to the Eastern Himalayan Syntaxis than Chuan-Dian Block so that more faults cut these two blocks into numerous fragments and accumulated differential rotation to accommodate the southward crustal extrusion related to the indentation of India into Eurasia (Fig. 3.5). Differential magnitude of the rotational and latitudinal movement of each segment of the Chuan-Dian Block and peripheral tectonic units, such as the Indochina Block and the Sibumasu Terrane, led to the boundary faults and those small faults in these blocks experienced complex tectonic evolution and accumulated diverse displacement along these faults (Fig. 3.5).

3.1.3 Evolution of Regional Block Motions During the Cenozoic: A Reconstruction with GPlates

To build a regional tectonic evolution model, we first compiled a comprehensive paleomagnetic dataset of 177 paleomagnetic sites based on 17,179 paleomagnetic directions in 81 paleomagnetic studies from Jurassic and younger rocks from South China, Indochina, and the Sibumasu Block. Paleomagnetic data were not included if they: (1) came from rocks older than Triassic or Quaternary and younger; (2) came from studies that do not provide information on demagnetization procedures; (3) were not analyzed by principle component analysis; (4) were (likely) re-magnetized according to the original authors; and (5) if site k- or K-values [precision parameters on directions or poles, respectively, by Fisher (1953)] are lower than 7 for sediment sites and 50 for volcanic sites. Since many of the available paleomagnetic data are derived from Cretaceous rocks deposited during the Cretaceous normal superchron, the presence of reversals is not a general requirement. Re-magnetized directions with precise age constraints on magnetization acquisition (Kornfeld et al., 2014; Li et al., 2017; Tong et al., 2020a; Yan et al., 2019) and anomalous directions due to local tectonics as pointed out by the original authors, are presented but not included in the final discussion.

The compiled paleomagnetic database were then compared to the Apparent Polar Wander Path of Eurasia (APWP, Torsvik et al., 2012) to identify the rotation and latitudinal displacement. We computed the declinations and inclinations at reference points 30° N, 105° E for the South China Block, and 23° N, 100.8° E for Indochina (Fig. 3.6). Because the paleomagnetic declinations in Indochina Block exhibit systematic variation in different parts, we chose the known faults (Dien Bien Phu, Mengxing, and Nanting faults) in the interior of the Indochina Block as boundaries and separated the Indochina Block into four sub-terranes: Lanping Basin (Fig. 3.3c), Southern Indochina (Vietnam in Royden et al.,2008, Fig. 3.3d), Southern Simao (Fig. 3.3e), and Simao Basin (Fig. 3.3f). The averages of each sub-terrane are calculated based on site averages. Age estimates on sampled sedimentary sections are typically within a geologic period with age errors spanning an entire period. For example, Early Cretaceous (145–100 Ma) was assigned at 122.5 Ma with an error of 22.5 Ma. We also calculate independent APWPs for the South China Block, Southern Indochina, Southern Simao and Northern Simao based on available

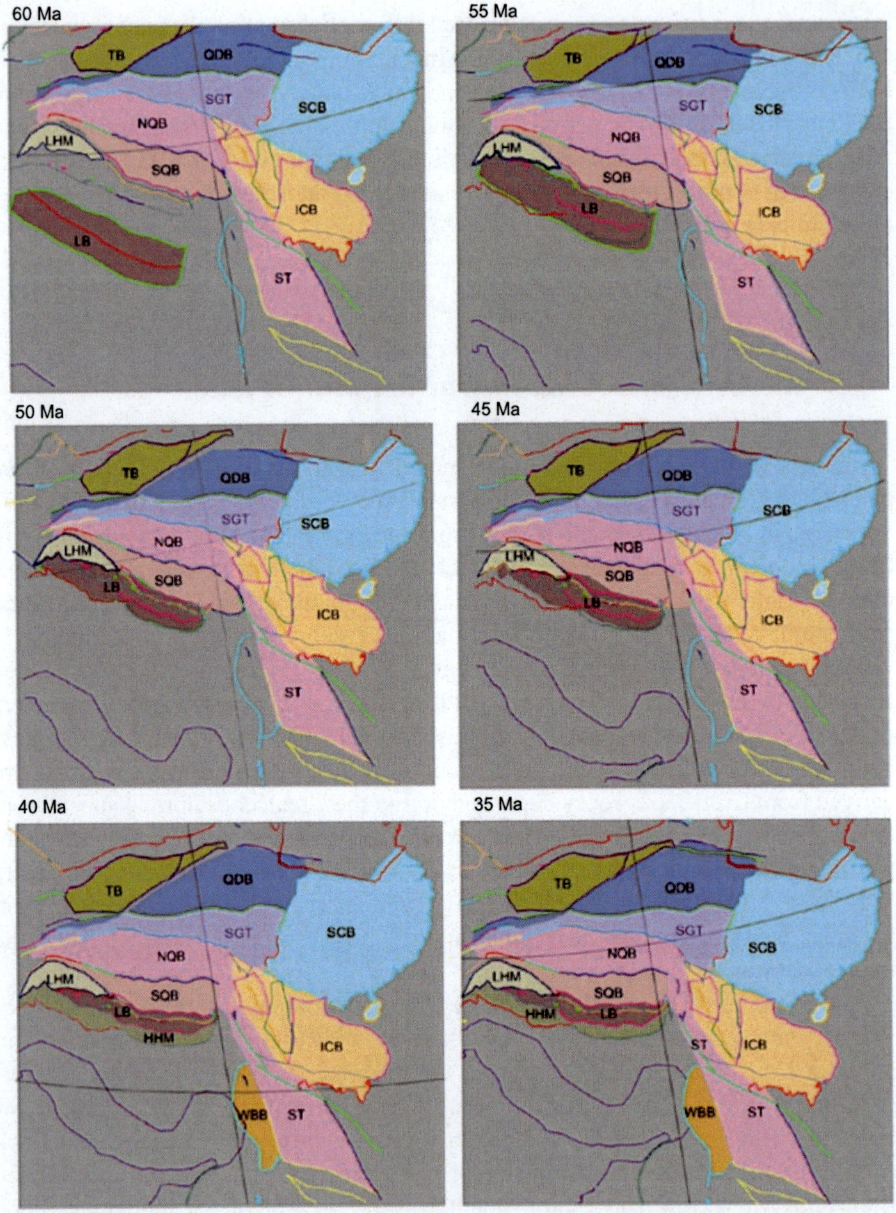

Fig. 3.6 Global plate reconstructions from 60 to 0 Ma in 5 million year time intervals. SCB: the South China Block, ICB: the Indochina Block, TB: the Tarim Basin, QDB: the Qaidam Basin, SGT: the Songpan-Ganze Terrane, NQB: the North Qiangtang Block, SQB: the South Qiangtang Block, ST: Sibumasu Terrane, LB: the Lhasa Block, HHM: the High Himalayan metamorphic rocks, LHM: Lesser Himalayan metasedimentary series, WBB: the West Burma Block. The different colored lines mark the boundaries between the different blocks and the sub-block

data, which we computed in 10 Myr time intervals from 180 to 40 Ma, using a 20 Myr sliding window which is the same approach as used in the reference APWP (Torsvik et al., 2012).

We designed an additional tool on the recent public online platform for paleomagnetic analysis Paleomagnetism.org (Koymans et al., 2016) that allows testing of kinematic restorations against paleomagnetic data. This tool allows to rotate the APWP whereby we use the version of Torsvik et al. (2012) into the coordinates of a reconstructed block if the Euler poles of this block are provided in 10 Myr intervals relative to South Africa. We have used this tool to test existing reconstructions of Indochina extrusion, and to develop an updated kinematic restoration of rotation and extrusion of Indochina using GPlates, a free plate reconstruction software (Boyden et al., 2011), (Fig. 3.6).

3.2 Active Block Models of the SE Tibetan Plateau

3.2.1 Characteristics of Major Block Boundary Faults of the Sichuan-Yunnan Block (also called Chuan-Dian Block)

As indicated by tectonic reconstruction scenarios in Fig. 3.6, the southeastern Tibetan Plateau experienced prolonged multiple-phase tectonic deformation over geological time scale. During the process, a myraid of active faults are formed, either by reactivating geologic inherited weakness or newly-forming structures in response to changed tectonic stress orientation (e.g., Replumaz & Tapponnier, 2003) (Fig. 3.7). The active fault network in the region consists mainly of conjugate NW- and NE-striking strike-slip faults mixed with normal or reverse component, depending on fault orientation and its angle with respect to the tectonic stress. In the northern part of the region, NW-trending faults are left-lateral strike-slip faults and NE-trending ones are right-lateral, whereas in the southern part, kinematics of the fault system is reversal, with NW-trending right-lateral and NE-trending left-lateral faults. These major faults together accommodate the southeastward extrusion of blocks or sub-blocks of varying size while the secondary faults with normal or reverse sense accommodate differential rotation between blocks or sub-blocks.

Under the frame of active blocks for China's mainland, the southeastern Tibetan Plateau consists of multiple level-II tectonic blocks. Among them, the Sichuan-Yunnan block is one of the most seismically active regions in China, and lies at the center of the targeted area of China Seismic Experimental Site (CSES). It is a rhomb-shaped block bounded by the Xianshuihe-Xiaojiang fault system in the east, the Red River fault in the southwest, and the Jinsha Jiang fault in the west (Figs. 3.8 and 3.9).

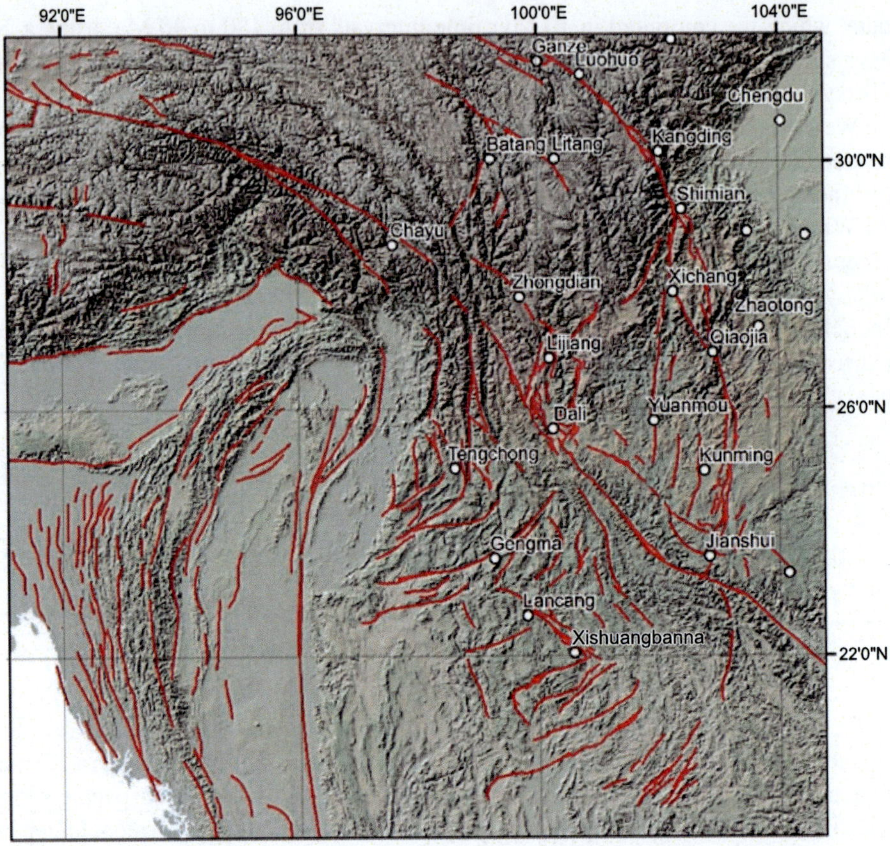

Fig. 3.7 Revised distribution of major active faults in southeast Tibetan Plateau

Yushu-Ganze-Xianshuihe-Xiaojiang Fault System

The 1600-km-long Yushu-Ganze-Xianshuihe-Xiaojiang fault system consists of the Xianshuihe, Anninghe, Zemuhe, and Xiaojiang faults, from northwest to southeast respectively (Fig. 3.9). Various block models, despite being different on block and geomtry, agree upon unaminously its role as a major block-bounding structure. It is a left-lateral strike-slip fault zone with a high slip rate up to 10–15 mm/yr, exhibiting numeous prominent offset landforms, hosting 17 historical earthquakes with magnitude of 7 and above. The northwest-trending Ganze-Yushu fault section is separated from the Xianshuihe section by the Ganze pull-apart basin. The left-lateral slip rate decrease slightly from ~ 13 to 10 mm/yr (Bai et al., 2018). The Xianshuihe fault consists of two sub-sections separated by the Kangding transpressional step-over, which also marks a change in fault strike from the NW-trending to NS-trending. The earthquake recurrence interval is ~ 200 years on average (Papadimitriou et al., 2004; Wen et al., 2008a, 2008b). The Anninghe fault section also consists of two sub-sections, marked by a slight change in fault strike near Mianning. The section to the north is estimated to slip at a rate of ~ 5 mm/yr (Wen, 2000; Wen et al.,

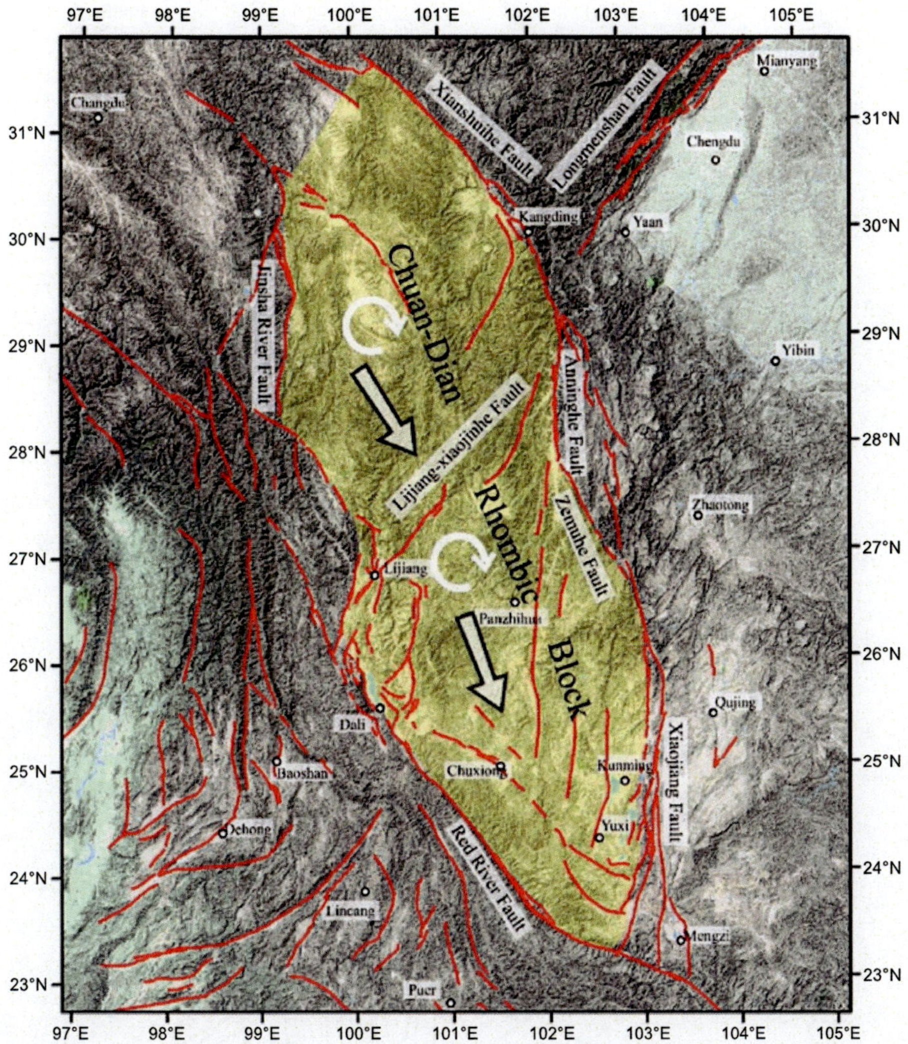

Fig. 3.8 The geometry and boundary faults of the Chuan-Dian block

2008a, 2008b); the last major earthquake occurred in A.D. 1485, with a magnitude of M ~ 7.5 (Wen, 2000; Wen et al., 2008a, 2008b). The section to the south, with the horizontal slip rate estimated to be ~ 6.5 mm/yr (Xu et al., 2003), experienced multiple historical earthquakes, including earthquakes M ~ 6¾ in A.D. 1489, M ~ 7.5 in A.D. 1536, and M ~ 6¾ in A.D. 1952. The Zemuhe fault is separated from the Anninghe fault by a sharp change in strike near Xicang, having a left-lateral slip rate of ~ 6.5 mm/yr and similar to the latter. Historical earthauakes occurred along the Zemuhe fault include the M ~ 6¾ earthquake in A.D. 1732, and the M ~ 7.5 earthquake in A.D. 1850.

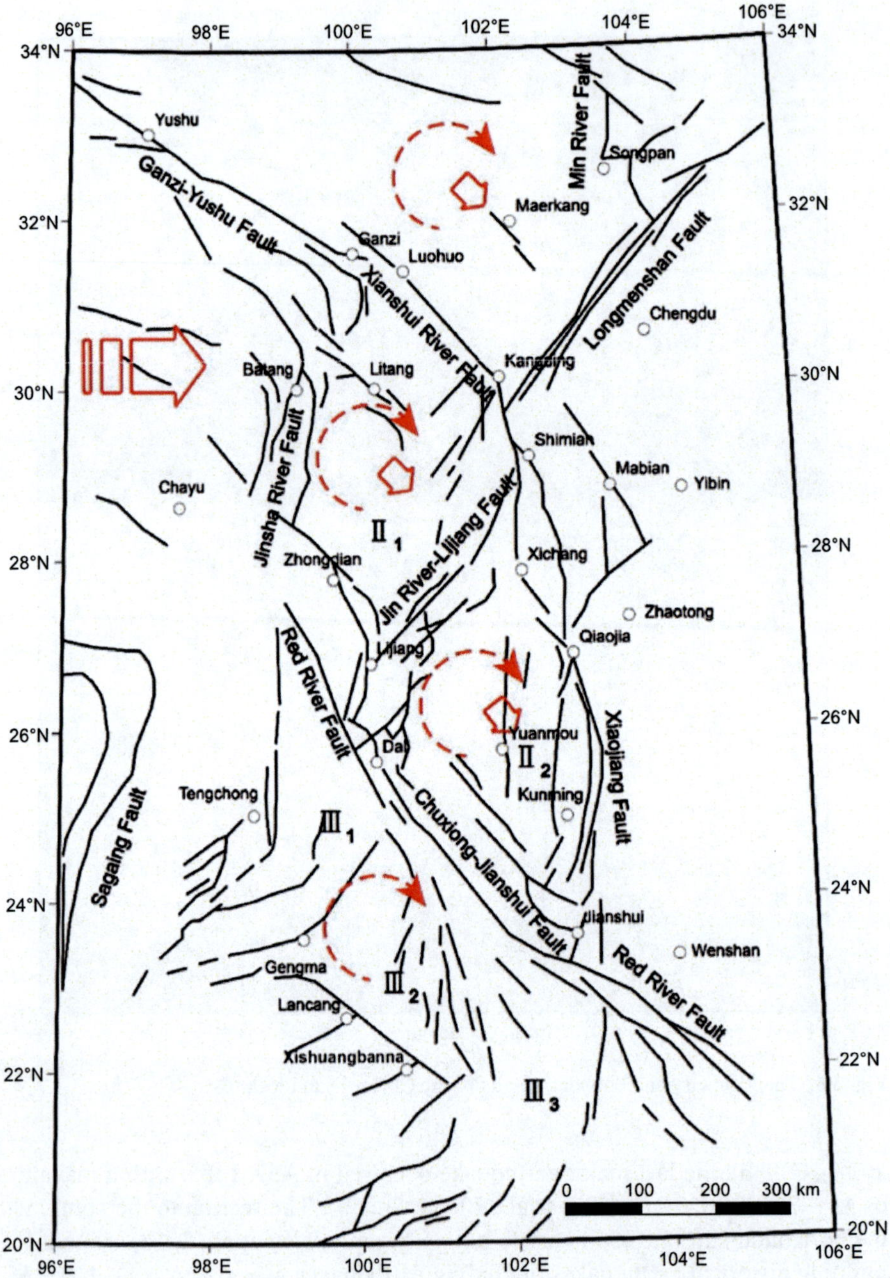

Fig. 3.9 The hierarchy and components of active blocks for the southeastern Tibetan Plateau, showing second-order differentiation of sub-blocks, their rotation, and the corresponding block-bounding faults

The 400 km-long NS-trending Xiaojiang fault is the southeast section of the Xianshuihe-Xiaojiang fault system (Figs. 3.8 and 3.9). Based on the characteristics of internal structure and geometry, the Xiaojiang fault zone is divided into the northern, middle, and southern sections. The northern section is a single fault with relatively simple geometry. The middle and southern sections consist of the eastern and the western branches, 12–16 km apart from each other. In the south, the branches crosscut or sub-parallel to each other, entwined in the plan-view. According to the GPS observations, the present-day slip rate of the Xiaojiang fault is ~ 10 mm/yr, with the slip rate gradually decreasing from north to south (Deng et al., 2002; King et al., 1997; Tang et al., 2006; Wang et al., 2001; Wen et al., 2011; Zhang et al., 2002, 2003). The average geologic slip rate on the Xiaojiang fault is 6–16.5 mm/yr during the Quaternary, of which the western branch accomodates 7.5–8.6 mm/yr and the eastern branch 4.5–5.1 mm/yr (Chen et al., 1988; He et al., 1993, 2002; Song et al., 1998). Since 1500, more than 20 earthquakes with magnitude of 6 and above occurred along the Xiaojiang fault zone (Ren, 2013; Wen et al., 2011), including Yiliang M ~ 7 earthquake in 1500s, Dongchuan M ~ 7.6 earthquake in 1733, the Huaning M ~ 7 earthquake in 1789, and the 1833 M ~ 8 Songming earthquake.

The Red-River Fault

The Red River fault divides the Indo-China (northernmost Sunda) block to the southwest and the South China block to the northeast, and has played a major role in the tectonic evolution of southeastern Tibetan Plateau and northern Southeast Asia (Allen et al., 1984; Lacassin et al., 1997; Tapponnier et al., 1982, 2001, 1990b; Tapponnier & Molnar, 1977) (Figs. 3.1, 3.5, 3.8). The fault can be considered being composed by three segments, a northern transtensional portion, and central and southern strike-slip segments, based on fault geometry and seismicity (Zhang, 2009).

The amount of sinistral slip on the RRF is widely accepted to be 500–1000 km (Leloup et al., 1995, 2007; Tapponnier et al., 1990a), although someone argue for the much less slip amount (e.g., Searle, 2006). A reversal from sinistral to dextral slip (Allen et al., 1984; Leloup et al., 1993) occurred along this fault between 13 and 5 Ma (Bergman et al., 1997; Cao et al., 2011; Leloup et al., 1993, 2001a, 2001b). This has been interpreted to accommodate a change in roles as the collision evolved (Lacassin et al., 1997; Tapponnier et al., 1982, 2001). The fault's subsequent dextral offsets of streams and Neogene basins range from meters to tens of kilometers (Allen et al., 1984; Chen, 2013; Guo et al., 2001; Replumaz et al., 2001; Schoenbohm et al., 2006; Trinh et al., 2012; Zuchiewicz et al., 2013), with maximum estimates of 50–70 km (Wang et al., 1998; Xiang et al., 2007).

Slip rates on the RRF's fault have been debated. The long term (millions of years) geological slip rate is estimated by dividing the largest dextral offsets based on inferred times of onset between 2 and 10 Ma, which yields 2–5 mm/yr along the central RRF zone (Allen et al., 1984; Replumaz et al., 2001; Schoenbohm et al., 2006) and 5–8 mm/yr along its southern segment in Vietnam (Zuchiewicz et al., 2013). Decadal strike-slip rates calculated from GPS profiles across the fault zone range between 0.2 and 2 mm/yr for all three segments (Cong & Feigl, 1999; Feigl et al., 2003; Hao et al., 2014; Shen et al., 2005; Wang et al., 2008; Zhao et al., 2012;

Zheng et al., 2017; Zuchiewicz & Cuong, 2009). Existing millennial strike-slip-rate estimates range from 1 to 6 mm/yr along the entire RRF zone in Vietnam (Chen, 2013; Trinh et al., 2012; Zuchiewicz et al., 2013), and 0.2–4 mm/year along the northern and central segments (Chen, 2013; Guo et al., 2001; Shi et al., 2018a, 2018b).

Since A.D. 886, there have been nine damaging earthquakes to the northwest, in the Nanjian-Dali-Eryuan region, but none along the lengthy, predominantly strike-slip segment from south of Nanjian to Hanoi (Hu et al., 2006; Zhang, 2009).

The Jinsha Jiang Fault

The 700 km-long Jinsha Jiang fault is considered to be the western boundary of the Sichuan-Yunnan block (Figs. 3.8 and 3.9). This fault has a long and diachronous activity, experiencing multiple phases of re-activation, first as a suture with prominent geologic indicators of compression and shortening structures (Xu et al., 1992). Overall, it strikes in NS-direction, but in detail the section north of Batang strikes NNW and the section to the south strikes NNE. In satellite images, the fault is marked by displaced geomorphic features such as linear though discontinuous troughs in hillslopes, offset ridge lines, and triangular facets, etc.

The kinematics of the Jinsha Jiang fault is disputed. Being a structure accommodating the southeastward extrusion of the southeastern Tibetan Plateau, the fault has experienced right-lateral strike slip motion since at least the Pliocene (Tang et al., 1993; Chang et al., 2014). Zhou et al. (2005) studied that the fault accommodated mostly EW shortening during the late Quaternary, and that the fault should not be considered as a major boundary for lateral extrusion. Someone considered the fault being inactive; the right-lateral shearing was intead taken up by a series of discontinous and distributed low-slip rate faults (Wang et al., 2003, 2008). Li et al. (2014) studied the kinematics of the fault by analyzing regional GPS data and concluded that the fault is right-lateral with a minor component of shortening, which is consistent with some geological studies of the fault (Xu et al., 2003; Tang et al., 1993; Chang et al., 2014). In an unpublished report, Chang (2014) gave an estimate of right-lateral slip rate of 3.3–4.1 mm/yr, and dip slip rate of 0.9–1.1 mm/a, which needs to be confirmed.

GPS velocity field show a prominent rotation of the velocity vectors from E, SE to SSE direction across the Jinsha Jiang fault (Fig. 3.10). To the west of the Jinsha Jiang fault, GPS velocity vectors are toward the east, which turns to SE in the vicinity of the fault, and then to SSE to the east of the fault. This pattern indicates a clockwise rotation toward SSE of the crust near the Jinsha Jiang fault. The rotation is more prominent in the southern section of the Jinsha Jiang fault, indicating that the horizontal motion in this area is strongly affected by the eastward extrusion of the Tibetan Plateau (Xu et al., 2020). The GPS horizontal velocity to the east of the Jinsha Jiang fault shows no obvious difference from that to the west, indicating that the fault does not absorb horizontal shortening of the crust, instead causes a change in the direction of crustal motion (Xu et al., 2020). GPS velocity vectors projected on profiles across the Jinshajiang fault indicate negligible horizontal slip rate in the northern section, but up to ~ 4.9 mm/yr in a right-lateral strike-slip fashion in the southern section. Further to the south, at the position of the Zhongdian fault, which

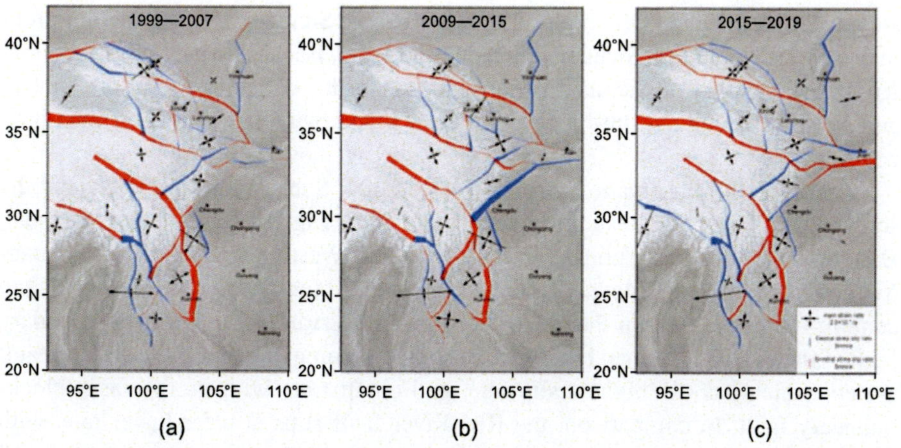

Fig. 3.10 Assuming a uniform elastic strain block model, the main strain rate and Euler rotational angular rate of the block boundaries are calculated for different periods, during **a** 1999–2007, **b** 2009–2015 and **c** 2015–2019

lies at the transition from Red River fault and Jinshajiang fault, GPS profile indicates a similar right-lateral rate of 5.5 mm/yr.

Strong historical earthquakes on the Jinshajiang fault include events with *M* ~ 6 in 1722, *M* ~ 7¼ in 1870, *M* ~ 6½ in 1923, and *M* ~ 6.7 earthquake swarms in 1989, respectively. However, the southern section south of Batang, for a 200 km reach, due to no strong earthquakes occurred on it, during the last 100 years, thus was considered as a "seismic gap" (*M*7 working group, 2012). Figure 3.10 shows also that the GPS velocity field in the eastern Tibetan Plateau is considerably sensitive to the coseismic and postseismic deformation of large earthquakes in the region, inferring the kinemaics of active blocks and their boundary faults. Therefore, the effect of large historical earthquake should be corrected or considered in the GPS analysis.

3.2.2 Various Block Models Based on Geologic Versus Geodetic Constraints

Replumaz and Tapponnier (2003) synthesized the data of the tectonic deformation of the Tibetan Plateau since the Indo-Asian collision, and reconstructed the relative positions and euler poles of various blocks through geological times. The reconstruction of the activity history of the Cenozoic microplates in the mainland of China is mainly based on the kinematics and slip rate of major faults. This kind of reconstruction target on the first-order scale micro-plate is like the block models with simplified geometry. For instance, the Qiangtang block is the largest block in size, and bounded by the Karakoram-Jiali fault system in the south, the Red River fault in the southeast, the Xianshuihe fault in the north and the Xiaojiang fault zone in the northeast. As a step toward refinement, Xu et al. (2003) divided the Sichuan-Yunnan area into four major blocks, including Sichuan-Yunnan, Maerkang, Baoshan-Puer and Myitkyina-Ximeng (Fig. 3.9). The Sichuan-Yunnan block is further divided into Middle Yunnan and Northwest Sichuan sub-blocks, being separated by the Lijiang-XiaoJinhe fault zone.

Based on GPS observations from 1998 to 2004, Shen et al. (2005) divided the southeastern Tibetan Plateau into five tectonic units (Fig. 3.11a): Northern Sichuan, Eastern Tibet, Sichuan-Yunnan, South China and Western Yunnan. The Sichuan-Yunnan block can be further divided into three sub-blocks: Yajiang, Shangrila and central Yunnan. Based on the current geodetic observations, the eastern boundary fault of the Sichuan-Yunnan block is Xianshuihe-Anninghe-Zemuhe-Xiaojiang fault system which has the highest slip rate, and compatible with its role as a block-boundary fault. In comparison, the Red River fault slips at a far lower rate, with both dextral and extensional rates of 2 ± 2 mm/yr in the northwest segment, and

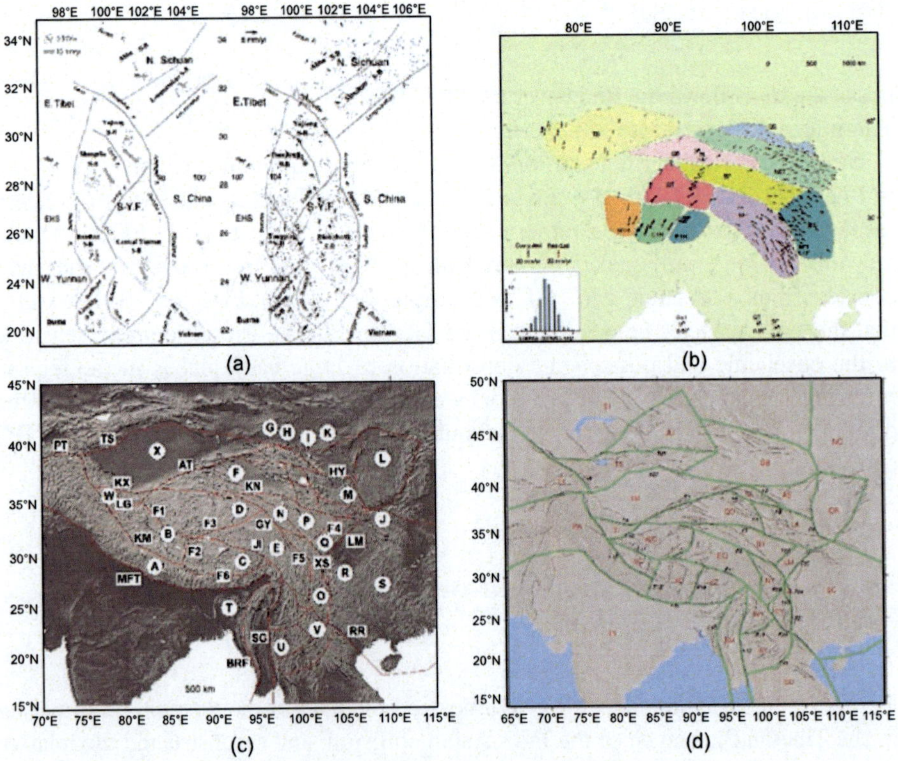

Fig. 3.11 Various versions of block models of the Tibetan Plateau, based on modern-day geodetic constraints. **a** Delineation of deformation blocks and their motions with respect to the Eurasia plate, and GPS velocity postfit residuals of the block motion model (Shen et al., 2005), **b** observed velocity field (black arrows) and block model of Tibet (Thatcher, 2007). **c** Active faults (black lines; Taylor & Yin, 2009) and reference model block geometry (red lines, with blocks identified by circled letters) (Loveless & Meade, 2011). **d** Active faults and block boundaries (green lines) for the Tibetan Plateau and surrounding areas (Wang et al., 2017)

dextral rate of 1 ± 2 mm/yr in the middle segment. Thus, the GPS results suggest that at present the Red River fault does not behave like a dominant intracontinental transform fault in the region as it used to be.

Loveless and Meade (2011) later combined geodetic observations and geologic fault slip rates, and proposed a more refined block model (Fig. 3.11c) to simultaneously estimate micro-plate rotation rates, interseismic elastic strain accumulation, fault slip rates and strain rates. This block model consists of 24 tectonic micro-plates which is built on the basis of Meade's model (2007), refines and improves the definition of block boundary and the location of boundary faults. For instance, the Red River fault zone is included in the block boundary, and a more detailed block division is given for the southeast margin of the Plateau.

Using 1854 GPS velocity vector data and considering the effects of block rotation and elastic strain, Wang et al. (2017) constructed a model with 30 blocks to describe the current deformation in western China (Fig. 3.11d). Based on the new velocity field, the block modelling approach is employed to obtain the regional kinematic characteristics and the slip rate of faults. The Sichuan-Yunnan region is divided into East Qiangtang (EQ), North Yunnan (NY), East Yunnan (EY), West Yunnan (WY), South Yunnan (SY) and Daliangshan (DL) blocks. In this model, blocks having significant residuals (less compatible) are located in the southern and southeastern Tibetan Plateau, suggesting complex tectonic settings in these regions (Wang et al., 2017), and the block model cannot fully explain the current crustal deformation in these regions.

3.2.3 Reappraisal of Block Models for the Sichuan-Yunnan Region

Block model can explain most of the geologic, geodetic, seismic activity observation data, suggesting that the forces acting on the microplate drive the activity of the fault system in the India-Eurasia collision region on a timescale of several decades to Quaternary, manifested by the local deformation characteristics near the boundaries of the blocks (Loveless & Meade, 2011; Meade, 2007; Replumaz & Tapponnier, 2003; Wang et al., 2017; Zhang et al., 2005).

However, block model solutions, or the division of active blocks and sub-blocks, are usually non-unique. Different studies proposed different sets of blocks and/or with varying block boundaries. Block models based on geologic data can be very different from those based on geodetic data as shown in Figs. 3.9 and 3.10 (Loveless & Meade, 2011; Meade, 2007; Replumaz & Tapponnier, 2003; Shen et al., 2005; Wang et al., 2017; Xu et al., 2003; Zhang et al., 2003, 2005). But none of the published block models can explain all observations so that it is difficult to evaluate which one is better than the others.

The biggest problem for the Sichuan-Yunnan block is perhaps the kinematic compatibility of all boundary faults. For the eastern boundary, the slip rate of

Xianshuihe-Anninghe-Zemuhe-Xiaojiang fault system reached about 10 mm/yr or even higher, consistently among different avenues of estimation, either by geodetic results (Jiang et al., 2015; King et al., 1997; Shen et al., 2005; Wang et al., 2009), geological observations (Allen et al., 1991; Bai et al., 2018; Chen et al., 2016; Papadimitriou et al., 2004; Zhang et al., 2016), or the block model inversion (Loveless & Meade, 2011; Meade, 2007; Wang et al., 2017). On the other hand, for the western boundary, the slip rate of Red River fault is much lower. GPS observation shows that the dextral slip rate of the Red River fault is about 2 mm/yr (Shen et al., 2005) (Fig. 3.10). Block model inversion indicates a rate ranging from 2.6 ± 0.3 to 4.0 ± 0.5 mm/yr, and the dextral slip rate of the southern section is only 0.8 ± 0.3 mm/yr (Wang et al., 2017). Millennial scale slip rate is estimated to be 1.1 ± 0.4 mm/yr for the middle section (Shi et al., 2018a). In addition, the northwest boundary of the Chuan-Dian block is still disputed. Some studies suggest that the Jinsha Jiang fault zone is the northwest boundary of the Chuan-Dian block (Xu et al., 2003; Loveless & Mead, 2010; Wang et al., 2017); others considered that the Jinsha Jiang fault zone is not active enough as a major block boundary and that the Chuan-Dian block is part of the bigger Qiangtang block to the northwest (Replumaz & Tapponnier, 2003; Meade, 2007; Thatcher, 2007). Thus, the Chuan-Dian block may not move as a porminent and separate block.

In summary, existing GPS observation data and the regional kinematics characteristics obtained by block models are usually associated with large residual errors in the southeastern Tibetan Plateau, suggesting a complex deformation pattern not easily being described by simple mechanism. Indeed, the seismotectonic boundary setting of the Sichuan-Yunnan region is of a high degree of complexity, being influenced by the eastward subduction of Myanmar arc in the west, northward subduction of the Indian plate and clockwise rotation of east and southeast Tibetan Plateau, gravitational effect of over-thickened crust of the central Tibetan Plateau, and slab rollback of Sumatra. This tectonic boundary condition is thus far more complex than California and the San Andreas fault system, where SCEC earthquake-related research is based. The combined effect of various dynamic forcing at multiple boundaries results in a high spatial and lateral change in block motions and a rotation in the major compressive stress field (Fig. 3.10). The changing plate configuration over geological time scale also suggests that a temporally evolving role of individual faults playing in accomodating regional deformation is both magnitude and kinematic polarity. Teconic inversion should be taken into account in block model schemes; block configuration based on geological observation is not necessarily be consistent with that based on modern-day geodetic data. Nonetheless, to fully understand the temporal evolution, we need more data from both geological and geodetic studies. After all, the Sichuan-Yunnan region, the test ground of the Chinese earthquake prediction experiment, is a vast region that hosts several dozens of faults, many of which are poorly understood. Previous efforts have been mainly poured on to the Xianshuihe-Xiaojiang fault system—itself with a length almost comparable to the San Andreas fault. Coordinated investigation is much needed for many under-studied faults in the region.

Acknowledgements This study received financial support from the China Earthquake Science Experiment (2017CESE0102), National Natural Science Foundation (41761144065, 41702223, U1839203, 42030305), National Key Research and Development Program of China (2018YFC1504104). We are grateful to Prof. Yonggang Li for his constructive suggestions and helps that significantly improved this manuscript. The data of block reconstruction can be downloaded at the CSES webpage (http://www.cses.ac.cn/).

References

Allen, C. R., Gilllespie, A., Yuan, H., et al. (1984). Red River and associated faults, Yunnan Province, China: Quaternary geology, slip rates, and seismic hazard. *Geological Society of America Bulletin, 95*(6), 686–700.

Allen, C. R., Luo, Z. L., Qian, H., et al. (1991). Field study of a highly active fault zone: The Xianshuihe fault of Southwestern China. *Geological Society of America Bulletin, 103*, 1178–1199. https://doi.org/10.1130/0016-7606

Bai, M. K., Chevalier, M. L., Pan, J. W., et al. (2018). Southeastward increase of the late Quaternary slip-rate of the Xianshuihe fault, eastern Tibet. Geodynamic and seismic hazard implications. *Earth and Planetary Science Letters, 485*, 19–31.

Bergman, S. C., Leloup, P. H., Tapponnier, P., Schärer, U., & O'Sullivan, P. (1997, Mar 23–27). Apatite fission track thermal history of the Ailao Shan–Red River shear zone. Paper presented at the European Union of Geoscience meeting, Strasbourg, France.

Bertrand, G., & Rangin, C. (2003). Tectonics of the western margin of the Shan plateau (central Myanmar): Implication for the India-Indochina oblique convergence since the Oligocene. *Journal of Asian Earth Sciences, 21*(10), 1139–1157. https://doi.org/10.1016/S1367-9120(02)00183-9

Boyden, J. A., Müller, R. D., Gurnis, M., Torsvik, T. H., Clark, J. A., Turner, M., Ivey-Law, H., Watson, R. J., & Cannon, J. S. (2011). Next-generation plate-tectonic reconstructions using GPlates. In Geoinformatics Cyberinfrastructure for the Solid Earth Sciences (pp. 95–113).

Burchfiel, B. C., & Chen, Z. (Eds.). (2012). Tectonics of the southeastern Tibetan Plateau and its adjacent foreland. *Geological Society of America, 210.*

Bureau of Geology Mineral Resources of Yunnan Province (BGMRY). (1990). *Regional geology of Yunnan province.* Geological Publishing House.

Cao, S., Neubauer, F., Liu, J., Genser, J., & Leiss, B. (2011). Exhumation of the Diancang Shan metamorphic complex along the Ailao Shan-Red River belt, southwestern Yunnan, China: Evidence from 40Ar/39Ar thermochronology. *Journal of Asian Earth Sciences, 42*(3), 525–550. https://doi.org/10.1016/j.jseaes.2011.04.017

Chang, Z., Zhang, Y., Li, J., et al. (2014). The geological and geomorphic characteristic of late Quaternary activity of the Deqin-Zhongdian-Daju fault. *Journal of Seismological Research, 37*(1), 46–52. (in Chinese).

Chen, J. (2013). Response of channel offsets to dextral-slip movement of the Red River fault zone [M.S. Thesis]. (M.S.), Zhejiang University, Hangzhou.

Chen, G. H., Xu, X. W., Wen, X. Z., et al. (2016). Late Quaternary slip-rates and slip partitioning on the Southeastern Xianshuihe fault system. *Eastern Tibetan Plateau. Acta Geologica Sinica, 90*(2), 537–554.

Chen, R., & Li, P. (1988). Slip rates and earthquake recurrence intervals of the western branch of the Xiaojiang fault zone. *Seismology and Geology, 10*(2), 1–13. (in Chinese).

Cong, D. C., & Feigl, K. L. (1999). Geodetic measurement of horizontal strain across the Red River fault near Thac Ba, Vietnam, 1963–1994. *Journal of Geodesy, 73*(6), 298–310.

Deng, Q. D., Zhang, P. Z., Ran, Y. K., et al. (2002). Basic characteristics of active tectonics of China. *Science in China. Science in China (series d), 32*(12), 1020–1031. (in Chinese).

Feigl, K. L., Cong, D. C., Becker, M., To, T. D., Neumann, K., & Xuyen, N. Q. (2003). Insignificant horizontal strain across the Red River fault near Thac Ba, Vietnam from GPS measurements 1994–2000. Paper presented at the EGS-AGU-EUG Joint Assembly, Nice, France.

Fisher, R. A. (1953). Dispersion on a sphere. *Proceedings of the Royal Society of London. Series A. Mathematical and Physical Sciences, 217*(1130), 295-305.

Funahara, S., Nishiwaki, N., Miki, M., Murata, F., Otofuji, Y.-I., & Wang, Y. Z. (1992). Paleomagnetic study of cretaceous rocks from the Yangtze block, central Yunnan, China: Implications for the India-Asia collision. *Earth and Planetary Science Letters, 113*(1–2), 77–91. https://doi.org/10.1016/0012-821X(92)90212-E

Gao, L., Yang, Z. Y., Tong, Y. B., Wang, H., An, C. Z., & Zhang, H. F. (2017). Cenozoic clockwise rotation of the Chuan Dian fragment, southeastern edge of the Tibetan Plateau: Evidence from a new paleomagnetic study. *Journal of Geodynamics, 112*, 46–57. https://doi.org/10.1016/j.jog.2017.10.001

Guo, S., Ji, F., Xiang, H., Dong, X., Yan, F., Zhang, S., Li, X., & Zhang, W. (2001). The Honghe active fault zone. Ocean Press. (in Chinese with English abstract).

Hao, M., Wang, Q., Shen, Z., Cui, D., Ji, L., Li, Y., & Qin, S. (2014). Present day crustal vertical movement inferred from precise leveling data in eastern margin of Tibetan Plateau. *Tectonophysics, 632*, 281–292. https://doi.org/10.1016/j.tecto.2014.06.016

He, H. L., Fang, Z. J., & Li, P. (1993). A preliminary approach to the fault activity of southern segment on Xiaojiang west branch fault. *Journal of Seismological Research, 6*(3), 291–298. (in Chinese).

He, H. L., Ikeda, Y., Song, F. M., et al. (2002). Late Quaternary slip rate of the Xiaojiang fault and its implication. *Seismology and Geology, 24*(1), 14–26. (in Chinese).

Hu, X., Zhang, J., Li, Z., & Lu, J. (2006). Relocation results of moderate-small earthquakes in the Red River fault zone of Yunnan and its relation with geologic structure. *Journal of Seismological Research, 29*(4), 349–354. (in Chinese with English abstract).

Huang, B., Yan, Y., Piper, J. D., Zhang, D., Yi, Z., Yu, S., & Zhou, T. (2018). Paleomagnetic constraints on the paleogeography of the East Asian blocks during Late Paleozoic and Early Mesozoic times. *Earth-Science Reviews, 186*, 8–36. https://doi.org/10.1016/j.earscirev.2018.02.004

Jiang, G., Xu, X., Chen, G., et al. (2015). Geodetic imaging of potential seismogenic asperities on the Xianshuihe-Anninghe-Zemuhe fault system, Southwest China, with a new 3-D viscoelastic interseismic coupling model. *Journal of Geophysical Research: Solid Earth, 120*(3), 1855–1873.

King, R. W., Shen, F., Burchfiel, B. C., et al. (1997). Geodetic measurement of crustal motion in Southwest China. *Geology, 25*, 179–182.

Kornfeld, D., Eckert, S., Appel, E., Ratschbacher, L., Pfänder, J., Liu, D., & Ding, L. (2014). Clockwise rotation of the Baoshan Block due to southeastward tectonic escape of Tibetan crust since the Oligocene. *Geophysical Journal International, 197*(1), 149–163. https://doi.org/10.1093/gji/ggu009

Koymans, M. R., Langereis, C. G., Pastor-Galán, D., & van Hinsbergen, D. J. (2016). Paleomagnetism.org: An online multi-platform open source environment for paleomagnetic data analysis. *Computers & Geosciences, 93*, 127–137.

Lacassin, R., Maluski, H., Leloup, P. H., Tapponnier, P., Hinthong, C., Siribhakdi, K., & Charoenravat, A. (1997). Tertiary diachronic extrusion and deformation of western Indochina: Structural and 40Ar/39Ar evidence from NW Thailand. *Journal of Geophysical Research: Solid Earth, 102*(B5), 10013–10037. https://doi.org/10.1029/96JB03831

Lacassin, R., Replumaz, A., & Hervé Leloup, P. (1998). Hairpin river loops and slip-sense inversion on southeast Asian strike-slip faults. *Geology, 26*(8), 703–706. https://doi.org/10.1130/0091-7613(1998)026%3c0703:HRLASS%3e2.3.CO;2

Leloup, P. H., Lacassin, R., Tapponnier, P., & Harrison, T. M. (2001a). Comment on "Onset timing of left-lateral movement along the Ailao Shan-Red river shear zone: Ar-40/Ar-39 dating constraint from the Nam Dinh area, northeastern Vietnam"; Wang et al., (2000). *Journal of Asian Earth Sciences 18*, 281–292; Discussion. *Journal of Asian Earth Sciences, 20*(1), 95–99.

Leloup, P. H., Arnaud, N., Lacassin, R., Kienast, J. R., Harrison, T. M., Trong, T. T. P., Replumaz, A., & Tapponnier, P. (2001b). New constraints on the structure, thermochronology, and timing of the Ailao Shan-Red River shear zone, SE Asia. *Journal of Geophysical Research—Solid Earth, 106*(B4), 6683–6732.

Leloup, P. H., Tapponnier, P., Lacassin, R., & Searle, M. P. (2007). Discussion on the role of the Red River shear zone, Yunnan and Vietnam, in the continental extrusion of SE Asia Journal, Vol. 163, 2006, 1025–1036. *Journal of the Geological Society, 164*(6), 1253–1260.https://doi.org/10.1144/0016-76492007-065

Leloup, P. H., Harrison, T. M., Ryerson, F. J., Wenji, C., Qi, L., Tapponnier, P., & Lacassin, R. (1993). Structural, petrological and thermal evolution of a tertiary ductile strike-slip shear zone, Diancang Shan. *Yunnan Journal of Geophysics Research, 98*(B4), 6715–6743. https://doi.org/10.1029/92jb02791

Leloup, P. H., Lacassin, R., Tapponnier, P., Schärer, U., Zhong, D., Liu, X., & Trinh, P. T. (1995). The Ailao Shan-Red River shear zone (Yunnan, China), tertiary transform boundary of Indochina. *Tectonophysics, 251*(1–4), 3–84. https://doi.org/10.1016/0040-1951(95)00070-4

Li, Y., Hao, M., Ji, L., et al. (2014). Fault slip rate and seismic moment deficit on major active faults in mid and south part of the eastern margin of Tibet Plateau. *Chinese Journal of Geophysics, 57*(4), 1062–1078. (in Chinese).

Li, S. H., Advokaat, E. L., van Hinsbergen, D. J., Koymans, M., Deng, C. L., & Zhu, R. X. (2017). Paleomagnetic constraints on the Mesozoic-Cenozoic paleolatitudinal and rotational history of Indochina and South China: Review and updated kinematic reconstruction. *Earth-Science Reviews, 171*, 58–77. https://doi.org/10.1016/j.earscirev.2017.05.007

Li, S. H., Deng, C. L., Yao, H. T., Huang, S., Liu, C. Y., He, H. Y., Pan, Y. X., & Zhu, R. X. (2013). Magnetostratigraphy of the Dali Basin in Yunnan and implications for late Neogene rotation of the southeast margin of the Tibetan Plateau. *Journal of Geophysical Research: Solid Earth, 118*(3), 791–807. https://doi.org/10.1002/jgrb.50129

Li, S., van Hinsbergen, D. J., Deng, C., Advokaat, E. L., & Zhu, R. (2018). Paleomagnetic constraints from the Baoshan Area on the deformation of the Qiangtang-Sibumasu Terrane around the Eastern Himalayan Syntaxis. *Journal of Geophysical Research: Solid Earth, 123*(2), 977–997. https://doi.org/10.1002/2017JB015112

Loveless, J. P., & Meade, B. J. (2011). Partitioning of localized and diffuse deformation in the Tibetan Plateau from joint inversions of geologic and geodetic observations. *Earth and Planetary Science Letters, 303*(1), 11–24.

Maurin, T., Masson, F., Rangin, C., Min, U. T., & Collard, P. (2010). First global positioning system results in northern Myanmar: Constant and localized slip rate along the Sagaing fault. *Geology, 38*(7), 591–594. https://doi.org/10.1130/G30872.1

Meade, B. J. (2007). Present-day kinematics at the India-Asia collision zone. *Geology, 35*, 81–84.

Metcalfe, I. (2013). Gondwana dispersion and Asian accretion: Tectonic and palaeogeographic evolution of eastern Tethys. *Journal of Asian Earth Sciences, 66*, 1–33. https://doi.org/10.1016/j.jseaes.2012.12.020

Otofuji, Y.-I., Liu, Y., Yokoyam, M., Tamai, M., & Yin, J. (1998). Tectonic deformation of the southwestern part of the Yangtze craton inferred from paleomagnetism. *Earth and Planetary Science Letters, 156*(1–2), 47–60. https://doi.org/10.1016/S0012-821X(98)00009-0

Pan, G. T., Wang, P. S., Xu, Y. R., Jiao, S. P., & Xiang, T. X. (1990). Cenozoic tectonic evolution of Qinghai-Xizang Plateau. *PR China Ministry of Geology and Mineral Resource Geology Mem. Series, 5*(9).

Papadimitriou, E., Wen, X. Z., Karakostas, V., et al. (2004). Earthquake triggering along the Xianshuihe fault zone of Western Sichuan, China. *Pure and Applied Geophysics, 161*(8), 1683–1707. https://doi.org/10.1007/s00024-003-2471-4

Ren, Z. K. (2013). Geometry and deformation features of the most recent co-seismic surfaceruptures along the Xiaojiang Fault and its tectonic implications for the Tibetan Plateau. *Journal of Asian Earth Sciences, 77*, 21–30.

Replumaz, A., Lacassin, R., Tapponnier, P., et al. (2001). Large river offsets and Plio-Quatemary dextral slip rate on the Red River fault (Yunnan, China). *Journal of Geophysical Research, 106*(B1), 819–836.

Replumaz, A., & Tapponnier, P. (2003). Reconstruction of the deformed collision zone between India and Asia by backward motion of lithospheric blocks. *Journal of Geophysical Research, 108*(B6), 2285. https://doi.org/10.1029/2001JB000661

Royden, L. H., Burchfiel, B. C., van der Hilst, R. D. (2008). The geological evolution of the Tibetan Plateau. *Science, 321*(5892), 1054-1058. https://doi.org/10.1126/science.1155371

Schoenbohm, L. M., Burchfiel, B. C., Liangzhong, C., & Jiyun, Y. (2006). Miocene to present activity along the Red River fault, China, in the context of continental extrusion, upper-crustal rotation, and lower-crustal flow. *Geological Society of America Bulletin, 118*(5–6), 672–688. https://doi.org/10.1130/b25816.1

Searle, M. P. (2006). Role of the Red River Shear zone, Yunnan and Vietnam, in the continental extrusion of SE Asia. *Journal of the Geological Society, 163*(6), 1025–1036. https://doi.org/10.1144/0016-76492005-144

Shen, Z. K., Lu, J. N., Wang, M., et al. (2005). Contemporary crustal deformation around the southeast borderland of the Tibetan Plateau. *Journal of Geophysical Research: Solid Earth, 110*, B11409. https://doi.org/10.1029/2004JB003421

Shi, X. H., Sieh, K., Weldon, R., et al. (2018a). Slip rate and rare large prehistoric earthquakes of the Red River fault, Southwestern China. *Geochemistry, Geophysics, Geosystems, 19*, 2014–2031.

Shi, X., Wang, Y., Sieh, K., Weldon, R., Feng, L., Chan, C.-H., & Liu-Zeng, J. (2018b). Fault slip and GPS velocities across the Shan Plateau define a curved southwestward crustal motion around the eastern Himalayan syntaxis. *Journal of Geophysical Research: Solid Earth, 123*(3), 2502–2518. https://doi.org/10.1002/2017jb015206

Song, F. M., Wang, Y. P., Yu, W. X., et al. (1998). Xiaojiang active fault zone. Seismological Press. (in Chinese).

Simons, W. J. F., et al. (2007), A decade of GPS in Southeast Asia: Resolving Sundaland motion and boundaries, *Journal of Geophysical Research*, 112, B06420, https://doi.org/10.1029/2005JB003868

Tamai, M., Liu, Y., Lu, L. Z., Yokoyama, M., Halim, N., Zaman, H., & Otofuji, Y. I. (2004). Palaeomagnetic evidence for southward displacement of the Chuan Dian fragment of the Yangtze Block. *Geophysical Journal International, 158*(1), 297–309.https://doi.org/10.1111/j.1365-246X.2004.02108.x

Tang, R. C., & Han, W. B. (Eds.). (1993). *Active faults and earthquakes in Sichuan Province.* Seismological Press. (in Chinese).

Tang, W. Q., Liu, Y. P., Chen, Z. L., et al. (2006). Modern activity in the southern and central parts of the Xiaojiang fault in Yunnan. *Sedimentary Geology and Tethyan Geology, 26*(2), 21–24. (in Chinese).

Tapponnier, P., Lacassin, R., & Leloup, P. H. (1990a). The Ailao Shan Red River metamorphic bilt: Tertiary left-lateral shear between Indochina and South China. *Nature, 343*, 431–437.

Tapponnier, P., Meyer, B., Avouac, J. P., Peltzer, G., Gaudemer, Y., Guo, S., & Dai, H. (1990b). Active thrusting and folding in the Qilian Shan, and decoupling between upper crust and mantle in northeastern Tibet. *Earth and Planetary Science Letters, 97*(3–4), 382–403. https://doi.org/10.1016/0012-821x(90)90053-z

Tapponnier, P., & Molnar, P. (1977). Active faulting and tectonics in China. *Journal of Geophysical Research, 82*(20), 2905–2930. https://doi.org/10.1029/JB082i020p02905

Tapponnier, P., Peltzer, G., & Armijo, R. (1986). On the mechanics of the collision between India and Asia. *Geological Society, London, Special Publications, 19*(1), 113–157. https://doi.org/10.1144/GSL.SP.1986.019.01.07

Tapponnier, P., Peltzer, G., Le Dain, A. Y., Armijo, R., & Cobbold, P. (1982). Propagating extrusion tectonics in Asia: New insights from simple experiments with plasticine. *Geology, 10*(12), 611–616.

Tapponnier, P., Zhiqin, X., Roger, F., Meyer, B., Arnaud, N., Wittlinger, G., & Jingsui, Y. (2001). Oblique stepwise rise and growth of the Tibet Plateau. *Science, 294*(5547), 1671–1677. https://doi.org/10.1126/science.105978

Thatcher, W. (2007). Microplate model for the present-day deformation of Tibet, *Journal of Geophysical Research,* 112, B01401, https://doi.org/10.1029/2005JB004244

Taylor, M., Yin, A., 2009. Active structures of the Himalayan-Tibetan orogen and their relationships to earthquake distribution, contemporary strain field, and Cenozoic volcanism. *Geosphere 5*(3), 199–214.https://doi.org/10.1130/GES00217.1

Tong, Y., Yang, Z., Pei, J., Wang, H., Wu, Z., & Li, J. (2020a). Crustal clockwise rotation of the southeastern edge of the Tibetan Plateau since the late Oligocene. *Journal of Geophysical Research: Solid Earth,* e2020JB020153. https://doi.org/10.1029/2020JB020153

Tong, Y., Wu, Z., Sun, Y., Yang, Z., Pei, J., Yang, X., Li, J., & Wang, C. (2020b). The interaction of the eastward extrusion of the Songpan-Ganzi Terrane and the crustal rotational movement of the Sichuan Basin since the late Paleogene: Evidence from cretaceous and paleogene paleomagnetic data sets of the Sichuan Basin. *Tectonics, 39*(2), e2019TC005784. https://doi.org/10.1029/2019TC005784

Tong, Y., Sun, Y., Wu, Z., Mao, C., Pei, J., Yang, Z., Pu, Z., Zhao, Y., & Xu, H. (2019). Passive crustal clockwise rotational deformation of the Sichuan Basin since the Miocene and its relationship with the tectonic evolution of the fault systems on the eastern edge of the Tibetan Plateau. *GSA Bulletin, 131*(1–2), 175–190. https://doi.org/10.1130/B31965.1

Tong, Y., Yang, Z., Jing, X., Zhao, Y., Li, C., Huang, D., & Zhang, X. (2016). New insights into the Cenozoic lateral extrusion of crustal blocks on the southeastern edge of Tibetan Plateau: Evidence from paleomagnetic results from Paleogene sedimentary strata of the Baoshan Terrane. *Tectonics, 35*(11), 2494–2514. https://doi.org/10.1002/2016TC004221

Tong, Y., Yang, Z., Mao, C., Pei, J., Pu, Z., & Xu, Y. (2017). Paleomagnetism of Eocene red-beds in the eastern part of the Qiangtang Terrane and its implications for uplift and southward crustal extrusion in the southeastern edge of the Tibetan Plateau. *Earth and Planetary Science Letters, 475*, 1–14. https://doi.org/10.1016/j.epsl.2017.07.026

Tong, Y. B., Yang, Z. Y., Wang, H., Gao, L., An, C. Z., Zhang, X. D., & Xu, Y. C. (2015). The Cenozoic rotational extrusion of the Chuan Dian Fragment: New paleomagnetic results from Paleogene red-beds on the southeastern edge of the Tibetan Plateau. *Tectonophysics, 658*, 46–60. https://doi.org/10.1016/j.tecto.2015.07.007

Torsvik, T. H., Van der Voo, R., Preeden, U., Mac Niocaill, C., Steinberger, B., Doubrovine, P. V., Van Hinsbergen, D. J., Domeier, M., Gaina, C., & Tohver, E. (2012). Phanerozoic polar wander, palaeogeography and dynamics. *Earth-Science Reviews, 114*(3–4), 325–368. https://doi.org/10.1016/j.earscirev.2012.06.007

Trinh, P. T., Liem, N. V., Huong, N. V., Vinh, H. Q., Thom, B. V., Thao, B. T., Tan, M. T., & Hoang, N. (2012). Late Quaternary tectonics and seismotectonics along the Red River fault zone. *North Vietnam. Earth-Science Reviews, 114*(3–4), 224–235. https://doi.org/10.1016/j.earscirev.2012.06.008

Wang, Y., Shen, J., & Wang, Q., et al. (2003). On the lateral extrusion of Sichuan-Yunnan Block. *Earth Science Frontiers, 10*(s1), 188–192. (in Chinese).

Wang, Y., Wang, E., Shen, Z.-K., Wang, M., Gan, W., Qiao, X., & Yang, Y. (2008). GPS-constrained inversion of present-day slip rates along major faults of the Sichuan-Yunnan region, China. *Science in China Series D: Earth Sciences, 51*(9), 1267–1283. (in Chinese), *38*(5), 582–597.

Wang, E., & Burchfiel, B. C. (1997). Interpretation of Cenozoic tectonics in the right-lateral accommodation zone between the ailao shan shear zone and the Eastern Himalayan syntaxis. *International Geology Review, 39*(3), 191–219. https://doi.org/10.1080/00206819709465267

Wang, E., Burchfiel, B. C., Royden, L. H., Chen, L., Chen, J., Li, W., & Chen, Z. (1998). Late Cenozoic Xianshuihe-Xiaojiang, Red River, and Dali fault systems of Southwestern Sichuan and Central Yunnan, China. *Geological Society of America Special Papers, 327*, 1–108. https://doi.org/10.1130/0-8137-2327-2.1

Wang, H., Tong, Y. B., & Gao, L. (2015). Paleomagnetlc analysis of Paleocene sedimentary strata in Chuan-Dian fragment on the southeastern margin of the Tibetan Plateau and its tectonic implication. *Geological Bulletin of China, 34*(1), 45–56.

Wang, H., Wright, T. J., & Biggs, J. (2009). Interseismic slip rate of the northwestern Xianshuihe fault from InSAR data. *Geophysical Research Letters, 36*, L03302. https://doi.org/10.1029/2008GL036560

Wang, H., & Yang, Z. Y. (2019). Differential rotation in the western Sichuan-Yunnan Block and its geological implications: New palaeomagnetic evidence from the Cretaceous red beds in the southeastern margin of the Tibetan Plateau. *Chinese Journal of Geophysics-Chinese Edition, 62*(5), 1789–1808. https://doi.org/10.6038/cjg2019L0786

Wang, H., Yang, Z. Y., Tong, Y. B., Gao, L., Jing, X. Q., & Zhang, H. F. (2016). Palaeomagnetic results from Palaeogene red beds of the Chuan-Dian Fragment, southeastern margin of the Tibetan Plateau: Implications for the displacement on the Xianshuihe-Xiaojiang fault systems. *International Geology Review, 58*(11), 1363–1381. https://doi.org/10.1080/00206814.2016.1157710

Wang, Q., Zhang, P. Z., Freymueller, J. T., et al. (2001). Present-day crustal deformation in China constrained by global positioning system measurements. *Science, 294*(5542), 574–577.

Wang, W., Qiao, X. J., Yang, S. M., et al. (2017). Present-day velocity field and block kinematics of Tibetan Plateau from GPS measurements. *Geophysical Journal International, 208*, 1088–1102.

Wen, X. Z. (2000). Character of rupture segmentation of the Xianshuihe-Anninghe-Zemuhe fault zone, Western Sichuan. *Seismology and Geology, 22*(3), 239–249 (in Chinese).

Wen, X. Z., Du, F., Long, F., et al. (2011). Tectonic dynamics and correlation of major earthquake sequences of the Xiaojiang and Qujiang-Shiping fault systems, Yunnan, China. *Science China Earth Sciences, 41*(5), 713–724. (in Chinese).

Wen, X. Z., Fan, J., Yi, G. X., et al. (2008a). A seismic gap on the Anninghe fault in western Sichuan. *Science in China D Series: Earth Sciences, 51*(10), 1375–1387. (in Chinese).

Wen, X. Z., Ma, S. L., Xu, X. W., et al. (2008b). Historical pattern and behavior of earthquake ruptures along the eastern boundary of the Sichuan-Yunnan faulted-block, southwestern China. *Physics of the Earth Planetary Interiors, 168*(1–2), 16–36. https://doi.org/10.1016/j.pepi.2008.04.013

Working Group of M7. 2012. Study on the mid- to long-term potential of large earthquakes on the Chinese continent. Seismological Press. (in Chinese).

Xiang, H.-F., Guo, S.-M., Zhang, W.-X., Han, Z.-J., Zhang, B.-L., Wan, J.-L., Dong, X., & Chen, L.-C. (2007). Quantitative study on the large scale dextral strike-slip offset in the southern segment of the Red River fault since Miocene. *Seismology and Geology, 29*(1), 52–65. (in Chinese with English abstract).

Xu, X. X., Ji, L. Y., Jiang, F. Y., et al. (2020). Study on current activity features of Jinshajiang fault zone based on GPS small earthquakes. *Journal of Geodesy and Geodynamics, 40*(10), 1062–1067. (in Chinese).

Xu, X. W., Wen, X. Z., Zheng, R. Z., et al. (2003). Pattern of latest tectonic motion and its dynamics for active blocks in Sichuan-Yunnan region, China. *Science in China, 46*(S2), 210–226.

Xu, Z., Hou, L., & Wang, Z. (1992). *Orogenic process beneath the Songpan-Garze orogenic belt*. Geology Press. (in Chinese).

Yan, Y., Duan, L., Liang, S., Wang, J., Huang, B., Zheng, W., & Zhang, P. (2020). Paleomagnetic constraint on the Carboniferous paleoposition of Indochina and its implications for the evolution of eastern Paleo-Tethys Ocean. *Tectonics, 39*(7), e2020TC006168. https://doi.org/10.1029/2020TC006168

Yan, Y. G., Huang, B. C., Zhao, J., Zhang, D. H., Liu, X. H., Charusiri, P., & Veeravinantanakul, A. (2017). Large southward motion and clockwise rotation of Indochina throughout the Mesozoic: Paleomagnetic and detrital zircon U-Pb geochronological constraints. *Earth and Planetary Science Letters, 459*, 264–278. https://doi.org/10.1016/j.epsl.2016.11.035

Yan, Y., Zhao, Q., Zhang, D., Charusiri, P., Huang, B., & Zhang, P. (2019). Palaeomagnetism of Late Triassic volcanic rocks from the western margin of Khorat Basin, Thailand and its implication for ambiguous inclination shallowing in Mesozoic sediments of Indochina. *Geophysical Journal International, 219*(2), 897–910. https://doi.org/10.1093/gji/ggz331

Yang, Z. Y., Sun, Z. M., Ma, X. H., Yin, J. Y., & Otofuji, Y. (2001). Palaeomagnetic study of the Early Tertiary on both sides of the Red River Fault and its geological implications. *Acta Geologica Sinica, 75*(1), 35–44.

Yin, A., & Harrison, T. M. (2000). Geologic evolution of the Himalayan-Tibetan orogen. *Annual Review of Earth and Planetary Sciences, 28*(1), 211–280. https://doi.org/10.1146/annurev.earth.28.1.211

Zhang, P. Z., Shen, Z. K., Wang, M., Gan, W. J., Bürgmann, R., Molnar, P., Wang, Q., Niu, Z. J., Sun, J. Z., & Wu, J. C. (2004). Continuous deformation of the Tibetan Plateau from global positioning system data. *Geology, 32*(9), 809–812.https://doi.org/10.1130/G20554.1

Zhang, J.-G. (2009). Study on the activity of Red river fault in China and Vietnam [Ph.D. thesis]. (Ph.D.), University of Science and Technology of China, Hefei. (in Chinese with English abstract).

Zhang, G. M., Ma, H. S., Wang, H., et al. (2005). Boundaries between active-tectonic blocks and strong earthquakes in the China's mainland. *Chinese Journal of Geophysics, 48*(3), 602–610. (in Chinese).

Zhang, P. Z., Deng, Q. D., Zhang, G. M., et al. (2003). Active tectonic blocks and strong earthquakes in the continent of China. *Science in China (series d), 46*(S2), 13–24.

Zhang, P. Z., Wang, Q., & Ma, Z. J. (2002). GPS velocity field and active crustal blocks of contemporary tectonic deformation in continental China. *Earth Science Frontiers, 9*(2), 430–441. (in Chinese).

Zhang, Y. S., Yao, X., Yu, K., et al. (2016). Late-Quaternary slip rate and seismic activity of the Xianshuihe fault zone in Southwest China. *Acta Geologica Sinica, 90*(2), 525–536.

Zhao, J., Huang, B., Yan, Y., & Zhang, D. (2015). Late Triassic paleomagnetic result from the Baoshan Terrane, West Yunnan of China: Implication for orientation of the East Paleotethys suture zone and timing of the Sibumasu-Indochina collision. *Journal of Asian Earth Sciences, 111*, 350–364. https://doi.org/10.1016/j.jseaes.2015.06.033

Zhao, J., Jiang, Z., Wu, Y., Liu, X., & Wei, W. (2012). Characteristics of block strain and fault movement in the Sichuan-Yunnan region before and after Wenchuan earthquake. *Geodesy and Geodynamics, 3*(2), 27–33.

Zheng, G., Wang, H., Wright, T. J., et al. (2017). Crustal Deformation in the India-Eurasia Collision Zone From 25 Years of GPS Measurements. *Journal of Geophysical Research Solid Earth, 122*(11), 9290–9312.

Zhou, R., Chen, G., Li, Y., et al. (2005). Research on active faults in Litang-Batang region, western Sichuan province, and the seismogenic structures of the 1989 Batang M6.7 earthquake swarm. *Seismology and Geology, 27*(1), 31–43. (in Chinese).

Zhu, R. X., Potts, R., Pan, Y. X., Lü, L. Q., Yao, H. T., Deng, C. L., & Qin, H. F. (2008). Paleomagnetism of the Yuanmou Basin near the southeastern margin of the Tibetan Plateau and its constraints on late Neogene sedimentation and tectonic rotation. *Earth and Planetary Science Letters, 272*(1–2), 97–104. https://doi.org/10.1016/j.epsl.2008.04.016

Zuchiewicz, W., & Cuong, N. Q. (2009). Quaternary tectonics of the Red River fault zone in Vietnam-A morphotectonic approach. *Geologia/Akademia Górniczo-Hutnicza im. Stanisława Staszica w Krakowie, 35*(2/1), 367–374.

Zuchiewicz, W., Cu'ò'ng, N. Q., Zasadni, J., & Yêm, N. T. (2013). Late Cenozoic tectonics of the Red River Fault Zone, Vietnam, in the light of geomorphic studies. *Journal of Geodynamics, 69*, 11–30.https://doi.org/10.1016/j.jog.2011.10.008

Chapter 4
CSES Community Velocity Models in Southwest China

Huajian Yao, Ying Liu, and Zhiqi Zhang

Abstract Southwest China, located at the southeastern margin of the Tibetan Plateau, plays an important role for the plateau growth and its material extrusion. It has complicated tectonic environment and severe earthquake hazards. There have been numerous geophysical studies in Southwest China. However, a community velocity model (CVM) in this region is still not available, which makes it difficult to have a consistent catalogue of earthquake locations and focal mechanisms, and a consistent velocity model for simulating strong ground motions and evaluating earthquake hazards. In this study, we first propose an overall strategy of building multi-scale CVMs in Southwest China. Then we build a high-resolution community velocity model (both Vp and Vs) of the crust and uppermost mantle in Southwest China along with earthquake locations by joint inversion of massive body-wave and surface-wave traveltime datasets. A multigrid strategy is adopted in the joint inversion. The spatial resolution of both Vp and Vs models can reach up to 0.5° horizontally and 10 km vertically according to checkerboard tests. The comparison of our inverted Vp and Vs models with those from other studies show general consistency in large-scale features of velocity anomalies. The inverted models are further validated using various seismic datasets. In general, our velocity models show two low Vp and Vs anomaly zones in the middle-lower crust and a prominent high velocity anomaly region in central Yunnan in between them. Our new models have been serve as the first CVMs in Southwest China (i.e., SWChinaCVM-1.0) for future seismological studies. In addition to the construction of the regional scale CVM in southwest China, other studies have obtained high-resolution velocity models of urban, fault zone, and fault scales using dense array seismic data in some regions in Southwest China. Finally, we discuss some perspective studies related to the construction of multi-scale CVMs in Southwest China as well as community models of other physical parameters.

Keywords Southwest China · Crust and Upper Mantle · Community Velocity Model · Joint Inversion

---

H. Yao (✉) · Y. Liu · Z. Zhang
Laboratory of Seismology and Physics of Earth's Interior, School of Earth and Space Sciences, University of Science and Technology of China, Hefei 230026, Anhui, China
e-mail: hjyao@ustc.edu.cn

© Higher Education Press 2022

Y.-G. Li et al. (eds.), *China Seismic Experimental Site*,
https://doi.org/10.1007/978-981-16-8607-8_4

4.1 Introduction

Due to the collision between the Indian plate and the Eurasian plate in the early Cenozoic, the Tibetan Plateau has experienced intense surface uplift, crustal shortening and thickening, and complex crustal and upper mantle deformation (Molnar & Tapponnier, 1975; Yin & Harrison, 2000). Southwest China, which mainly includes Sichuan and Yunnan provinces and is located at the southeastern margin of the Tibetan Plateau, is an important window for transferring crust and upper mantle materials that were extruded from the Tibetan Plateau to Southeast Asia. There are many active faults in this region, including the Longmenshan fault (LMSF), the Xianshuihe fault (XSHF), the Xiaojiang fault (XJF), the Lijiang-Xiaojinhe fault (LXJF), and the Red River fault (RRF) (Burchfiel & Chen, 2012; Royden et al., 1997; Wang, 1998). These fault systems divide Southwest China into different tectonic units, such as the Songpan-Ganze block (SPGZB), the Qiangtang block (QTB), the diamond shaped Chuan-Dian block (bounded by XSHF, XJF and RRF), the Yangtze Craton (YZC), the Shan Thai block (STB), and the Youjiang block (YJB) (Fig. 4.1).

Due to very active tectonic activities and complex lithospheric structures and deformations, Southwest China is one of the most seismically active regions in China, which is located in the southern part of the South-North Seismic Belt of China, with more than one-third large earthquakes of $M_w > 7$ that had occurred in history (Zhang, 2013). For example, the most devastating 2008 $M_w7.9$ Wenchuan earthquake and the 2013 $M_w6.6$ Lushan earthquake occurred along the LMSF, which resulted in severe causalities and property losses. Therefore, earthquake monitoring and forecasting as well as earthquake hazard prevention and mitigation is always an important task in Southwest China. At present time, the studies on seismicity and seismic hazard distribution, the structural and tectonic background of earthquake generation, earthquake source physics, earthquake early warning and prediction, etc., are important fields of earthquake studies in China. The construction of high resolution and multi-scale regional 3-D community velocity models (CVMs) is fundamental for these researches.

The concept of CVM was first proposed by the Southern California Earthquake Center (SCEC) in 1996. The SCEC Unified Structural Representation (USR), which is a 3-D description of crustal and upper mantle structure in California, integrates different model results derived from different data types and inversion methods. Two versions of the CVM have been released, that is, CVM-S and CVM-H. The latest versions are CVM-S4.26 (Lee et al., 2014) and CVM-H15.10 (Shaw et al., 2015). The CVM of SCEC USR has served a broad range of scientific and hazards assessment efforts, including fault systems analysis, strong ground motion prediction, and Probabilistic Seismic Hazard Analysis. Supported by the China Seismic Experiment Site (CSES) of China Earthquake Administration (CEA) and also inspired by the SCEC CVM strategies, our goal is to build a high-resolution and multi-scale CVM of the crust and uppermost mantle in Southwest China with multiple data types and joint inversion approaches.

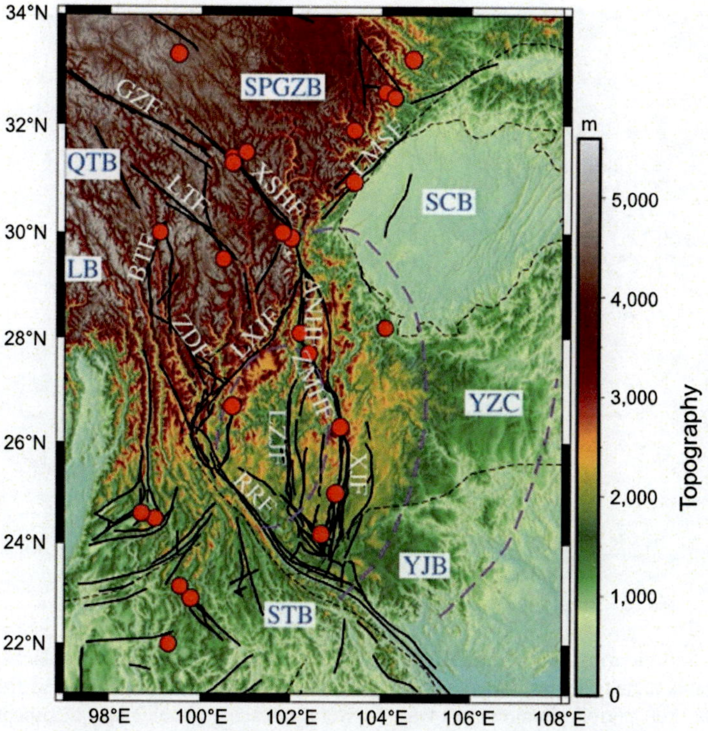

Fig. 4.1 Topography map of southwest China with the main tectonic features and major fault systems in this region. The thin dashed black lines indicate tectonic block boundaries. QTB, Qiangtang block; SPGZB, Songpan-Ganze block; SCB, Sichuan Basin; LB, Lhasa block; YZC, Yangtze craton; STB, Shan Thai block; YJB, Youjiang block. Thick solid black lines show major faults. GZF, Ganze fault; LTF, Litang fault; BTF, Batang fault; ZDF, Zhongdian fault; XSHF, Xianshuihe fault; LXJF, Lijiang-Xiaojinhe fault; ANHF, Anninghe fault; ZMHF, Zemuhe fault; LZJF, Lüzhijiang fault; RRF, Red River fault; XJF, Xiaojiang fault; LMSF, Longmenshan fault. The epicentre of large historical earthquakes ($M_w \geq 7.0$) are plotted in red filled circles. The inner, intermediate and outer zone boundaries of the Emeishan large igneous province (ELIP) (Xu et al., 2004) are denoted by purple dashed lines (Figure from Liu et al., 2021)

4.2 The Current Status of Building the CVM in Southwest China

Compared with southern California of the United States, the construction of the CVM in Southwest China started late. In 2016, the CSES of CEA began to organize the systematic investigation and research work on the construction of the CVM in the Sichuan-Yunnan region, Southwest China. Prior to this, the density and quality of permanent and transportable seismic networks in Sichuan and Yunnan have improved substantially in recent decades (Fig. 4.2a). There have been many research publications related to crust and upper mantle models in Southwest China using seismic tomography or inversion. However, the lateral scales of the published models are complex and diverse, and the data types and inversion methods used are also diverse.

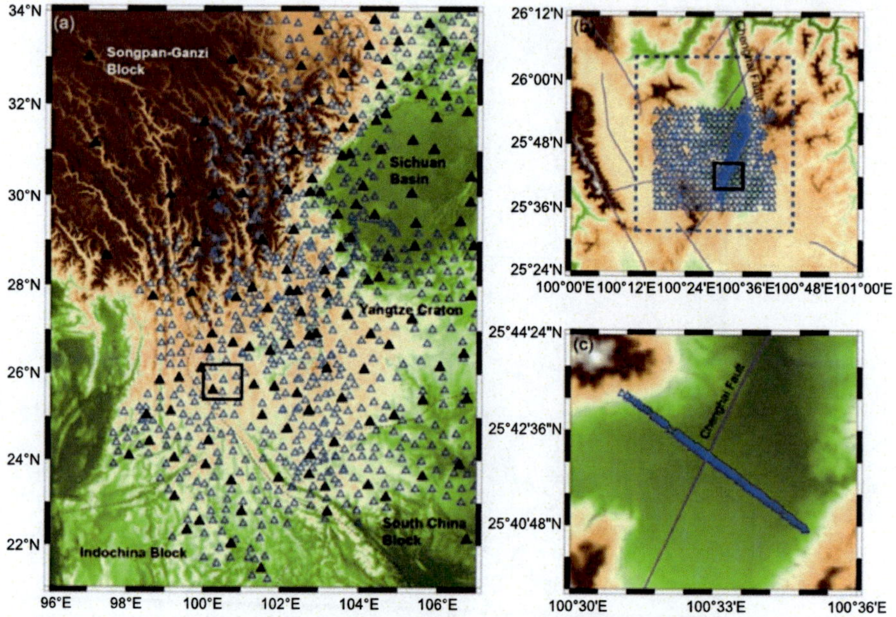

Fig. 4.2 Seismic arrays of different scales in Sichuan and Yunnan: **a** permanent (black triangles) and transportable (blue triangles) broadband stations; **b** dense short-period planar array (blue triangles; about 400 short-period seismometers with an average spacing about 2 km) in Binchuan Basin in western Yunnan; the graphic area is shown as the black box in (**a**); **c** the ~ 8 km long dense short-period linear array consisting of 150 short-period seismometers (blue triangles) with an average spacing of 50 m across the Chenghai fault in western Yunnan. The graphic area is shown as the black box in (**b**) (Figure from Yao, 2020)

For instance, the crust and upper mantle structures in Southwest China have been obtained by many different seismic tomographic methods, such as body wave travel time tomography (Huang et al., 2002; Huang et al., 2015, 2019; Wang et al., 2003, 2015; Wang et al., 2010; Xu & Song, 2010; Wei et al., 2013; Xin et al., 2019), surface wave travel time tomography (Yao et al., 2006, 2008; Li et al., 2009; Huang et al., 2010; Yang et al., 2012; Zhang et al., 2020b), receiver functions (Zhang et al., 2010; Wang et al., 2017; Hu et al., 2018), and joint inversion of different types of seismic data (Liu et al., 2014; Bao et al., 2015; Yang et al., 2020b).

Due to inherent disparate model uncertainties and model sensitivities of different data types, there exist considerable differences among various models. Even the models obtained from the same type of data show different features due to the differences in inversion methods, model parameterization and regularization approaches. Overall, the crustal velocity models from previous seismic tomography studies have limited resolutions of ~ 50–100 km horizontally and 10–20 km vertically. Thus, it is difficult to study detailed crustal structures and delineate fine features from previous velocity models. For example, previous studies have observed low velocity zones in the middle to lower crust (Huang et al., 2002, 2010, 2015; Wang et al., 2003, 2007, 2017; Yao et al., 2006, 2008; Li et al., 2009; Yang et al., 2012; Hu et al., 2018; Yang et al., 2020b; Zhang et al., 2020b) and high velocity anomalies in central Yunnan and its adjacent areas (Liu et al., 2001; Qiao et al., 2018; Xu et al., 2015; Yao et al., 2008; Zheng et al., 2016), but their depth extent and shape of these velocity anomalies are not consistent among them. Till now, a well-recognized and widely used CVM in Southwest China still does not exist. Therefore, it is necessary to construct a consistent CVM from different datasets.

At present, there is almost no research work to evaluate the accuracy, applicability and updateability of these velocity models with different resolutions and different scales. In addition, the accuracy of the earthquake catalog information (such as location, focal mechanism, etc.) used to construct velocity models needs to be further improved. The lag or lack of these fundamental work will greatly restrict the in-depth development of other research works in Southwest China, such as the construction of the community fault model (CFM), simulation of strong ground motion, earthquake hazard prediction, and earthquake dynamics simulation.

4.3 The Overall Strategy of Building Multi-scale CVMs

With the support of the CSES of CEA, and on the basis of many investigations and discussions, an overall strategy for the construction of CVMs in Sichuan and Yunnan has been formed (Fig. 4.3a) (Yao, 2020). Based mainly on seismological data, we carried out a systematic research work by the route of three steps: "multi-scale velocity model construction—earthquake source parameter determination—velocity model and source parameter evaluation" in the Sichuan-Yunnan area. The three research units support each other for better improving the reliability and accuracy of CVM. According to the needs of earthquake science studies and earthquake engineering applications, based on the waveform data of seismic arrays with different lateral scales and station spacing (Fig. 4.2) and methods of joint seismic tomography or full waveform inversion, we need to construct CVMs of four different scales in Southwest China (Fig. 4.3b), namely the regional block scale, fault zone scale, fault scale, and urban scale, with its lateral resolution about 10–50, 1–5, 0.1–0.5 and 1 km, respectively.

The CVM in regional block scale can be obtained by inversion of data from permanent and transportable broadband stations with station spacing about 10–50 km in the Sichuan-Yunnan region (Fig. 4.2a). This type of models is of great significance for understanding scientific issues related to differences in the distribution and physical properties of major blocks, tectonic characteristics and geodynamic problems, inversion of regional earthquake source parameters, and the structural and tectonic background for moderate and strong earthquakes. The velocity models in fault zone scale, fault scale and urban scale are mainly obtained by inversion of data from dense planar and linear transportable arrays (Fig. 4.2b, c). In the Sichuan-Yunnan area, some large active fault zones are composed of multiple faults, such as the Longmenshan fault zone and the Xiaojiang fault zone. In order to better obtain the physical characteristics and spatial interaction between different faults, we need to

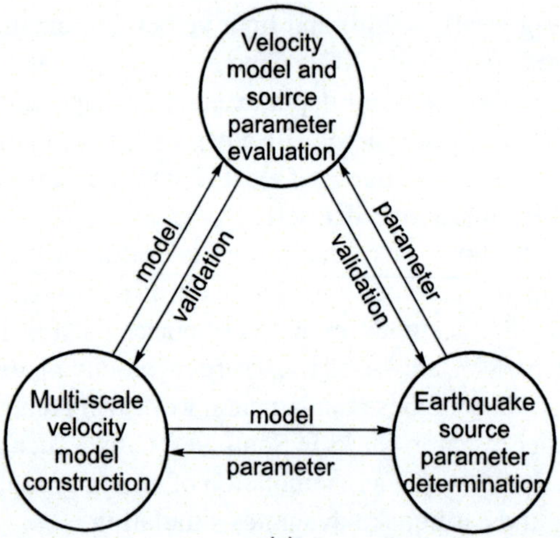

(a)

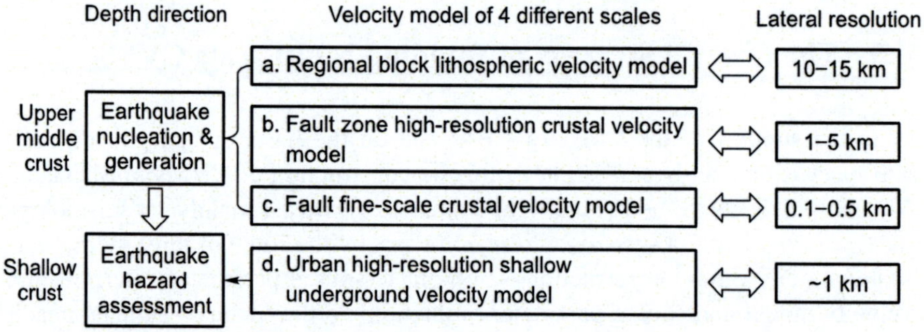

(b)

Fig. 4.3 **a** The overall strategy of constructing CVM in the Sichuan-Yunnan region and **b** the strategy of constructing velocity models of four different scales (Figure from Yao, 2020)

have the velocity model of fault zone scale with a lateral resolution of 1–5 km or higher. The nucleation of earthquakes often starts from a certain active fault, and then the rupture propagates along the fault, and even jumps to other adjacent faults. Therefore, the fine-scale velocity structure of the fault itself (lateral resolution of 0.1–0.5 km or higher) is important for understanding earthquake physics, such as nucleation, rupture initiation, propagation, and stopping. For instance, to obtain the structure of fault fracture zones usually requires the deployment of ultra-dense linear or even planar arrays across the faults, with station spacing of only tens to hundreds of meters.

The models of regional block scale, fault zone scale and fault scale will provide essential velocity structure information for the upper and middle crust, where earthquakes mainly nucleate and occur. The densely populated urban areas are greatly affected by earthquake hazards. Therefore, obtaining the shallow crustal structure of major cities in Southwest China has very important earthquake engineering meanings, including the exploration of fault distribution in urban areas, simulation of strong ground motion, earthquake hazard assessment, urban planning, etc. Its lateral resolution needs to be at least kilometer level to reliably simulate higher frequency (1 Hz and higher) ground motion. Therefore, it is possible to achieve the required lateral resolution of the urban scale model by deploying planar dense arrays with station spacing in the order of kilometers in major urban areas. Since near surface velocity structures have significant effects on ground motion, it is also important to collect near surface seismic velocity measurements from boreholes or other seismic surveys, e.g., from active or passive surface wave studies.

4.4 Building an Initial Regional-Scale Reference Velocity Model in Southwest China

Since 2017, with the support of CSES, CEA, we have initiated to building a multi-scale CVM in the Sichuan-Yunnan region, Southwest China, and have made substantial progresses. For the regional block scale model, we combined data from ambient noise, teleseismic body and surface waves, and petroleum wells and seismic-reflection profiles to better constrain the crustal shear velocity structure in Southwest China. We jointly inverted the Rayleigh wave dispersion (5–40 s period), Rayleigh wave ZH ratio (20–60 s period), and P-wave receiver function for 114 permanent stations with a stepwise linearized joint inversion method (Zhang & Yao, 2017). And then we combined all the 1-D models beneath each station to form a 3-D shear wave velocity model of the crust and uppermost mantle in the Sichuan-Yunnan area with lateral resolution about 50–100 km (Yang et al., 2020b). Compared to previous tomography results, we observe higher shear wave velocity for the sedimentary rocks within the Sichuan Basin, which is consistent with sonic logging measurements. Our model reveals widespread low-velocity zones in the mid-lower crust, and their boundaries correlate well with major fault systems. Between two main mid-crustal low-velocity channels, a prominent high-velocity region surrounded by earthquakes is observed in the inner zone of the Emeishan large igneous province (ELIP) and around the Anninghe-Zemuhe fault zone. These observations are comparable to regional tomography results using very dense arrays.

Based on this initial reference velocity model (Yang et al., 2020b), we further integrated the interstation teleseismic and ambient noise surface wave phase velocity dispersion data of permanent stations (period range 5–150 s), and obtained the 3-D shear wave velocity model of the crust and upper mantle (depth of 0–300 km) in the Sichuan and Yunnan region (Zhang et al. 2020b) using the direct inversion method of surface wave travel times (Fang et al., 2015). The lateral resolution is about 50–100 km in the crust and about 100–150 km in the upper mantle. This model (Fig. 4.4) has better revealed the tectonic activities of different modes in the crust and upper mantle in the southeastern margin of the Tibetan Plateau.

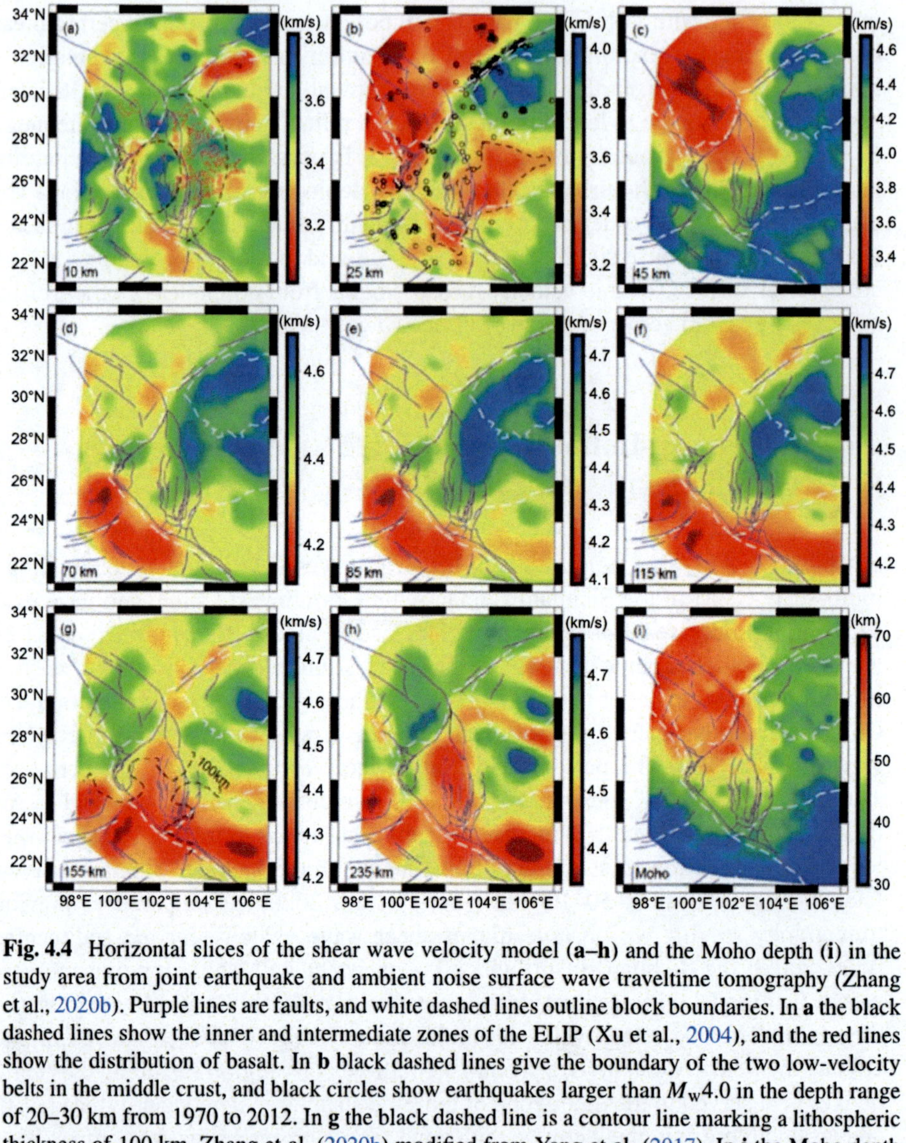

Fig. 4.4 Horizontal slices of the shear wave velocity model (**a–h**) and the Moho depth (**i**) in the study area from joint earthquake and ambient noise surface wave traveltime tomography (Zhang et al., 2020b). Purple lines are faults, and white dashed lines outline block boundaries. In **a** the black dashed lines show the inner and intermediate zones of the ELIP (Xu et al., 2004), and the red lines show the distribution of basalt. In **b** black dashed lines give the boundary of the two low-velocity belts in the middle crust, and black circles show earthquakes larger than M_W4.0 in the depth range of 20–30 km from 1970 to 2012. In **g** the black dashed line is a contour line marking a lithospheric thickness of 100 km, Zhang et al. (2020b) modified from Yang et al. (2017). In **i** the Moho depth data come from Wang et al. (2017) and He et al. (2014) (Figure from Zhang et al., 2020b)

4.5 Building the CVM V1.0 of Southwest China from Joint Seismic Tomography

The initial reference velocity model of Southwest China (Yang et al., 2020b) is based on the pointwise inversion of surface wave data and teleseismic receiver functions. However, the resolution is still limited by the spacing of the regional permanent stations, which is unevenly distributed. In addition, this model hasn't been validated by regional body wave traveltime data. In order to improve the quality and resolution of the regional reference model in Southwest China, we further built the CVM V1.0 (the first version of CVM) in Southwest China using a joint inversion method by combining a large amount of body wave travel times and surface wave dispersion data at most of the permanent stations and a number of temporary stations in Southwest China. Then, we applied a joint inversion method to simultaneously invert for 3-D P- and S-wave velocity structures of the crust and uppermost mantle in Southwest China, and also relocated all seismic events at the same time. Compared to previous studies, which mostly inverted body and surface wave data separately, the joint inversion method can take advantages of the complementary strengths of both body and surface wave data, thus obtaining a more unified velocity model. We illustrate the detailed data, method, tomographic model, model resolution, and model validation in the following subsections.

4.5.1 Data and Method

4.5.1.1 Body Wave Arrival Time Data

We assembled the catalog of P- and S-wave arrival time data from October 2008 to September 2016 recorded by the catalog of China Digital Seismic Network in Southwest China. During this period, nearly 939,000 P- and 918,000 S-wave arrival times from 201,751 earthquakes were recorded and manually picked at 230 permanent and temporary seismic stations (Dataset 0).

To avoid outliers of P- and S-wave picks in this catalog, we first removed travel time data that are away from the main trend of travel time curves. We also filtered out earthquake events which were recorded by fewer than 10 stations. After applying these two criteria, 69,767 earthquake events with 649,306 P- and 620,276 S-wave first arrivals were retained (Dataset 1). Among these 69,767 earthquakes, more than 50% (36,415) are located along the LMSF. This uneven distribution of body wave data could lead to an unstable inversion problem and result in unreliable features in the inverted velocity models. To avoid this potential problem, we divided our study region into small bins (1-km-thick cell with 3 km $\times$ 3 km area) and only used 1 randomly selected seismic event in each bin to construct our final body wave dataset (Dataset 2) to achieve a more homogenous data distribution. This results in 33,357 seismic events with 386,958 P- and 372,662 S-wave first arrivals (Fig. 4.5), among which about 38% are located along the LMSF. The differential arrival times of near-by events are calculated. Finally, we construct a body wave dataset with 386,958 P-wave first arrivals, 372,662 S-wave first arrivals and 1,381,763 event-pair differential arrival times.

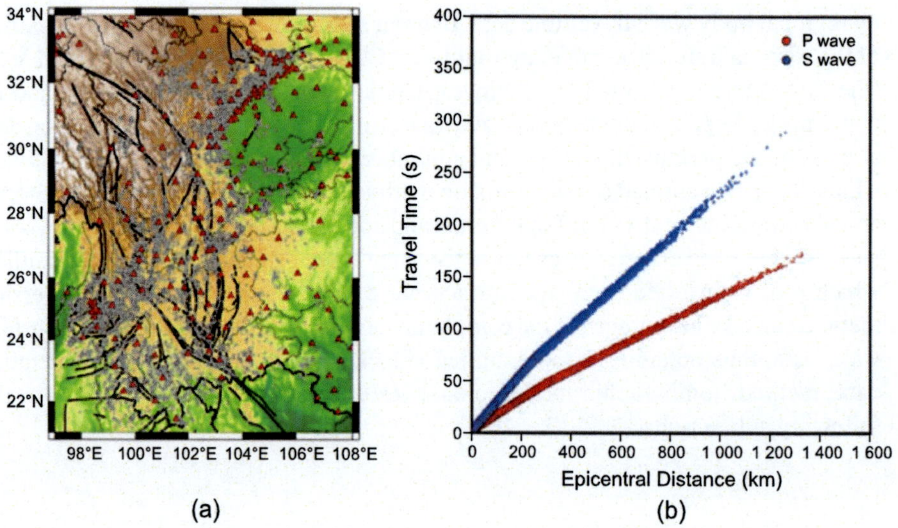

Fig. 4.5 **a** The distribution of selected seismic events (gray filled circles) and stations (red filled triangles) of body wave data used in joint inversion. **b** Travel-time picks of selected P- and S-wave used in joint tomography (Figure from Liu et al., 2021)

4.5.1.2 Surface Wave Dispersion Data

We also assembled nearly 8100 Rayleigh wave phase velocity dispersion curves with period range from 5 to 50 s (Fig. 4.6). This dataset consists of ambient noise Rayleigh wave phase velocity dispersion data from three studies: 6910 Rayleigh wave phase velocity dispersion curves (from 5 to 50 s) extracted by Yang et al. (2020b) from 124 permanent stations (deployed by the China National Seismic Network, Zheng et al., 2010) in Southwest China, 544 Rayleigh wave phase velocity dispersion curves (from 5 to 45 s) extracted by Yao et al. (2010) from 35 temporary stations (deployed by Massachusetts Institute of Technology and Lehigh University) in Southwest China, and 640 Rayleigh wave phase velocity dispersion curves (from 5 to 40 s) extracted by Qiao et al. (2018) from 16 temporary stations (deployed by the Institute of Earth Sciences, "Academia Sinica") in northern Vietnam (Nguyen et al., 2013) and 49 permanent stations in Yunnan Province.

4.5.1.3 Joint Body and Surface Wave Travel Time Tomography

In order to use both body wave travel times and surface wave dispersion data to constrain the velocity structures, we adopted the joint inversion algorithm developed by Fang et al. (2016) in this study. This algorithm combines the double-difference tomography method (Zhang & Thurber, 2003) for body wave travel times and the direct surface wave tomography method (Fang et al., 2015) for surface wave

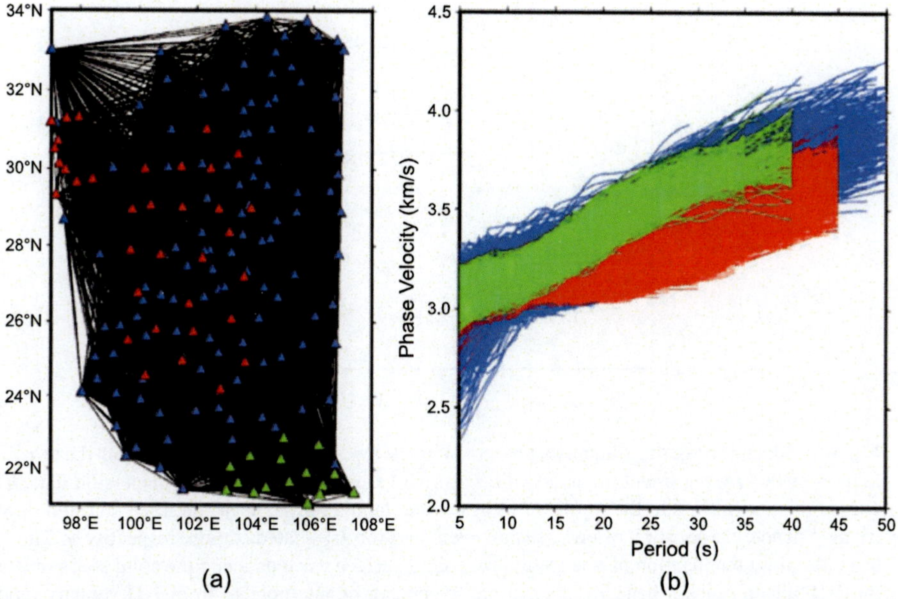

Fig. 4.6 **a** The distribution of path coverage of surface wave dispersion data used in the joint inversion. Red, green and blue filled triangles denote stations used to extract interstation ambient noise Rayleigh wave dispersion data in Yao et al. (2010), Qiao et al. (2018) and Yang et al. (2020b), respectively. **b** Rayleigh wave phase velocity dispersion curves, with color indicating different datasets (Figure from Liu et al., 2021)

travel time data, to directly invert for 3-D P- and S-wave velocity structures and simultaneously relocate the seismic events.

In this algorithm, the model is set in layers with depth relative to the mean sea level. To take into consideration of the topography effect in our study region, we set a layer at depth of − 5 km in the model parameterization. For the body wave travel time data, it traces the ray path between source and station using pseudo-bending method (Um & Thurber, 1987). For the surface wave dispersion data, the real ray path, which might be non-great circle due to heterogeneities in the structure, is traced using fast marching method (Rawlinson et al., 2004) at each period with the phase velocity. The depth sensitivity kernel is calculated using a layered model beneath each geographic point. Considering the large topographic variations in this region (from nearly 5 km in the Tibetan Plateau to less than 0.5 km in Sichuan Basin), it is not appropriate to use a constant layer thickness to calculate surface wave sensitivity kernel. Therefore, we slightly modified this algorithm by incorporating the true topography at each grid node (Fig. 4.7) to calculate the sensitivity kernels of surface wave dispersion data.

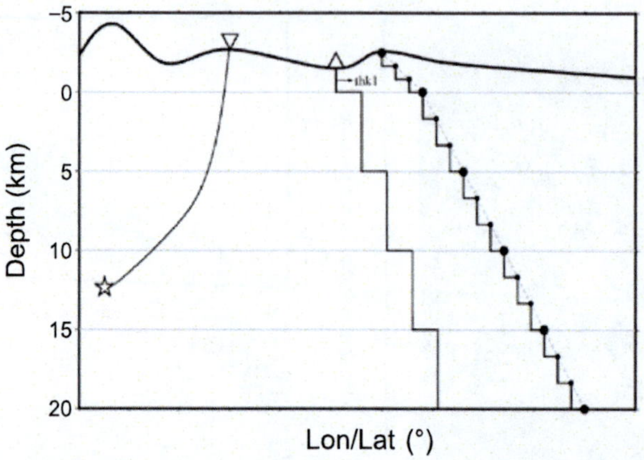

Fig. 4.7 Illustration of the scheme that we consider the topography for calculating both the sensitive kernels of body wave travel times and the sensitive kernels of surface wave dispersion data. The curved thick solid line in the top layer indicates the topography. The white filled star and down-triangle denote the location of one seismic event and the associated station, respectively. The up-triangle shows the location of one station recording surface wave data and the solid stairs indicate the 1-D velocity model beneath this station, which can be interpolated from 1-D velocity model of surrounded grids (black dots). The 1-D velocity model beneath one grid (shown in bigger black dots) is interpolated (smaller black dots and thick solid stairs) for calculating the depth sensitivity kernel of dispersion data. When we calculate the sensitive kernels of surface wave dispersion data, the thickness of the top layer above the mean sea level is set as the true topography after modification (e.g. thk1), instead of a constant value (e.g. 5 km) before modification (Figure from Liu et al., 2021)

4.5.2 *Tomographic Details and Model Resolution Tests*

4.5.2.1 Inversion Details

As we know, a good initial velocity model is helpful in linearized inversion. Thus, we incorporated the 3-D Vs model from Yang et al. (2020b) to construct the initial Vs model. To preserve the main features but avoid sharpness in the model of Yang et al. (2020b), we first applied Gaussian filtering to this model to build a smooth 3-D initial Vs model. We then converted the smooth initial Vs model to build the initial Vp model using the empirical relationship in Brocher (2005). Thus, a priori information from surface wave dispersion data, ZH ratios and receiver functions is implicitly included.

The multigrid inversion strategy was adopted to achieve multi-scale inversion. We first used a grid of 0.5° × 0.5°, and then a finer grid of 0.25° × 0.25° in the horizontal directions to perform the joint inversion. The vertical grid nodes were set relative to the mean sea level in depth, and the grid interval in the vertical direction varies from 5 km (−5 km to 50 km depth), to 10 km (50–70 km depth), and to 30 km (70–100 km depth), respectively. The inversion results of coarser grids (0.5° × 0.5°) were used as the initial model for the inversion of finer grids (0.25° × 0.25°). In this way, we can obtain higher resolution structures in the area with denser data coverage while still remaining stable features in the other areas.

To prevent the inversion results from being dominated by body wave or surface wave data in the joint inversion, an appropriate weight is chosen according to the variances of the two data types. The weight of body wave data is set to be 1.0 and the relative weight of surface wave data (α) is calculated as following:

$$\alpha = \sqrt{\frac{N_{BW}\sigma_{BW}^2}{N_{SW}\sigma_{SW}^2}} \tag{4.1}$$

where N_{BW} and N_{SW} are the numbers of body wave and surface wave travel times, σ_{BW} and σ_{SW} are the estimated uncertainties of body wave and surface wave travel times, respectively. Note that the relative weight (α) is calculated and updated before each iteration.

The root-mean-square travel time residuals of body- and surface-wave reduce to 0.4339 s (from 0.9285 s) and 2.062 s (from 4.161 s) after 11 iterations (Fig. 4.8), respectively. The body- and surface-wave travel time residuals of each single event-station pair is calculated using the initial and final inverted models, respectively. From the histogram distributions (Fig. 4.8), it can be seen that after the inversion,

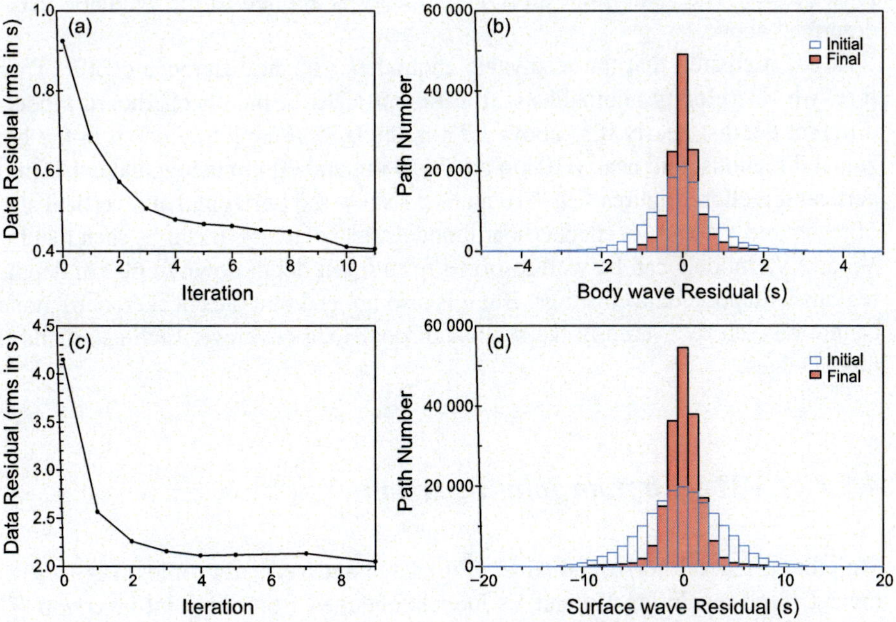

Fig. 4.8 **a** The evolution of body wave data residuals; **b** the histograms of body wave travel time residuals ($t_{obs} - t_{pre}$) before and after the inversion; **c** the evolution of surface wave data residuals; **d** the histograms of surface wave travel time residuals ($t_{obs} - t_{pre}$) before and after the inversion (Figure from Liu et al., 2021)

the travel time residuals of both body- and surface-wave are within a smaller range, indicating good fitness of the final inverted model with real data.

4.5.2.2 Checkerboard Resolution Tests

The checkerboard resolution tests were applied to investigate the resolution of the final inverted models by using the same distribution of events and stations with real data. Considering the insensitivity of surface wave dispersion data to sharp anomaly structures, here we constructed checkerboard model with a sinusoidal velocity variation in both horizontal and vertical directions with $\pm$ 5% perturbations of Vp and Vs relative to the reference velocity model. To better resolve the model resolution under the real ray path distribution, we chose our final inverted Vp and Vs models, instead of the initial velocity model, as the reference model in generating the checkerboard model. Then we computed the synthetic body wave travel times and surface wave dispersion data using the same distribution of sources and receivers as the real dataset by using pseudo-bending and fast marching methods, respectively. To simulate the manually picking errors, we add Gaussian distributed random noise with a standard deviation of 0.1 s to the synthetic body wave travel times. We also add Gaussian distributed random noise with a standard deviation of 2% to synthetic surface wave dispersion data, according to the analysis of phase velocity bias in Yao and van der Hilst (2009). The same joint inversion strategy is applied to the synthetic data as described above.

It is anticipated that the resolvable anomalies vary at difference depths. Therefore, we set velocity anomalies with different sizes in the checkerboard model at different depths: nearly 0.5° above 20 km and 0.75° below that depth in the horizontal directions, and nearly 10 km above 30 km and 20 km below that depth in the vertical direction. Figures 4.9, 4.10 and 4.11 show the horizontal and vertical slices of recovered Vp and Vs checkerboard models, respectively. It can be seen that both Vp and Vs models can be well resolved from 0 km depth down to 60 km depth in regions with good data coverage. But it is also noticed that the checkerboard models cannot be well recovered in the regions of poor data coverage, such as around the model edges.

4.5.3 CVM V1.0 from Joint Inversion

We plotted several horizontal slices (Fig. 4.12) and vertical profiles (Fig. 4.13) of the 3-D high-resolution Vp and Vs models obtained from the joint inversion. The areas with poor data coverage and poor checkerboard recovery were cut off in the depth slices of 5, 10, 20, 30, 40 and 50 km (Fig. 4.12). Four east–west vertical profiles (AA′, BB′, CC′ and DD′ in Fig. 4.13) and three south-north vertical profiles (EE′, FF′ and GG′ in Fig. 4.13) are mostly inside the well-resolved area. From both horizontal

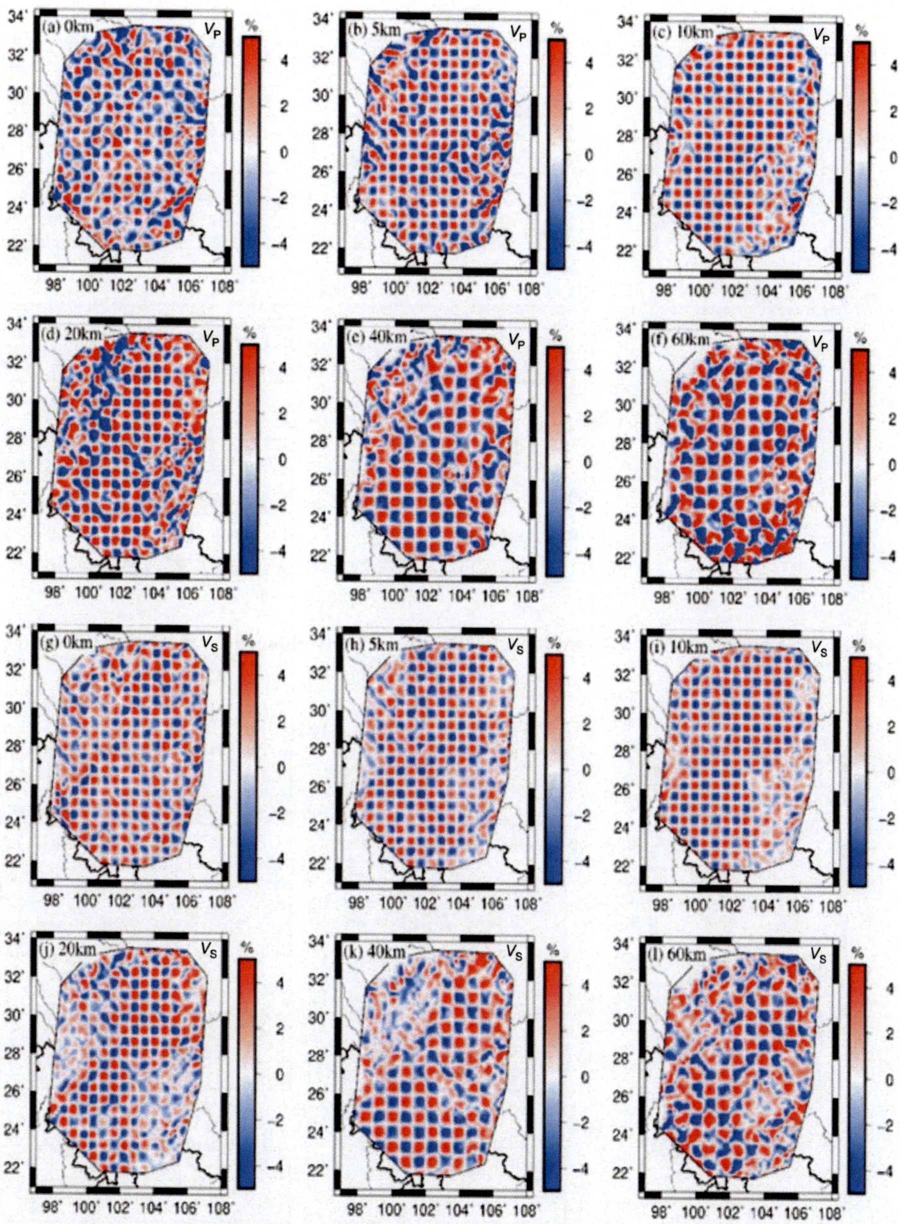

Fig. 4.9 Recovered checkerboard models of P-wave (**a–f**) and S-wave (**g–l**) velocity at selected depths of 0, 5, 10, 20, 40 and 60 km, respectively. The checkerboard size is nearly 0.5° above 20 km and 0.75° below that depth in the horizontal directions, and nearly 10 km above 30 km depth and 20 km below that depth in the vertical direction (Figure from Liu et al., 2021)

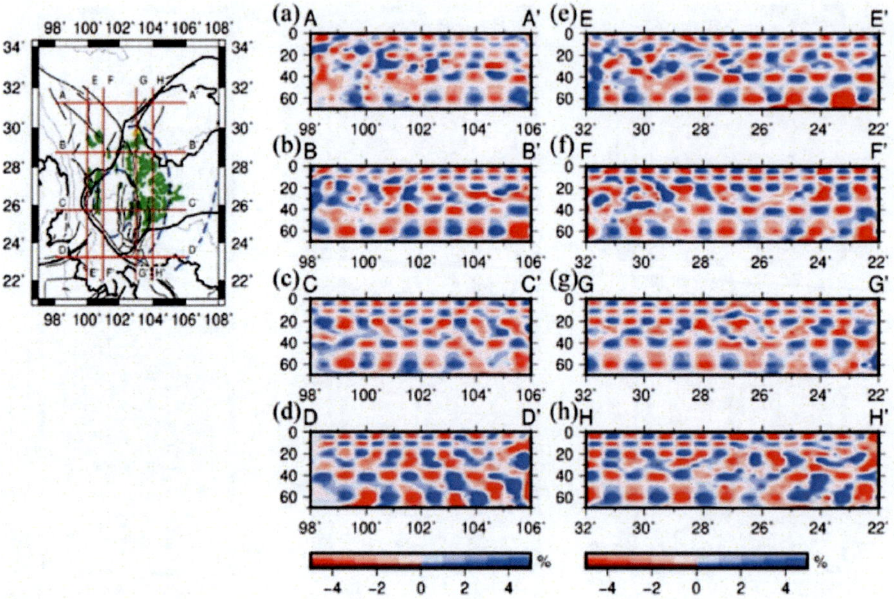

Fig. 4.10 Recovered checkerboard model of P-wave velocity along eight vertical profiles (Figure from Liu et al., 2021)

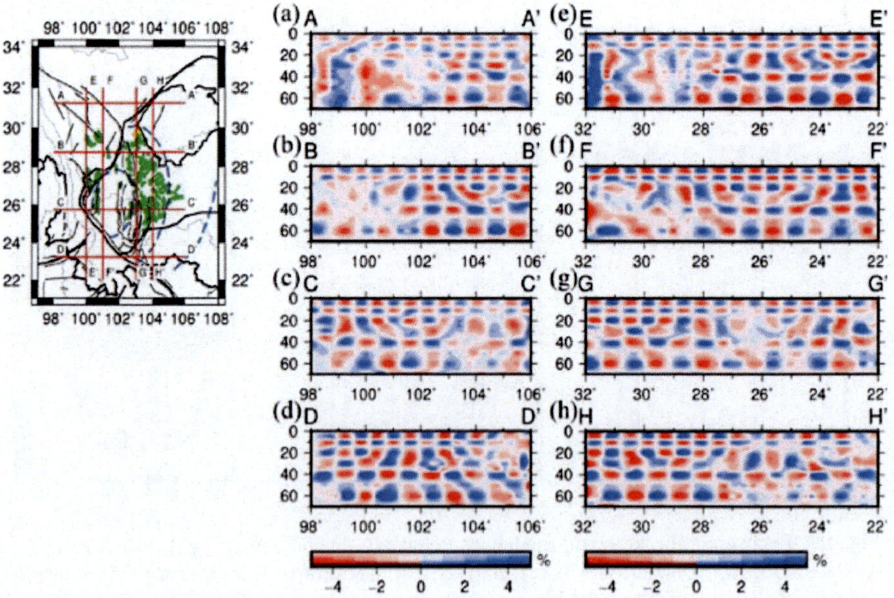

Fig. 4.11 Recovered checkerboard model of S-wave velocity along eight vertical profiles (Figure from Liu et al., 2021)

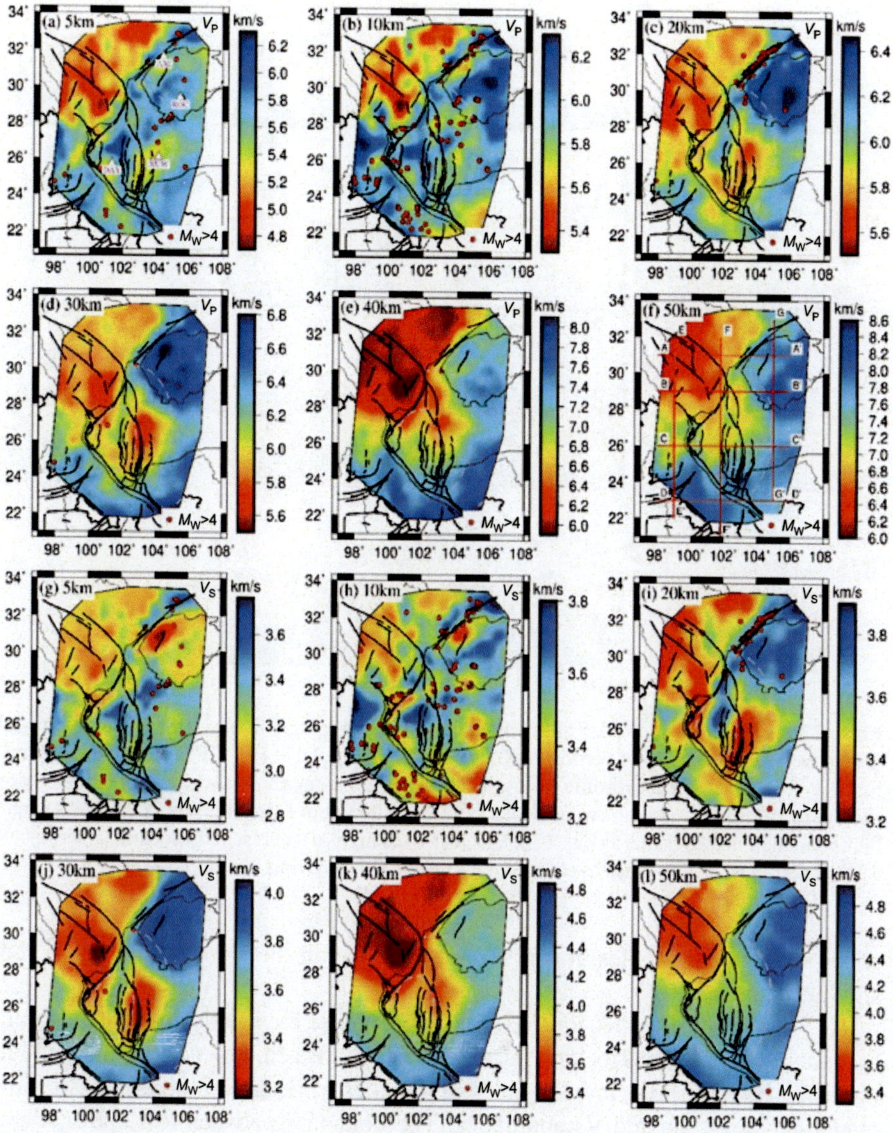

Fig. 4.12 Map view of the inverted P-wave (**a–f**) and S-wave (**g–l**) velocity models at selected depths of 5, 10, 20, 30, 40 and 50 km, respectively. The thin and thick solid black lines denote the block boundaries and major faults in our study region, respectively. The boundaries of the inner and intermediate zones of the ELIP are shown in grey dashed lines. The relocated seismic events with Mw ≥ 4.0 are plotted in red filled circles. Four selected stations, AXI, DAY, ROC and XUW, of which we compare 1D models, Rayleigh wave dispersion curves and ZH ratios in Sect. 5.4, are plotted as white filled triangles in (**a**). Note that velocity structures without good resolution are cut off according to the checkerboard test (Figure from Liu et al., 2021)

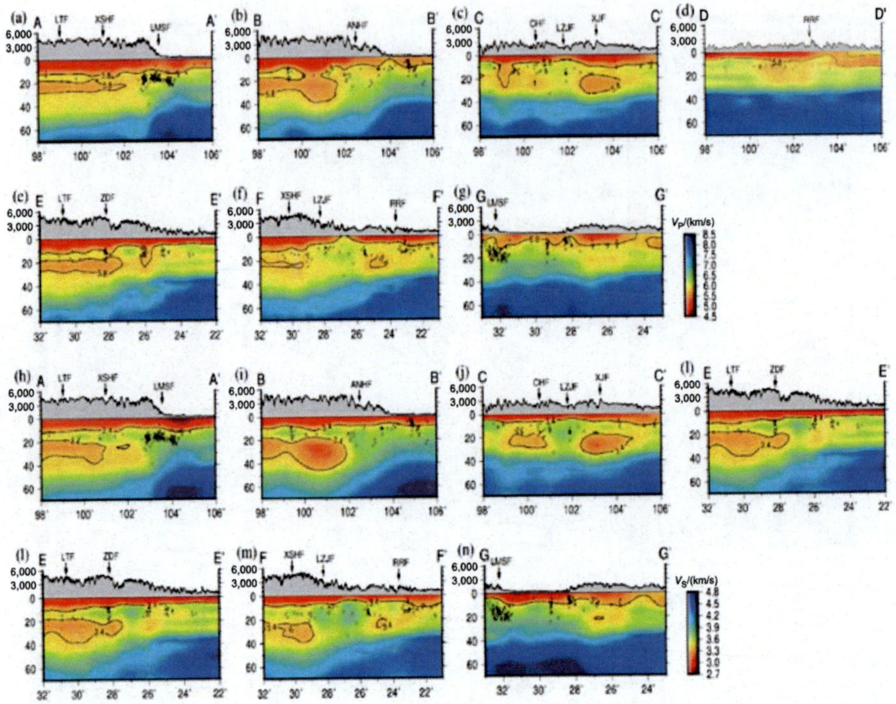

Fig. 4.13 Seven vertical profiles of the inverted P-wave (**a–g**) and S-wave (**h–n**) velocity models, respectively. The locations of vertical profiles AA', BB', CC', DD', EE', FF' and GG' are plotted in Fig. 4.12f. For each vertical profile (**a–n**), the grey shaded areas above show the topography along each vertical profile, with arrows and the associated text above denoting the locations of major faults and their names (same as Fig. 4.1). Black crosses in (**a–n**) represent the locations of relocated earthquakes (Mw $\geq$ 4.0) close to each vertical profile (Figure from Liu et al., 2021)

slices and vertical profiles, it can be seen that overall the Vp and Vs models show similar features.

At 5- and 10-km depths, low Vp and Vs anomalies are mostly associated with the Sichuan Basin and the Tibetan Plateau. In Sichuan Basin, it is noticed that only the northwestern part shows low Vp and Vs anomalies while the southeastern part shows moderately high Vp and Vs anomalies. The boundary between low and moderate velocity anomalies is associated with the Longquanshan fault. Low Vp and Vs anomalies also appear near the Red-River Fault and the Xiaojiang Fault. Inside the Chuan-Dian diamond block, it mainly exhibits relatively high Vp and Vs anomalies. And the Emerishan large igneous province (ELIP) is associated with high Vp and Vs anomalies. Along the Longmenshan fault, Vp and Vs are relatively higher than both sides.

At 20-km depth, the entire Sichuan Basin starts to display relatively entire high Vp and Vs anomalies while the Tibetan Plateau still exhibits relatively low Vp and Vs anomalies. The Longmenshan fault clearly separates the velocity contrasts between the Tibetan plateau and the Sichuan Basin. Such features can also be clearly seen in vertical profiles AA′ and BB′ (Fig. 4.13b, c, i, j). The ELIP keeps high Vp and Vs anomalies at 20-km depth, which can also be observed from vertical profiles CC′ and FF′ (Fig. 4.13d, g, k, n). However, these high Vp and Vs anomalies associated with the ELIP disappear at 30-km depth. Compared to depth slices of 5 and 10 km, two predominant low velocity zones (LVZs) are observed at the depth of 30 km, which are mostly located to the west of the Longmenshan fault and Lijiang-Xiaojinghe fault and along the Xiaojiang fault, respectively. Such LVZs in the mid-lower crust are also clearly shown in most vertical profiles except for the southernmost (DD′) and easternmost (GG′) ones. Two mid-lower crustal LVZs are separated by the relatively high velocity anomaly corresponding to the ELIP near the Anninghe, Zemuhe and Lüzhijiang faults, which can be seen clearly at profile CC′. They are defined as two low-Vs channels in Bao et al. (2015) and are also revealed by Yang et al. (2020b).

At deeper depths (40–50 km), the Vp and Vs structures show very similar features. Both Vp and Vs are lower in the northwestern part but higher in the southeastern part of our study area, indicating the variation of crustal thickness in Southwest China (Wang et al., 2017).

The relocations of earthquakes with magnitude larger than 4 are also shown in Figs. 4.12 and 4.13. They are mostly located above 20-km depth and most of them are either distributed along faults or around the boundaries between high and low velocity anomalies. The relocations of more events ($M_w > 2.0$) are shown in Figs. 4.14, 4.15 and 4.16. Figure 4.14 shows the distributions of seismic events before and after relocation in horizontal view. Earthquakes with magnitude larger than 3.0 are mostly distributed along faults and the seismicity is more concentrated after relocation. We also plotted several profiles to show the vertical distributions of event locations with velocity structures in Figs. 4.15 and 4.16. It can be clearly seen that our relocations from joint inversion are more concentrated than initial locations in both horizontal and vertical directions. Generally, the relocated events are distributed in the relatively high velocity anomalies area above 30-km depth, although some seismic events are relocated in the shallow low velocity zones.

4.5.4 *Validation of CVM V1.0*

To further validate our jointly inverted models from joint inversion of body and surface wave data, we first compared them with existing models in the region that were obtained using different methods and different data in this region (Shen et al., 2016; Xin et al., 2019; Yang et al., 2020b) with 1-D profiles. Then we compared the simulated Rayleigh wave dispersion data and ZH ratio data from different models with the observed data. P wave travel times from active-source seismic experiments, which were not used for inversion, were also used to compare with the predicted travel times simulated by different models.

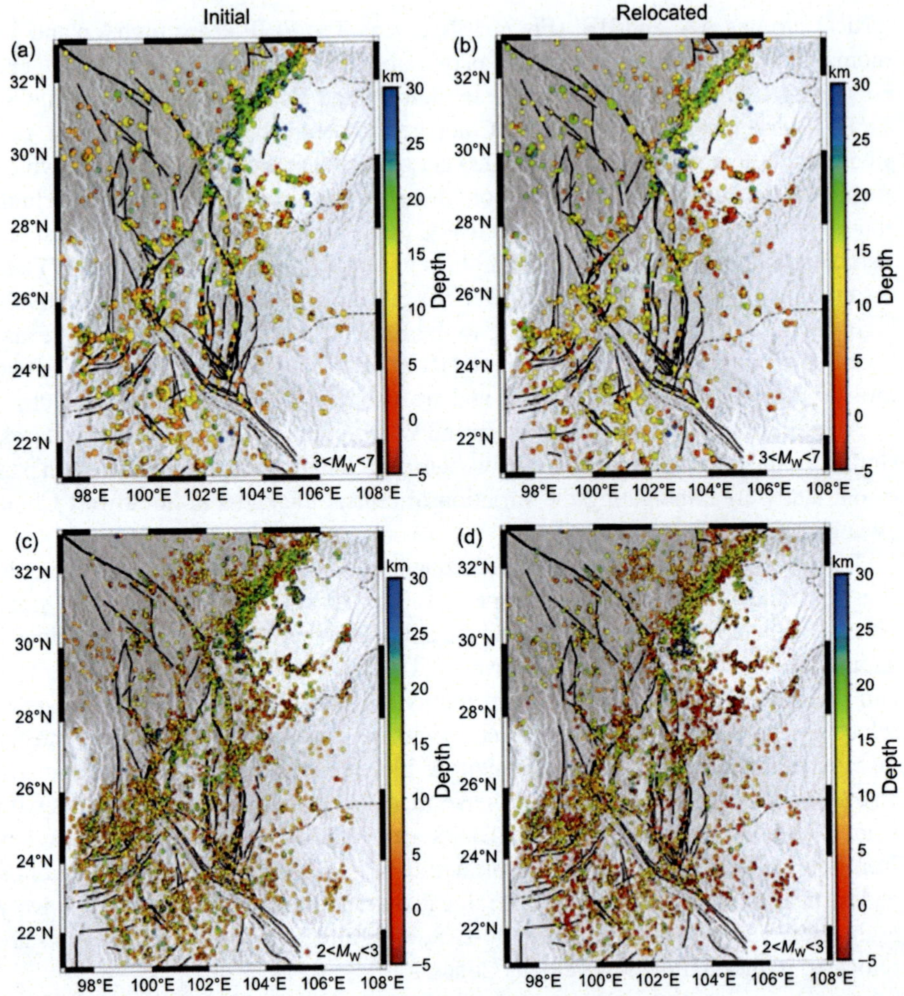

Fig. 4.14 The horizontal distributions of seismic events with initial and relocated locations. The events are colored by depth and scaled by magnitude. **a–d** show events with $M_w > 3$ and $2 < M_w < 3$, respectively (Figure from Liu et al., 2021)

4.5.4.1 Comparison of 1-D Profile with Other Models

Three models from other studies using different data and methods are chosen for comparison: (1) a reference crust and uppermost mantle Vs model of China from surface wave tomography by Shen et al. (2016), (2) USTClitho1.0, new lithospheric Vp and Vs models of continental China from body wave travel time data by Xin et al. (2019), (3) SWChinaVs_2018, a new crustal Vs model of Southwest China from joint inversion by Yang et al. (2020b). For comparison, 1D velocity profiles beneath each seismic station are obtained by interpolating them from 3-D models. We only compare our Vp model with the model of Xin et al. (2019) since Shen

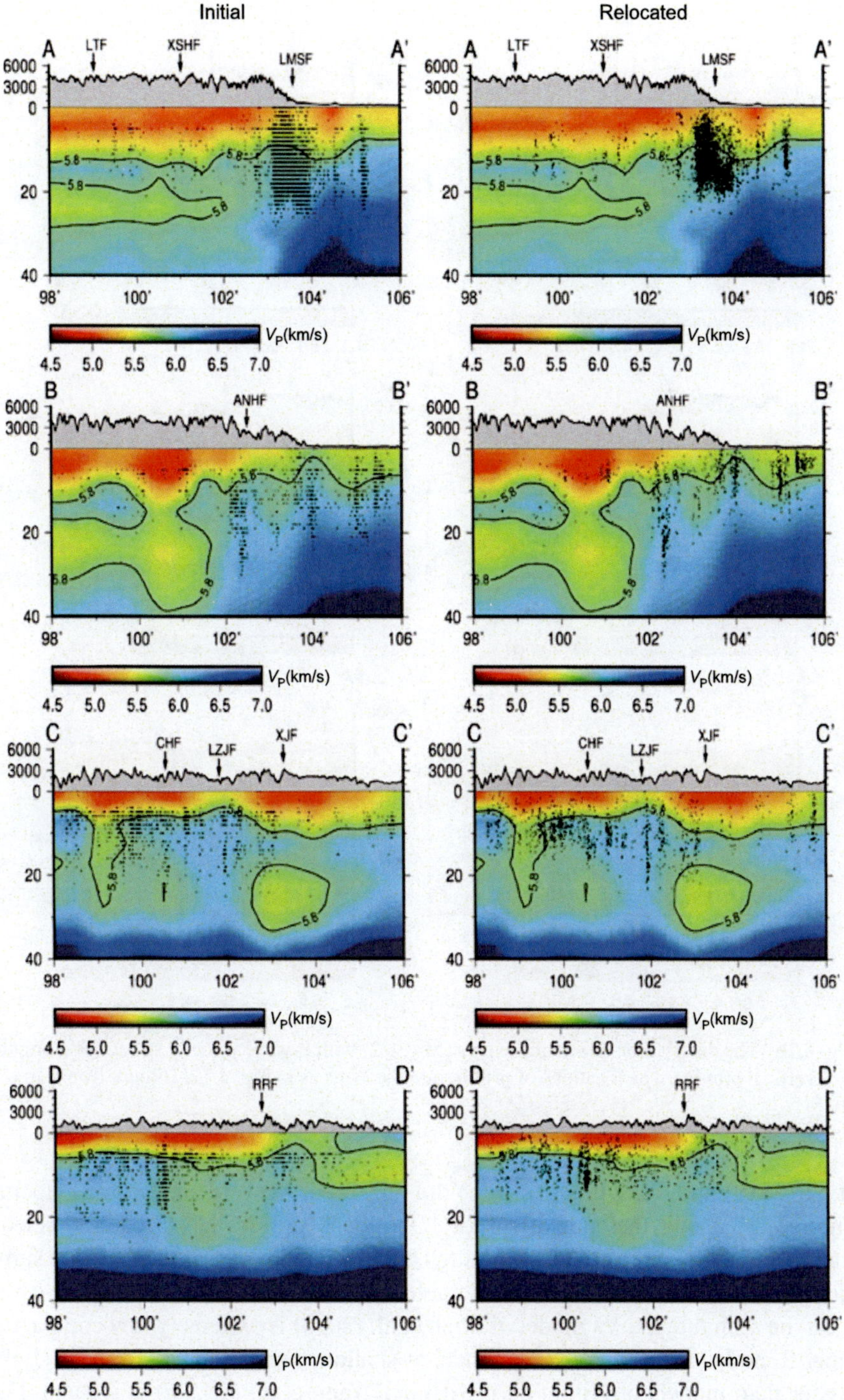

Fig. 4.15 The distribution of seismic events ($M_w \geq 2$) with P-wave velocity structures along four E-W vertical profiles. The locations of profiles are the same as in Fig. 4.12f (Figure from Liu et al., 2021)

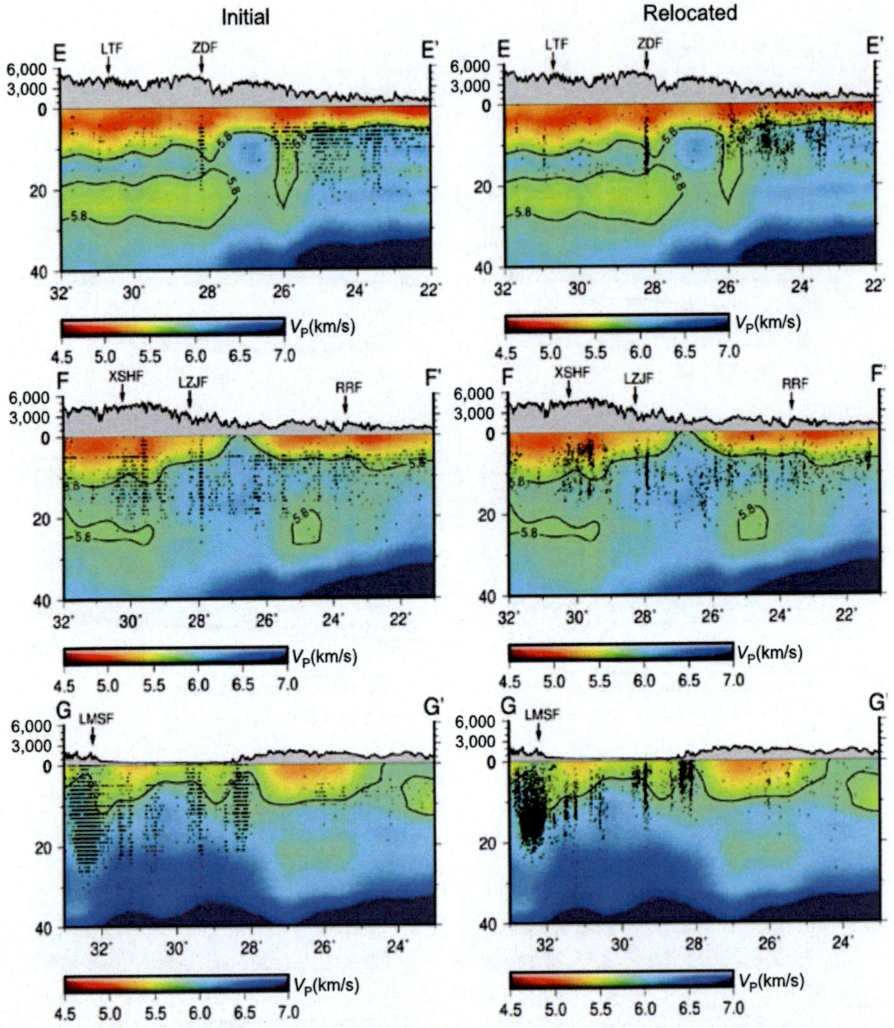

Fig. 4.16 The distribution of seismic events (Mw ≥ 2) with P-wave velocity structures along three N-S vertical profiles. The locations of profiles are the same as in Fig. 4.12f (Figure from Liu et al., 2021)

et al. (2016) and Yang et al. (2020b) did not directly invert for P-wave velocities. Figure 4.17 shows the comparison of 1D velocity profiles at 4 randomly selected stations (marked in Fig. 4.12a). Although the main trends of these models are similar along depth but there still exist considerable differences. For Vs, at depths above 5 km, it can be seen that the Vs model of Shen et al. (2016) is extremely low compared to other 3 models. Our inverted Vs model at shallow depths (above 5 km) is slightly higher than models of Xin et al. (2019) and Yang et al. (2020b) at stations DAY, ROC and XUW. From about 5 to 40 km depth, Vs models from our study, models of Shen et al. (2016) and Xin et al. (2019) are very similar to each other and change

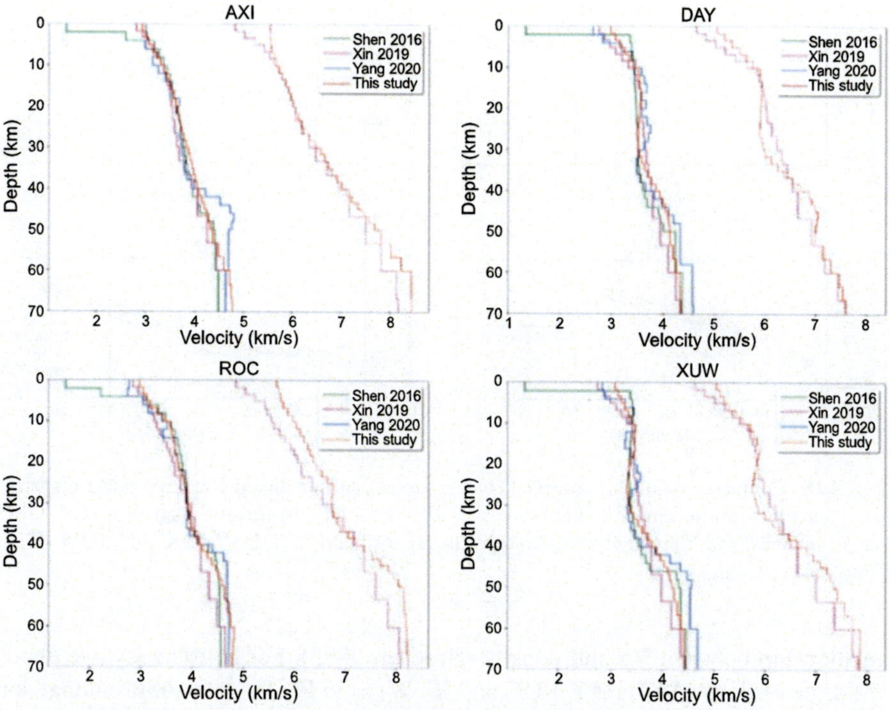

Fig. 4.17 Comparisons of our 1D velocity models (red lines) above 70-km depth below the sea level with Shen et al. (2016) (green lines), Xin et al. (2019) (magenta lines) and Yang et al. (2020b) (blue lines) at station AXI, DAY, ROC and XUW. Solid and dashed lines represent Vs and Vp model, respectively. The locations of stations can be found in Fig. 4.12a (Figure from Liu et al., 2021)

smoothly with depth, while the Vs model of Yang et al. (2020b) indicates larger anomalies along depth and varies more abruptly at depths. Below 40 km depth, the trends of these 4 Vs models are similar but the values differ at all stations. For Vp, the discrepancies between our inverted Vp model and the Vp model of Xin et al. (2019) at shallow depths can reach up to almost 1 km/s at station AXI and ROC. Note that station DAY and XUW are located above the two LVZs, respectively (Fig. 4.12), and low Vp at depths of 20–30 km can be seen in the 1D profiles of our inverted Vp model but not in that of Xin et al. (2019).

4.5.4.2 Comparison with Rayleigh Wave Dispersion Data

We evaluate the reliability of our inverted Vp and Vs models by comparing the predicted and observed Rayleigh wave dispersion data. The Rayleigh wave dispersion data at each station obtained by Yang et al. (2020b) are chosen as observations. We interpolate our 3-D velocity models to build 1D Vp and Vs profiles beneath each station. Then, 1-D density profile beneath each station is derived according to the

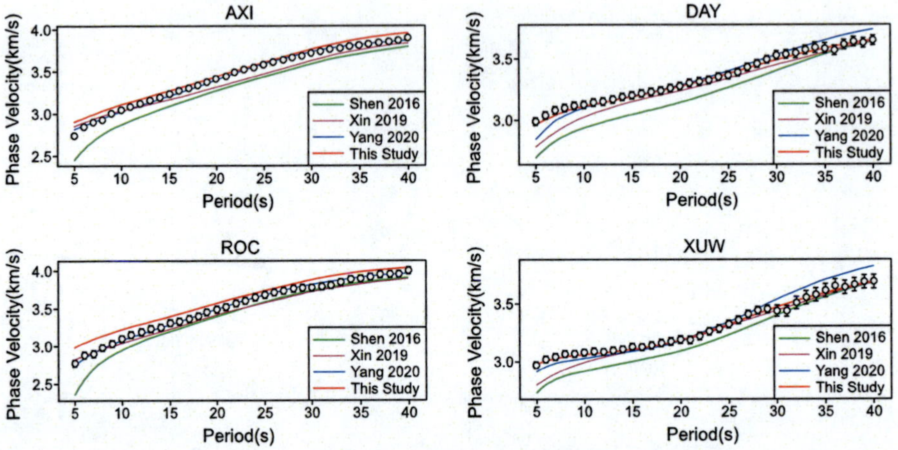

Fig. 4.18 Comparisons of the observed (black circles) and predicted Rayleigh wave dispersion data using our inverted models, the models of Shen et al. (2016) (green lines), Xin et al. (2019) (magenta lines) and Yang et al. (2020b) (blue lines) at station AXI, DAY, ROC and XUW (Figure from Liu et al., 2021)

relationship between Vp and density (Brocher, 2005). For consistency, we choose the same stations (AXI, DAY, ROC and XUW) as in Fig. 4.17 to show comparisons between the observed and predicted Rayleigh wave dispersion data using different velocity models (Fig. 4.18). In general, the predicted Rayleigh wave dispersion data using our inverted models can better fit the observed data than using models of Shen et al. (2016) and Xin et al. (2019). For station DAY and XUW, our predicted Rayleigh wave dispersion data are closer to the observed data than Yang et al. (2020b) at short (5–10 s) and long (> 30 s) periods. For station AXI and ROC, the predicted Rayleigh wave dispersion data of Yang et al. (2020b) are closer to the observed data than our inverted models at short (5–10 s) and long (> 30 s) periods. For intermediate periods (10–30 s), the predicted Rayleigh wave dispersion data of our inverted models and Yang et al. (2020b) are both close to the observed data.

4.5.4.3 Comparison with ZH Ratio Data

Similar to Sect. 4.5.4.2, we also calculate the Rayleigh wave ZH ratios at each station using different models, and compare with observations obtained by Yang et al. (2020b). Figure 4.19 displays the observed and predicted Rayleigh wave ZH ratios at station AXI, DAY, ROC and XUW using different velocity models. As expected, the predicted Rayleigh wave ZH ratios using models of Yang et al. (2020b) can better fit the observed data. The predicted Rayleigh wave ZH ratios of Shen et al. (2016) have large deviations from the observed data, especially at short periods. But it is hard to tell whether the predicted Rayleigh wave ZH ratios using our inverted models or models of Xin et al. (2019) have better data fitness at these stations. Back to our predicted Rayleigh wave ZH ratios, it can be noticed that they are not close to the

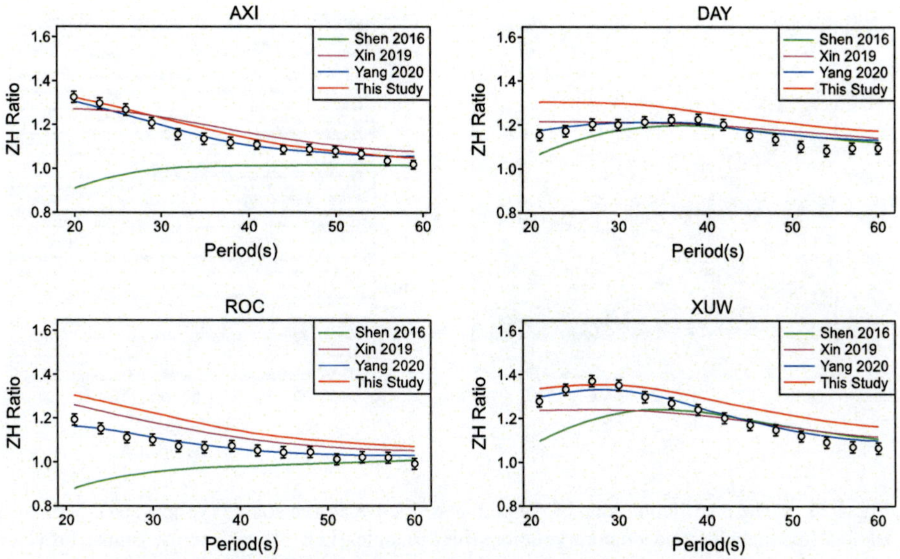

Fig. 4.19 Comparisons of the observed (black circles) and predicted Rayleigh wave ZH ratios using our inverted models, the models of Shen et al. (2016) (green lines), Xin et al. (2019) (magenta lines) and Yang et al. (2020b) (blue lines) at station AXI, DAY, ROC and XUW (Figure from Liu et al., 2021)

observed data except for station AXI (Fig. 4.19). At station XUW, our predicted Rayleigh wave ZH ratios are close to the observed data at 20–30 s periods, but are about 7% higher than the observed data at periods of 45–60 s. For stations DAY and ROC, the differences between the predicted and observed Rayleigh wave ZH ratios can reach to about 12% at short periods.

4.5.4.4 Comparison with P-Wave Travel Times from Active Sources

Considering that the Rayleigh wave dispersion data and ZH ratios are less sensitive to the P-wave velocities, we use an active-source dataset to evaluate our inverted Vp model more effectively and directly. We collect stacked signals simulated by airgun shot experiments at a fixed airgun source in Yunnan Province (Wang et al., 2012), recorded at 67 permanent stations and 271 temporary stations (Fig. 4.20). The P-wave first arrivals are first manually picked, then re-picked by applying a new automatic arrival time picking algorithm (Guo et al., 2018) based on Akaike Information Criteria. After that, 83 high-quality P-wave first arrivals are obtained among 114 manually picked ones. Predicted P-wave travel times are calculated using our inverted Vp model, and the Vp models of Xin et al. (2019), Yang et al. (2020b) and Shen et al., (2016) and compared to the observed P wave arrival times, respectively. Note that Vp models of Yang et al. (2020b) and Shen et al. (2016) are converted from Vs models following the empirical relationship of Brocher (2005).

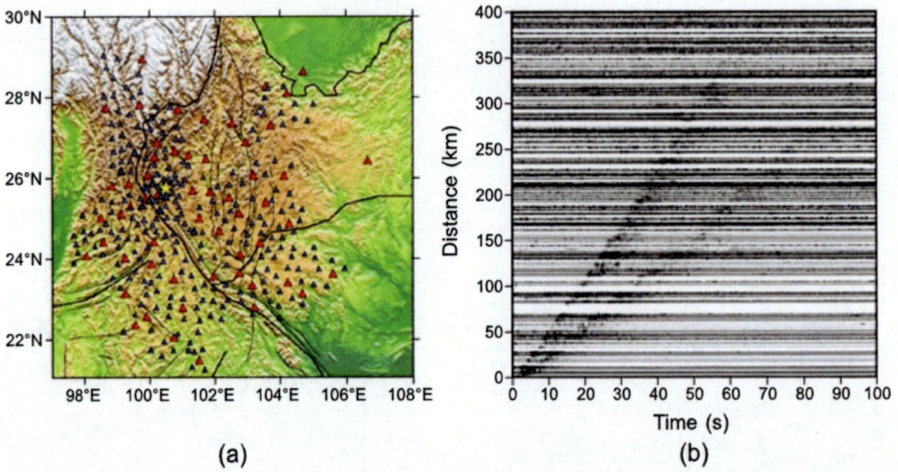

Fig. 4.20 **a** The distribution map of the permanent active airgun source (yellow star), permanent stations (red triangles) and temporary stations (blue triangles) in Binchuan county, Yunnan Province. **b** Stacked waveforms of airgun source signals (Figure from Liu et al., 2021)

Figure 4.21 shows histograms of absolute and relative travel time residuals between the observed and predicted P-wave first arrival travel times using four different models, respectively. Generally, the absolute and relative travel time residuals of our inverted models are mostly within [−2 s, 2 s] and [−10%, 10%] range, respectively. However, there are still a few P-wave arrival times that have 30–40% relative travel time residuals (Fig. 4.21b). These large relative travel time residuals are associated with stations close to the active source and have very small arrival times (about 1.0 s). It is likely that these large relative travel time residuals are caused by either picking errors or inaccuracy of the very shallow structures or both.

The data fitness of different models does not show a very significant difference. Quantitatively, the standard deviations of the absolute and relative P-wave travel time residuals using our inverted models are smaller. However, the medians of the absolute and relative P-wave travel time residuals using models of Xin et al. (2019) are closer to zero. In addition, we also predict synthetic P- and S-wave travel times of the body wave dataset used in the real data inversion, which indicates that our inverted models have better real data fitness.

4.6 Discussion of the CVM V1.0

4.6.1 Error Analysis of the Jointly Inverted Models

We adopt bootstrapping analysis, which is a statistical resampling method, to estimate the uncertainties of the jointly inverted models (the final models). First, we randomly

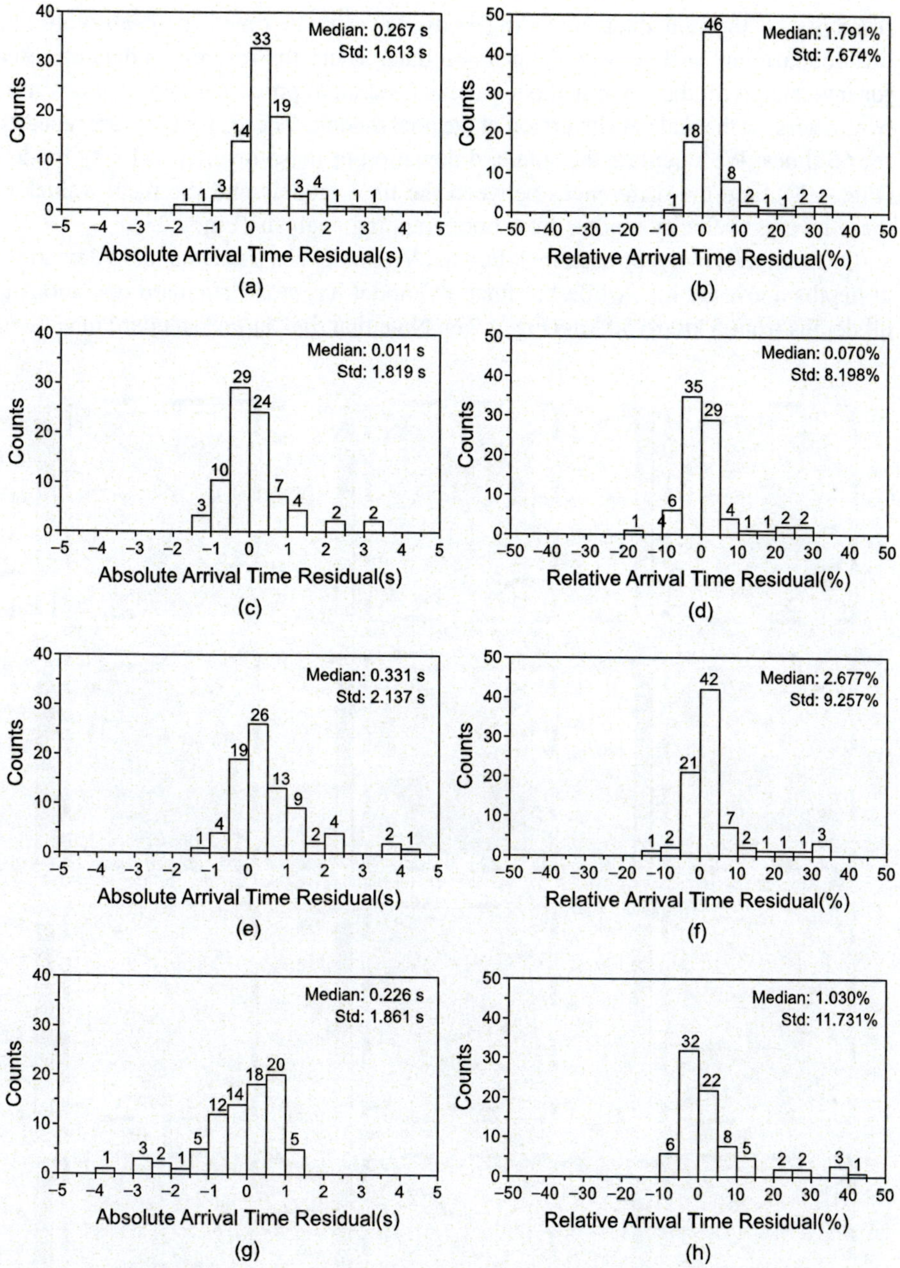

Fig. 4.21 Histograms of absolute residuals ($t_{obs} - t_{pre}$, left column) and relative residuals ($(t_{obs} - t_{pre})/t_{obs} \times 100$, right column) between the observed and predicted airgun P-wave arrival times generated by different Vp models: our inverted Vp model (**a**, **b**), the Vp model of Xin et al. (2019) (**c**, **d**), Yang et al. (2020b) (**e**, **f**) and Shen et al. (2016) (**g**, **h**). Note that Vp models of Yang et al. (2020b) and Shen et al. (2016) are converted from their inverted Vs models. The median and standard deviations are texted in the top right corner of each panel (Figure from Liu et al., 2021)

resample all the real data, including body wave travel times (both absolute and differential), and surface wave dispersion data. Then, the resampled data are used for inversion with the same inversion details (iterative processes and regularization parameters) as the real data inversion of the final models. These processes are repeated for 60 times. We calculate the standard deviation of these inverted velocity models (Fig. 4.22), and the differences between the final models and the mean values of these inverted velocity models from bootstrapping analysis (Fig. 4.23).

Qualitatively, it can be seen that the final Vp model has small standard deviations at depths above 30 km, while the final Vs model has small standard deviations at all depths from 5 km to 50 km (Fig. 4.22). Note that the "large" standard deviations

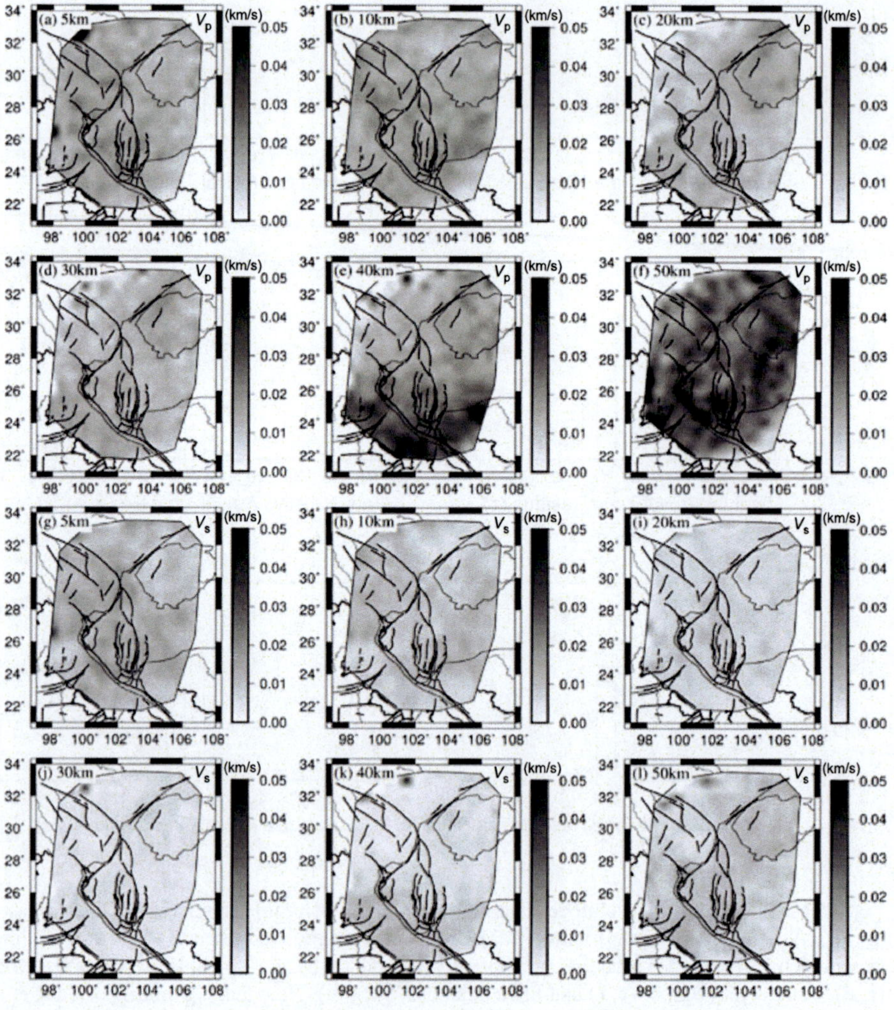

Fig. 4.22 Map view of the standard deviations of P-wave (**a–f**) and S-wave (**g–l**) velocity models from bootstrapping analysis. Other symbols are the same as in Fig. 4.12 (Figure from Liu et al., 2021)

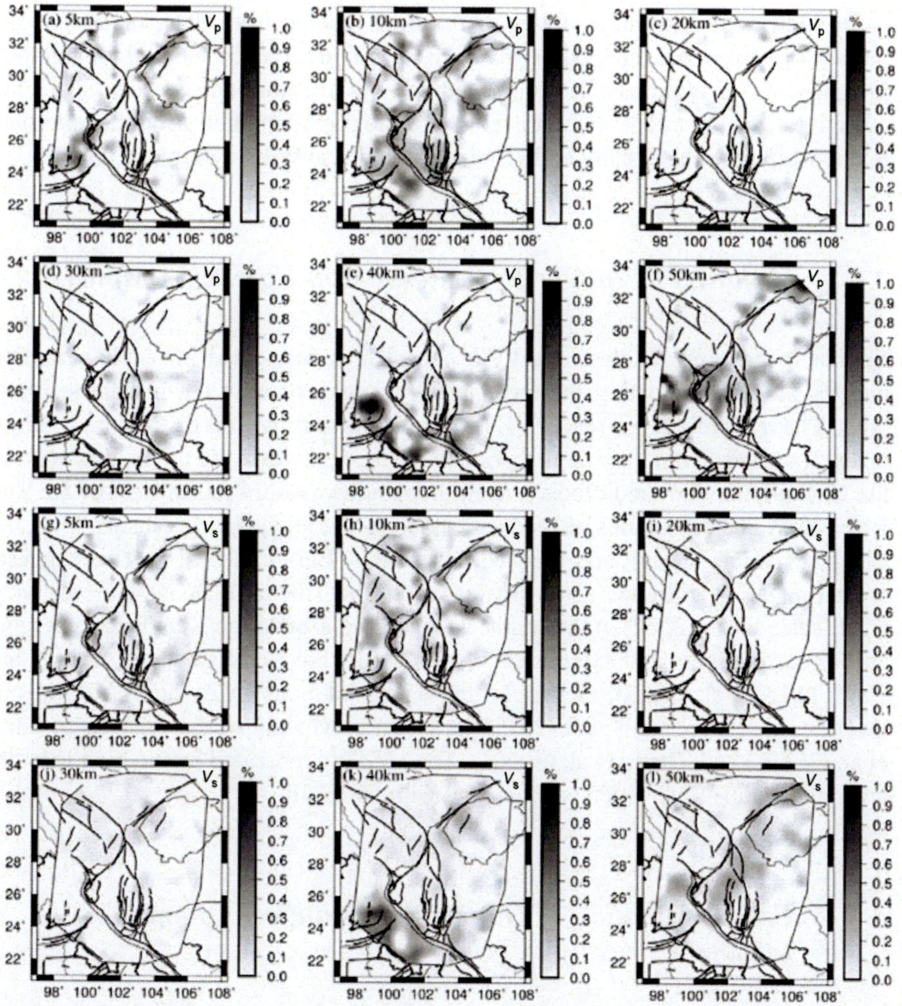

Fig. 4.23 Map view of model differences (in percentage) between the mean P-wave (**a–f**) and S-wave (**g–l**) velocity models from bootstrapping analysis and the final inverted velocity models (SWChinaCVM-1.0). Other symbols are the same as in Fig. 4.12 (Figure from Liu et al., 2021)

of the final Vp model at depths of 40 km and 50 km are ~ 0.05 km/s. Also, it can be seen that both mean Vp and Vs models from bootstrapping analysis have small differences (< 0.2%) with the final models in most regions at all depths (Fig. 4.23). Quantitatively, the standard deviations of the final Vp model are less than 0.05 km/s at 99.63% inversion grid nodes and the standard deviations of the final Vs model are less than 0.05 km/s at 99.94% inversion grid nodes. Compared with the final models, the mean Vp model has less than 0.5% differences at 99.10% inversion grid nodes and the mean Vs model has less than 0.5% differences at 99.55% inversion grid nodes.

In addition to the uncertainties of the final models from data, random errors might exist given that the models are inverted from a travel-time based tomography method with assumptions of isotropic medium. Also, depths of discontinuities are not taken into considerations in the travel time tomography method. And the empirical relationship between density and velocities could also bring systematic bias.

4.6.2 Major LVZs in Our High Resolution Vp and Vs Models

Similar to previous studies, two LVZs are clearly delineated in our inverted Vp and Vs models, indicating the mechanically weak zones in the middle to lower crust. The relatively large-scale LVZ (LVZ1) in the northwest is distributed continuously. It extends from the SongPan-Ganze block southwestward to the northwest of the Chuan-Dian diamond block, and further southwestward to the west of the Red River fault. The boundaries of LVZ1 are associated with the Longmenshan fault, the Lijiang-Xiaojinghe fault and the Red River fault (Bai et al., 2010). This suggests crustal materials extruded southeastward from the Tibetan Plateau. The relatively small-scale LVZ (LVZ2) in the southeast is located around the Xiaojiang fault zone and its eastern area, and extends southward across the Red River fault (Bao et al., 2015; Qiao et al., 2018). Generally, the depths of both LVZ1 and LVZ2 in our inverted models are shallower (20–30 km) than previously studies, similar to Yang et al. (2020b) and Zhang et al. (2020b). The LVZ2 is separated by the high velocity anomaly from LVZ1, which is associated with the ELIP.

4.6.3 Cross-Validation of Our High Resolution Vp and Vs Models

To cross-validate our inverted models, we first compare 1D velocity profiles (both Vp and Vs) beneath several stations with that of Shen et al. (2016), Xin et al. (2019) and Yang et al. (2020b) (Fig. 4.17). Then, we also compare the Rayleigh wave dispersion data (Fig. 4.18) and ZH ratio data (Fig. 4.19) predicted by these models with the observations. It can be clearly seen that these four Vs models and two Vp models are mainly consistent in the mid-lower crust (10–40 km) overall. But they still show non-negligible differences in the uppermost and lowermost crust, especially above 5 km depth. Those major differences are likely caused by the different constraints on velocity structures of different data types, especially at shallow and deep depth. For example, ZH ratios are very sensitive to very shallow structures in the crust while receiver functions are sensitive to interfaces in the crust and upper mantle. In addition to that, different inversion methods could also bring discrepancies into velocity structures even using the same dataset. In general, our inverted models show good consistency on large scale structures with previous studies but provide finer features, and can mostly fit the observed Rayleigh wave dispersion data at stations and P-wave travel times from the active source. Since both body wave travel times and surface wave dispersion data used in this study are lack of good constraints on velocity structures at very shallow depths, our inverted models cannot well fit the observed ZH ratio data. In this case, adding ZH ratio data would be very helpful to improve constraints on very shallow velocity structures (e.g., Yang et al. 2020b).

4.6.4 *Improvement of the Crust and Uppermost Structure in Southwest China*

Many tomographic studies have been carried out in Southwest China to determine its crustal velocity structures. However, most Vp and Vs models proposed recently were determined by only inverting a single type of data, such as body wave travel times (Huang et al., 2019; Xin et al., 2019) or surface wave dispersion data (Shen et al., 2016; Zhang et al., 2020b). Both velocity models of Xin et al. (2019) and Shen et al. (2016) are presenting the crust and upper mantle structures in continental China, thus only providing smooth constraints on the velocity structures in Southwest China with few details due to the limitations of resolution. Huang et al. (2019) presented 3-D Vp and Vs models of the crust and upper mantle in Southwest China by local and teleseismic P- and S-wave travel time tomography, while Zhang et al. (2020b) obtained a 3-D Vs model of the crust and upper mantle in Southwest China by directly inversion of Rayleigh wave dispersion data. These two studies have relatively higher resolution in the crust (about 100 km), but still cannot provide information for very fine crustal structures. Moreover, there are some Vs models presented by joint inversion of two or three types of data (Liu et al., 2014; Bao et al., 2015; Yang et al., 2020b). Study regions of Liu et al. (2014) and Bao et al. (2015) did not cover the whole Southwest China even though they used very dense arrays and revealed very fine structures. Yang et al. (2020b) proposed a new crustal Vs model in Southwest China (exactly the same region as this study) by joint inversion of surface wave dispersion data, ZH ratio and receiver function. Their new model has relatively higher resolution than previous models. However, it was obtained by a stepwise linearized joint inversion method (Zhang & Yao, 2017) at each station and then interpolated to a 3-D model. Thus, it is not a directly inverted 3-D model and some bias might be brought due to interpolation. Overall speaking, due to the limitation of data and method, there is not a high-resolution community velocity model of both P- and S-wave velocities of the crust and uppermost mantle in Southwest China to date.

It is known that a high-resolution velocity model is very important for studying earthquake physics, simulating strong ground motion and assessing seismic hazards. By incorporating a large quantity of body- and surface-wave travel time data with high quality and applying a directly joint inversion method, we have obtained 3-D Vp and Vs models of the crust and uppermost mantle in Southwest China and relocated seismic events simultaneously in this study. Through resolution tests, model uncertainty analysis, cross-validation and data fitting evaluation, we think that our inverted velocity models are with high resolution and stability. But it is still difficult to estimate the resolution or error of the Vp/Vs model directly determined by Vp and Vs as they usually have different quality. It is not suggested for users to directly use the calculated Vp/Vs. The Vp/Vs model will be improved in our future work. Although there are a lot of efforts to be made towards achieving a CVM such as in southern California, our inverted models can serve as the first version of community velocity model in Southwest China for future studies, including using, testing and improving this model.

4.7 Constructing Velocity Models of Other Scales in Southwest China

Besides the regional block scale CVM of both P- and S-wave velocities in the crust and uppermost mantle in Southwest China, other studies have been conducted to construct velocity models of other scales (Figs. 4.2 and 4.3) in Southwest China. For the construction of fault zone scale and urban scale models, Zhang et al. (2020a) conducted earthquake relocation and 3-D body wave travel time tomography mainly based on the dense 2-D Binchuan array data (composed of more than 400 short-period seismographs, with an average station spacing of 2 km; Fig. 4.2b) in Yunnan. They obtained a high-resolution 3-D upper crustal P-wave and S-wave velocity model (9 km depth above) of the Binchuan Basin and its adjacent area (Fig. 4.24; the area indicated by the blue dotted line in Fig. 4.2b). The lateral resolution of this model can reach 5 km. This model provides a good high-resolution 3-D shallow crustal velocity model for understanding the main 3-D structural characteristics of the regional fault zone in the Binchuan Basin. This high-resolution model also enables the simulation of strong ground motion and earthquake hazard estimation in the Binchuan basin and urban areas by considering a scenario earthquake occurred at the nearby Chenghai fault.

For the fault scale model, an ultra-dense linear array (Fig. 4.2c) was deployed across the Chenghai fault in Binchuan, Yunnan, which is composed of 150 short-period seismometers, approximately 8 km long, with an average station spacing of 50 m. An ultra-high-resolution 2-D shear wave velocity model to a depth of 2 km below the linear array (Fig. 4.25) is obtained using surface wave short-period dispersion data from ambient noise and the direct inversion method from surface wave travel times (Yang et al., 2020a). The lateral resolution is about 200–500 m, which reveals the fine features of the 3.4-km-wide sediments-controlling shallow low-velocity zone of the Chenghai fault, providing an important foundation for investigation of high-resolution depth-dependent fault zone properties and future simulation of earthquake rupture of this fault.

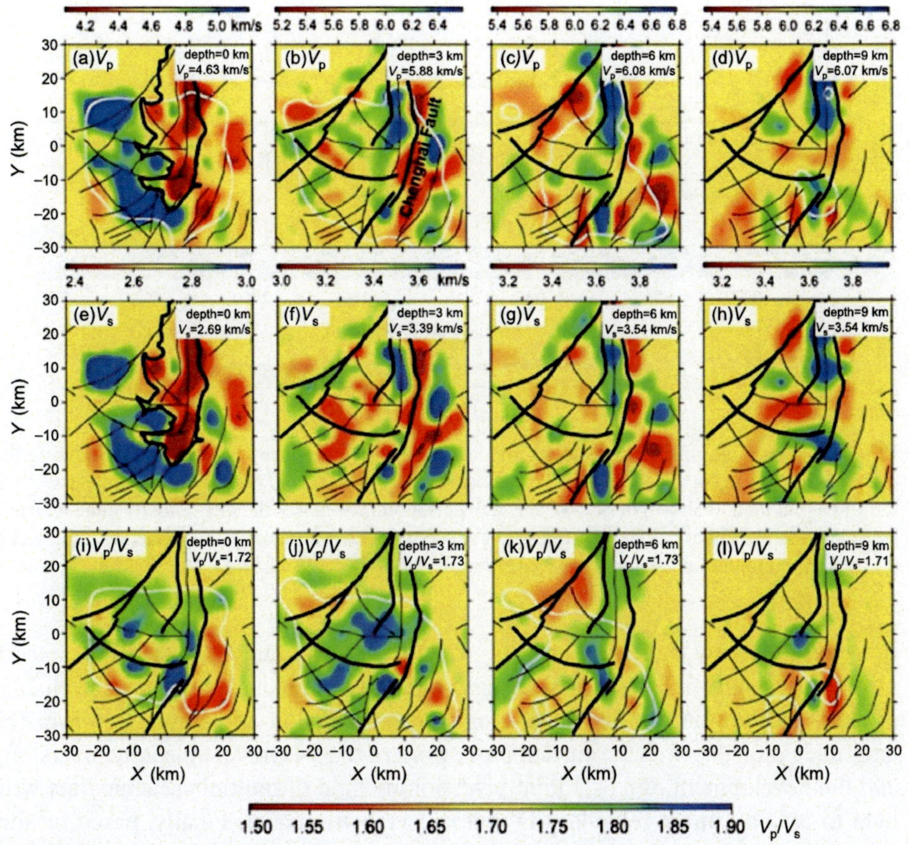

Fig. 4.24 High-resolution fault zone and urban scale Vp, Vs, and Vp/Vs models of Binchuan, Yunnan, from 3-D body wave travel time tomography using the dense array data (Fig. 4.2b) (Figure modified from Zhang et al., 2020a)

At present, a number of research teams have deployed planar and linear arrays of different scales and station spacing in some key earthquake monitoring regions, such as the Anninghe-Zemuhe-Xiaojiang fault zone, the Red-River fault zone, the Longmenshan fault zone, and the shale gas production area in the southeastern Sichuan Basin. These arrays will provide important data support for the construction of multi-scale velocity models in Southwest China.

4.8 Perspectives

At present, we have constructed the 1.0 version of the 3-D CVM of Sichuan and Yunnan, Southwest China, that is, SWChinaCVM-1.0, based on the joint body wave and surface wave travel time tomography, mainly using data from the permanent stations. We will further add body wave and surface wave travel time data of dense

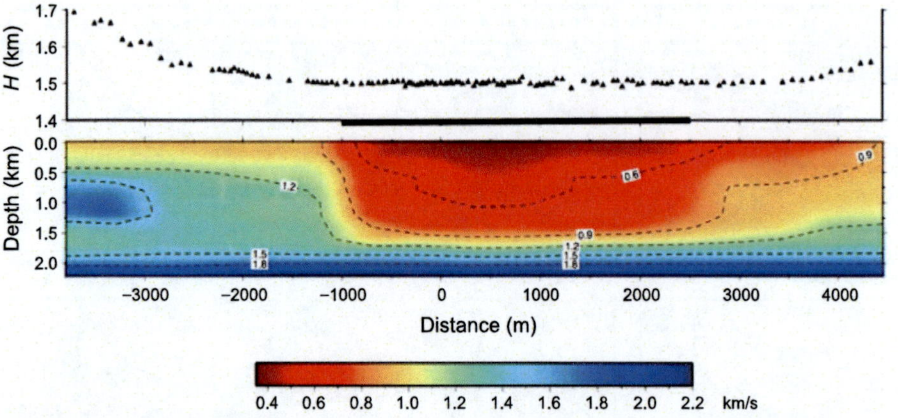

Fig. 4.25 The fault zone scale velocity model along the linear array across the Chenghai fault, Yunnan. The top panel shows the station altitudes and locations from northwest to southeast across the Chenghai fault as shown in Fig. 4.2c. Black horizontal bar marks the low-velocity zone inferred from teleseismic P waves. The bottom panel shows shear wave velocity model beneath the array derived from ambient noise tomography (Figure from Yang et al., 2020a)

transportable array stations (Fig. 4.2a) to build velocity models with higher lateral resolution (about 15–30 km), including azimuthally and radially anisotropic velocity models. In addition, the integration of receiver function and Rayleigh wave ZH amplitude ratio data into the current framework of the 3-D travel time joint inversion, and the development of a 3-D joint inversion method of multiple seismic data will help to obtain a more reliable 3-D regional velocity model. Finally, based on the waveform data of regional events and surface wave waveform data from ambient noise cross-correlation, a full waveform tomography or adjoint tomography method (such as Dong et al., 2019; Lee et al., 2014) will be used to construct a more accurate 3-D CVM. And on this basis, P-wave and S-wave receiver function migration imaging based on the 3-D CVM can be further used to construct better regional-scale 3-D community interface models, such as Moho interface, lithosphere-asthenosphere boundary, etc.

In addition to the combination of active and passive source travel time and waveform tomography methods for constructing structural models of fault zones, we also need to use waveform modeling, head wave and trapped wave analysis, frequency spectrum analysis and other methods using waveforms from small and moderate earthquakes to obtain high resolution velocity models of fault zones/faults and structural models of fault fracture zones. In recent years, the technology of environmentally-friendly active sources (such as air gun sources, methane sources, etc.) has developed rapidly. In the future, these sources can be excited within or around urban regions, which can better improve the resolution of structural images of urban areas in combination with ambient noise methods. In addition, for important active fault zone areas, the use of ambient noise and repeating sources (such as repeating earthquakes, active sources) to study temporal changes of 3-D velocity structures can better improve the research capabilities of earthquake physical processes as well as earthquake monitoring and prediction.

In the future, we need to construct multi-scale 3-D velocity models around important active fault zones and faults, large and medium-sized cities, and important infrastructure regions in Southwest China. The integration of these high-resolution models into the regional block scale 3-D CVMs of Southwest China will result in a true multi-scale CVM. In addition, we have a strong need to build an evaluation framework for multi-scale 3-D velocity models, and systematically perform the reliability and accuracy evaluation of the multi-scale CVMs.

Besides the CVMs, Southwest China also needs unified structure models. For example, the amplitude information of earthquake waveforms can be used to construct a 3-D community attenuation model, the joint inversion of gravity data and seismological data to construct a 3-D community density model, the inversion of magnetotelluric data to construct a 3-D community conductivity model, combination of geological and geophysical data to build a 3-D community fault model, and integration of velocity structure, density structure and rock physics data to build a 3-D community rheological model. These researches can better serve the need of earthquake science and earthquake engineering in Southwest China and also provide ideas and solutions for the construction of unified structure models in other regions of China.

Acknowledgements We are grateful to the researchers from China Earthquake Administration and many universities and research institutes in China for their valuable comments and suggestions on the construction of the community velocity model in the Sichuan-Yunnan region. The earthquake catalogue and body wave travel time data were provided by the China Earthquake Data Center (http://data.earthquake.cn, last accessed October 2018). The continuous waveform data were provided by Data Management Centre of China National Seismic Network at Institute of Geophysics, China Earthquake Administration (doi: https://doi.org/10.11998/SeisDmc/SN, http://www.seisdmc.ac.cn). Most plots were made using the Generic Mapping Tools (GMT; Wessel and Smith, 1998). This research was funded by the National Key R&D Program (2018YFC1503400) and the China Seismic Experiment Site Project (CSES20180101) and. The community velocity model SWChinaCVM-1.0 obtained in this study can be downloaded at https://github.com/liuyingustc/SWChinaCVM.

References

Bao, X., Sun, X., Xu, M., Eaton, D. W., Song, X., Wang, L., Ding, Z., Li, M., Li, H., & Huang, Z. (2015). Two crustal low-velocity channels beneath SE Tibet revealed by joint inversion of Rayleigh wave dispersion and receiver functions. *Earth and Planetary Science Letters, 415*, 16–24.

Brocher, T. M. (2005). Empirical relations between elastic wavespeeds and density in the Earth's crust. *Bulletin of the Seismological Society of America, 95*(6), 2081–2092.

Burchfiel, B. C., & Chen, Z. (2012). Tectonics of the southeastern Tibetan Plateau and its adjacent foreland. *Geological Society of America, 210*. https://doi.org/10.1130/2012.1210

Bai, D., Unsworth, M. J., Meju, M. A., Ma, X., Teng, J., Kong, X., Sun, Y., Sun,J., Wang, L., Jiang, C. et al. (2010). Crustal deformation of the eastern Tibetan plateau revealed by magnetotelluric imaging, *Nature Geoscience, 3*, 358–362. https://doi.org/10.1038/ngeo830

Dong, X., Yang, D., & Niu, F. (2019). Passive adjoint tomography of the crustal and upper mantle beneath eastern Tibet with a W2-Norm misfit function. *Geophysical Research Letters, 46*, 12986–12995.

Fang, H., Yao, H., Zhang, H., Huang, Y. C., & van der Hilst, R. D. (2015). Direct inversion of surface wave dispersion for three-dimensional shallow crustal structure based on ray tracing: Methodology and application. *Geophysical Journal International, 201*(3), 1251–1263.

Fang, H., Zhang, H., Yao, H., Allam, A., Zigone, D., Ben-Zion, Y., Thurber, C. H., & van der Hilst, R. D. (2016). A new algorithm for three-dimensional joint inversion of body wave and surface wave data and its application to the Southern California plate boundary region. *Journal of Geophysical Research: Solid Earth, 121*(5), 3557–3569.

Guo, H., Zhang, H., & Froment, B. (2018). Structural control on earthquake behaviors revealed by high-resolution Vp/Vs imaging along the Gofar transform fault, East Pacific Rise. *Earth and Planetary Science Letters, 499*, 243–255.

He, R., Shang, X., Yu, C., Zhang, H., Van der Hilst, R. D. (2014). A unified map of Moho depth and Vp/Vs ratio of continental China by receiver function analysis. *Geophysical Journal International, 199*, 1910–1918.

Hu, J., Badal, J., Yang, H., Li, G., & Peng, H. (2018). Comprehensive crustal structure and seismological evidence for lower crustal flow in the southeastern margin of Tibet revealed by receiver functions. *Gondwana Research, 55*, 42–59.

Huang, H., Yao, H., & van der Hilst, R. D. (2010). Radial anisotropy in the crust of SE Tibet and SW China from ambient noise interferometry. *Geophysical Research Letters, 37*(21). https://doi.org/10.1029/2010GL044981

Huang, J., Zhao, D., & Zheng, S. (2002). Lithospheric structure and its relationship to seismic and volcanic activity in southwest China, *Journal of Geophysical Research: Solid Earth, 107*(B10), ESE-13.

Huang, Z., Wang, P., Xu, M., Wang, L., Ding, Z., Wu, Y., Xu, M., Mi, N., Yu, D., & Li, H. (2015). Mantle structure and dynamics beneath SE Tibet revealed by new seismic images. *Earth and Planetary Science Letters, 411*, 100–111.

Huang, Z., Wang, L., Xu, M., Zhao, D., Mi, N., & Yu, D. (2019). P and S wave tomography beneath the SE Tibetan Plateau: Evidence for lithospheric delamination. *Journal of Geophysical Research: Solid Earth, 124*(10), 10292–10308.

Lee, E. J., Chen, P., Jordan, T. H., Maechling, P. B., Denolle, M. A., & Beroza, G. C. (2014). Full-3-D tomography for crustal structure in southern California based on the scattering-integral and the adjoint-wavefield methods. *Journal of Geophysical Research: Solid Earth, 119*(8), 6421–6451.

Li, H., Su, W., Wang, C. Y., & Huang, Z. (2009). Ambient noise Rayleigh wave tomography in western Sichuan and eastern Tibet. *Earth and Planetary Science Letters, 282*(1–4), 201–211.

Liu, J., Liu, F., He, J., Chen, H., & You, Q. (2001). Study of seismic tomography in Panxi paleorift area of southwestern China. *Science in China Series d: Earth Sciences, 44*(3), 277–288.

Liu, Q. Y., Van Der Hilst, R. D., Li, Y., Yao, H. J., Chen, J. H., Guo, B., Qi, S. H., Wang, J., Huang, H., & Li, S. C. (2014). Eastward expansion of the Tibetan Plateau by crustal flow and strain partitioning across faults. *Nature Geoscience, 7*(5), 361–365.

Liu, Y., Yao, H., Zhang, H., & Fang, H. (2021). The community velocity model V1.0 of southwest China, constructed from joint body- and surface-wave traveltime tomography. *Seismological Research Letters*. https://doi.org/10.1785/0220200318

Molnar, P., & Tapponnier, P. (1975). Cenozoic tectonics of Asia: effects of a continental collision. *Science, 189*(4201), 419–426.

Nguyen, V. D., Huang, B. S., Le, T. S., Dinh, V. T., Zhu, L., & Wen, K. L. (2013). Constraints on the crustal structure of northern Vietnam based on analysis of teleseismic converted waves. *Tectonophysics, 601*, 87–97.

Qiao, L., Yao, H., Lai, Y. C., Huang, B. S., & Zhang, P. (2018). Crustal structure of southwest China and northern Vietnam from ambient noise tomography: Implication for the large-scale material transport model in SE Tibet. *Tectonics, 37*(5), 1492–1506.

Rawlinson, N., & Sambridge, M. (2004). Wave front evolution in strongly heterogeneous layered media using the fast marching method. *Geophysical Journal International, 156*(3), 631–647.

Royden, L. H., Burchfiel, B. C., King, R. W., Wang, E., Chen, Z., Shen, F., & Liu, Y. (1997). Surface deformation and lower crustal flow in eastern Tibet. *Science, 276*(5313), 788–790.

Shaw, J. H., Plesch, A., Tape, C., Suess, M. P., Jordan, T. H., Ely, G., Hauksson, E., Tromp, J., Tanimoto, T., Graves, R., & Olsen, K. (2015). Unified structural representation of the southern California crust and upper mantle. *Earth and Planetary Science Letters, 415*, 1–15.

Shen, W., Ritzwoller, M. H., Kang, D., Kim, Y., Lin, F. C., Ning, J., Wang, W., Zheng, Y., & Zhou, L. (2016). A seismic reference model for the crust and uppermost mantle beneath China from surface wave dispersion. *Geophysical Journal International, 206*(2), 954–979.

Um, J., & Thurber, C. H. (1987). A fast algorithm for two-point seismic ray tracing. *Bulletin of the Seismological Society of America, 77*(3), 972–986.

Wang, B., Ge, H., Yang, W., Wang, W., Wang, B., Wu, G., & Su, Y. (2012). Transmitting seismic station monitors fault zone at depth. *Eos, Transactions American Geophysical Union, 93*(5), 49–50.

Wang, C. Y., Chan, W. W., & Mooney, W. D. (2003). Three-dimensional velocity structure of crust and upper mantle in southwestern China and its tectonic implications. *Journal of Geophysical Research: Solid Earth, 108*(B9). https://doi.org/10.1029/2002JB001973

Wang, C. Y., Han, W. B., Wu, J. P., Lou, H., & Chan, W. W. (2007). Crustal structure beneath the eastern margin of the Tibetan Plateau and its tectonic implications. *Journal of Geophysical Research: Solid Earth, 112*(B7). https://doi.org/10.1029/2005JB003873

Wang, C. Y., Wen-Cai, Y., Jian-Ping, W., & Zhi-Feng, D. (2015). Study on the lithospheric structure and earthquakes in North-South Tectonic Belt. *Chinese Journal of Geophysics-Chinese Edition, 58*(11), 3867–3901.

Wang, E. (1998). Late Cenozoic Xianshuihe-Xiaojiang Red River, and Dali fault systems of Southwestern Sichuan and Central Yunnan, China. *Geological Society of America, 327*. https://doi.org/10.1130/0-8137-2327-2.1

Wang, W., Wu, J., Fang, L., Lai, G., & Cai, Y. (2017). Crustal thickness and Poisson's ratio in southwest China based on data from dense seismic arrays. *Journal of Geophysical Research: Solid Earth, 122*(9), 7219–7235.

Wang, Z., Zhao, D., & Wang, J. (2010). Deep structure and seismogenesis of the north-south seismic zone in southwest China. *Journal of Geophysical Research: Solid Earth, 115*(B12). https://doi.org/10.1029/2010JB007797

Wei, W., Zhao, D., & Xu, J. (2013). P-wave anisotropic tomography in southeast Tibet: New insight into the lower crustal flow and seismotectonics. *Physics of the Earth and Planetary Interiors, 222*, 47–57.

Wessel, P., & Smith, W. H. (1998). New, improved version of Generic Mapping Tools released. *Eos, Transactions American Geophysical Union, 79*(47), 579–579.

Xin, H., Zhang, H., Kang, M., He, R., Gao, L., & Gao, J. (2019). High-resolution lithospheric velocity structure of continental China by double-difference seismic travel-time tomography. *Seismological Research Letters, 90*(1), 229–241.

Xu, Y. G., He, B., Chung, S. L., Menzies, M. A., & Frey, F. A. (2004). Geologic, geochemical, and geophysical consequences of plume involvement in the Emeishan flood-basalt province. *Geology, 32*(10), 917–920.

Xu, Z. J., & Song, X. (2010). Joint inversion for crustal and Pn velocities and Moho depth in Eastern Margin of the Tibetan Plateau. *Tectonophysics, 491*(1–4), 185–193.

Xu, T., Zhang, Z., Liu, B., Chen, Y., Zhang, M., Tian, X., Xu, Y., & Teng, J. (2015). Crustal velocity structure in the Emeishan large igneous province and evidence of the Permian mantle plume activity. *Science China Earth Sciences, 58*(7), 1133–1147.

Yang, H., Duan, Y., Song, J., Jiang, X., Tian, X., Yang, W., et al. (2020a). Fine structure of the Chenghai fault zone, Yunnan, China, constrained from teleseismic travel time and ambient noise tomography. *Journal of Geophysical Research: Solid Earth, 125*, e2020JB019565. https://doi.org/10.1029/2020JB019565

Yang, Y., Yao, H., Wu, H., Zhang, P., & Wang, M. (2020b). A new crustal shear-velocity model in Southwest China from joint seismological inversion and its implications for regional crustal dynamics. *Geophysical Journal International, 220*(2), 1379–1393.

Yang, Y., Ritzwoller, M. H., Zheng, Y., Shen, W., Levshin, A. L., & Xie, Z. (2012). A synoptic view of the distribution and connectivity of the mid-crustal low velocity zone beneath Tibet. *Journal of Geophysical Research: Solid Earth, 117*(B4). https://doi.org/10.1029/2011JB008810

Yang, H., Peng, H., Hu, J. (2017). The lithospheric structure beneath southeast Tibet revealed by P and S receiver functions. *Journal of Asian Earth Sciences, 138*, 62–71.

Yao, H., Beghein, C., & van der Hilst, R. D. (2008). Surface wave array tomography in SE Tibet from ambient seismic noise and two-station analysis-II crustal and upper-mantle structure. *Geophysical Journal International, 173*, 205–219.

Yao, H., van der Hilst, R. D., & De Hoop, M. V. (2006). Surface-wave array tomography in SE Tibet from ambient seismic noise and two-station analysis—I phase velocity maps. *Geophysical Journal International, 166*(2), 732–744.

Yao, H., & van der Hilst, R. D. (2009). Analysis of ambient noise energy distribution and phase velocity bias in ambient noise tomography, with application to SE Tibet. *Geophysical Journal International, 179*(2), 1113–1132.

Yao, H., van der Hilst, R. D., & Montagner, J. P. (2010). Heterogeneity and anisotropy of the lithosphere of SE Tibet from surface wave array tomography. *Journal of Geophysical Research: Solid Earth, 115*(B12). https://doi.org/10.1029/2009JB007142

Yao, H. (2020). Building the community velocity model in the Sichuan-Yunnan region, China: Strategies and progresses, Science China Earth Sciences, *63*(9), 1425–1428. https://doi.org/10.1007/s11430-020-9645-3

Yin, A., & Harrison, T. M. (2000). Geologic evolution of the Himalayan-Tibetan orogen. *Annual Review of Earth and Planetary Sciences, 28*(1), 211–280.

Zhang, H., & Thurber, C. H. (2003). Double-difference tomography: The method and its application to the Hayward fault, California. *Bulletin of the Seismological Society of America, 93*(5), 1875–1889.

Zhang, P. (2013). A review on active tectonics and deep crustal processes of the Western Sichuan region, eastern margin of the Tibetan Plateau. *Tectonophysics, 584*, 7–22.

Zhang, P., & Yao, H. (2017). Stepwise joint inversion of surface wave dispersion, Rayleigh wave ZH ratio, and receiver function data for 1D crustal shear wave velocity structure. *Earthquake Science, 30*(5–6), 229–238.

Zhang, Y., Wang, B., Lin, G., Wang, W., Yang, W., & Wu, Z. (2020a). Upper crustal velocity structure of Binchuan, Yunnan revealed by dense array local seismic tomography. *Chinese Journal of Geophysics, 63*(9), 3292–3306. https://doi.org/10.6038/cjg2020N0455 (in Chinese).

Zhang, Z., Yao, H., & Yang, Y. (2020b). Shear wave velocity structure of the crust and upper mantle in Southeastern Tibet and its geodynamic implications. Science China Earth Sciences, *63*(9), 1278–1293.

Zheng, C., Ding, Z. F., & Song, X. (2016). Joint inversion of surface wave dispersion and receiver functions for crustal and uppermost mantle structure in southeast Tibetan Plateau. *Acta Geophysica Sinica, 59*(9), 3223–3236.

Zheng, X. F., Yao, Z. X., Liang, J. H., & Zheng, J. (2010). The role played and opportunities provided by IGP DMC of China National Seismic Network in Wenchuan earthquake disaster relief and researches. *Bulletin of the Seismological Society of America, 100*(5B), 2866–2872.

Chapter 5
Three-Dimensional Community Active Faults Models of CSES

Renqi Lu, Maomao Wang, Zhigang Li, and Jing Liu-Zeng

Abstract Accurate three-dimensional (3-D) geometry of active fault in the subsurface is vital to decoding the rupture process, kinematic behavior, and dynamical mechanism of earthquakes. Since the geometrical complexity appears to be an inherent characteristic of the fault structure, many studies have been conducted to explore the impact of the fault geometry on rupture dynamic processes. Scientists of the Southern California Earthquake Center have contributed to the development of a community-based three-dimensional fault model (CFM) to study active faulting and earthquake phenomena and to assess probabilistic seismic hazard, as well as to predict strong ground motions. It is important to construct the CFM in the CSES, though a difficult task giving the area large in extent and with complex tectonic background. Mapping 3-D representations of active faults is an important component in developing various projects on the CSES. In this study, we illustrate several typical active faults (eg. Xianshuihe fault, Longmenshan fault, Anninghe-Xiaojiang-Zemuhe faults, and Red River fault), and introduce the 3-D modeling methods and processes. Our results and the 3-D CFM could effectively support the scientific investigations of CSES in China.

R. Lu (✉)
Institute of Geology, China Earthquake Administration, Beijing 100029, China
e-mail: lurenqi@163.com

M. Wang
College of Oceanography, Hohai University, Nanjing 210098, China
e-mail: wangmm@hhu.edu.cn

Z. Li
School of Earth Sciences and Engineering, Sun Yat-Sen University, Guangzhou 510275, China
e-mail: lizhigang@mail.sysu.edu.cn

J. Liu-Zeng
Institute of Surface-School of Earth System Science, Tianjin University, Tianjin 300072, China
e-mail: liu_zeng@tju.edu.cn

© Higher Education Press 2022

Y.-G. Li et al. (eds.), *China Seismic Experimental Site*,
https://doi.org/10.1007/978-981-16-8607-8_5

5.1 Introduction

In the past, earthquake hazard assessments are largely based on inferences about the location and magnitude of past and future earthquakes, which often involves defining the activity and slip rates on faults using geologic, seismological, or geodetic observations (Plesch et al., 2020). However, for most faults, the fault plane is nonplanar with complex geometry and heterogeneity. The three-dimensional (3-D) fault architecture controls the dynamics of earthquake swarms (Ross et al., 2020), and more precise 3-D representations are needed to accurately define subsurface fault geometry (Plesch et al., 2007).

The level of accuracy in 3-D geometric model of seismogenic fault governs how well the community understands the mechanism for great earthquakes (Stockmeyer et al., 2014). 3-D structure and model of active fault is an important basis for studying earthquake and geological deformation (Hubbard et al., 2010; Yue et al., 2005). 3-D models of active fault allows for a comprehensive understanding of the potential size of past and future earthquakes, and the detailed subsurface fault geometry are critical for accurate seismic hazard assessment (Riesner et al., 2017). In addition, to map accurate 3-D fault representations is essential to analyze the complicate dynamic process of a devastating earthquake (Marshall et al., 2017).

Scientists of the Southern California Earthquake Center (SCEC) have contributed to the development of a community-based 3-D fault model to study active faulting and earthquake phenomena and to assess probabilistic seismic hazard, as well as to predict strong ground motions (Field et al., 2014; Plesch et al., 2007; Shaw et al., 2015). At present, SCEC collaborators from Harvard and Caltech have developed a new (Fig. 5.1), statistical way to rapidly identify the most likely fault (or sets of candidate faults) in the Community Fault Model (CFM) that generated an earthquake, including detailed 3-D representations or more than 300 active fault segments that are deemed capable of generating moderate- to-large earthquakes (Nicholson et al., 2019).

The CFM team of CSES collaborates closely with scientists at the Harvard University and SCEC, as well as the Houston University. We develop our own workflow and methods to study the active fault and 3-D modeling, because the background of CSES is quite different from the SCEC. The CSES is located in the western China with a total area of about 780,000 km^2 (Fig. 5.2), is much larger than southern California—the target region of SCEC. This area includes the Sichuan-Yunnan rhombic-shaped block near the eastern tectonic transition zone of the Qinghai-Tibet Plateau and its surroundings. The tectonic boundary condition is very complex, with various types of fault systems such as compressional, shearing, and extensional and various combinations. In the past 50 years, an average of 14 earthquakes with a magnitude of 6.0 or higher and 3 earthquakes with a magnitude of 7.0 occurred every 10 years, which caused serious casualties and property losses.

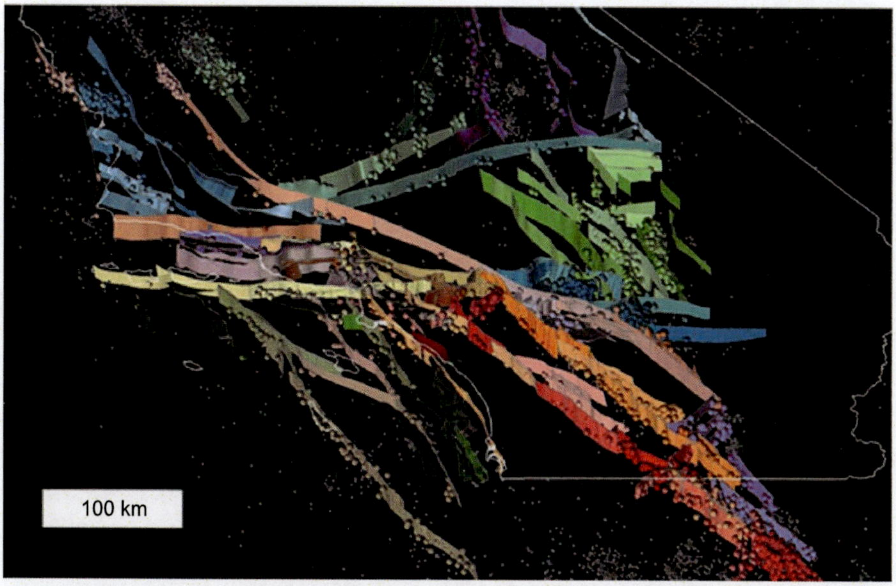

Fig. 5.1 SCEC CFM 5.2 Modeled Associations for the southern California earthquake catalog. Similarly-colored faults (planes) and earthquakes (spheres) have the highest association probability between them. Small gray symbols are not associated with a known fault (Evans et al., 2019)

5.2 Methods

The CFM is an object-oriented 3-D representation of more than 100 faults in the CSES (Fig. 5.2). The 3-D CFM could be defined by many different types of constraints, including surface trace of active faults, focal mechanism orientations, relocated earthquake hypocenters, seismic reflection profiles and geologic cross sections. Meanwhile, some advanced workflow enables us to control the weights assigned to different constraints, increasing the accuracy of 3-D fault models (Carena et al., 2004; Riesner et al., 2017). The 3-D modeling workflow consists of a series of defined steps that can be iterated and reproduced changes to input data and workflow parameters such as the weighting of constraints (Plesch et al., 2020).

In this study, we adopted a universal workflow to build the 3-D active fault model of the study region in the Sub-surface Knowledge United Approach (SKUA)-Geology Object Computer Aided Design (GOCAD) software, a powerful 3-D modeling platform. All available datasets were formatted to the Universal Transverse Mercator (UTM) grid system and input into one project, which was defined with the World Geodetic System (WGS) 84 and 48 zone coordinate system. The fault surface can be made with discrete smooth interpolation algorithm (Mallet, 2002). Faults in the CFM are represented as irregular triangulated surfaces (t-surfs), which consist of nodes or points (x, y, and z) associated with one another in triangular elements, and which usually are (or were) smoothed by Kriging interpolation. Triangulated

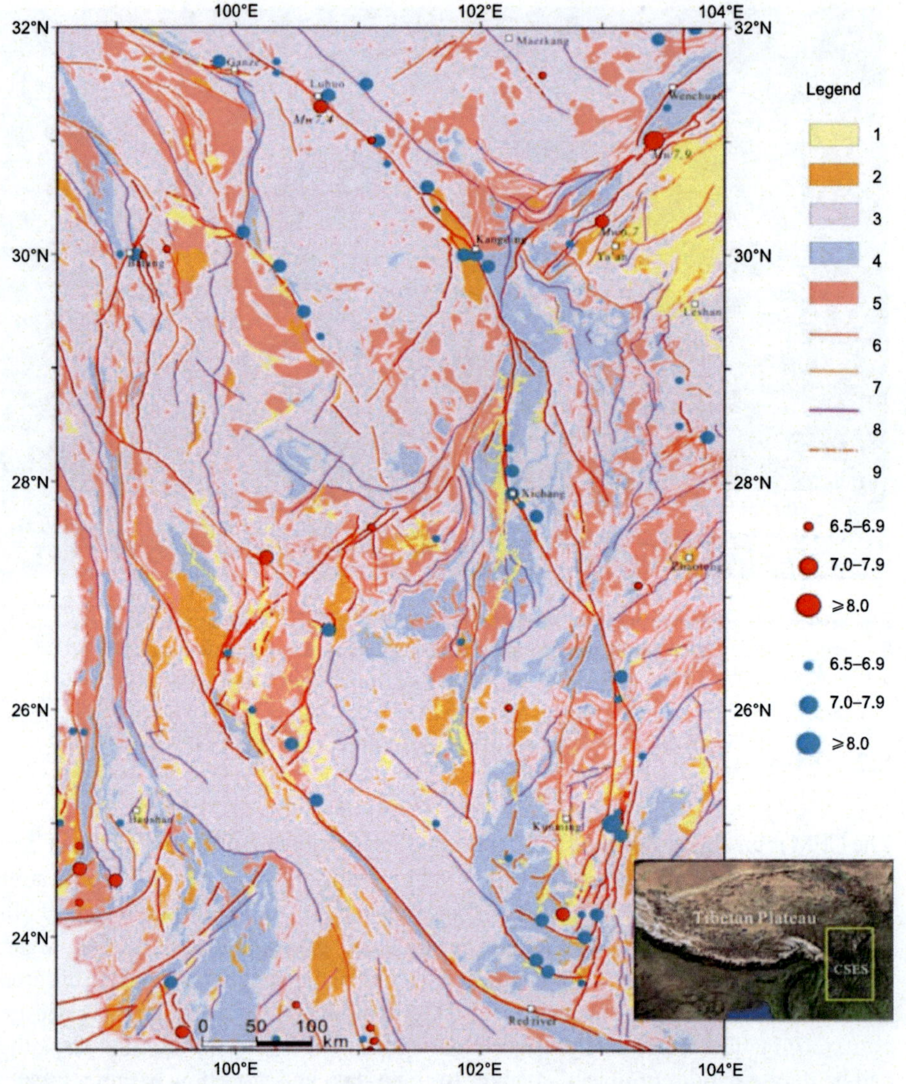

Fig. 5.2 Geological and structural background of the CSES (modified from Wu et al., 2018). Legend: 1—Quaternary strata; 2—Neogene and Paleogene strata; 3—Mesozoic and Paleozoic strata; 4—Precambrian strata; 5—Magmatic of Mesozoic and Paleozoic; 6—Holocene fault; 7—Late Pleistocene fault; 8—Early to middle Pleistocene fault; 9—Blind fault. Red circles show the earthquakes since 1970, and blue circles are the history earthquakes before 1970. F1: Xianshuihe fault; F2: Longmenshan fault; F3: Anninghe-Zemuhe faults; F4: Daliangshan fault; F5: Xiaojinhe fault; F6: Xiaojiang fault; F7: Red River fault

surfaces are advantageous because they accurately define complex surface geometries and topologies. Node (point) spacing is allowed to vary as a function of fault shape or data resolution (Plesch et al., 2007).

5.3 3-D Active Fault Model of the Bayan Har Block

The Bayan Har block in the southeastern Tibetan Plateau (Fig. 5.2), is an actively deforming (moving?) block and hosts several large earthquakes over the past two decades (Deng et al., 2010). The Xianshuihe fault (F1) and Longmenshan fault (F2) are two boundary faults of the Bayan Har block. Many large earthquakes occurred along the two faults, including the 1973 Luhuo M_w 7.4 earthquake in the Xianshuihe fault, the 2008 Wenchuan M_w 7.9 earthquake in the Longmenshan fault belt (Fig. 5.3).

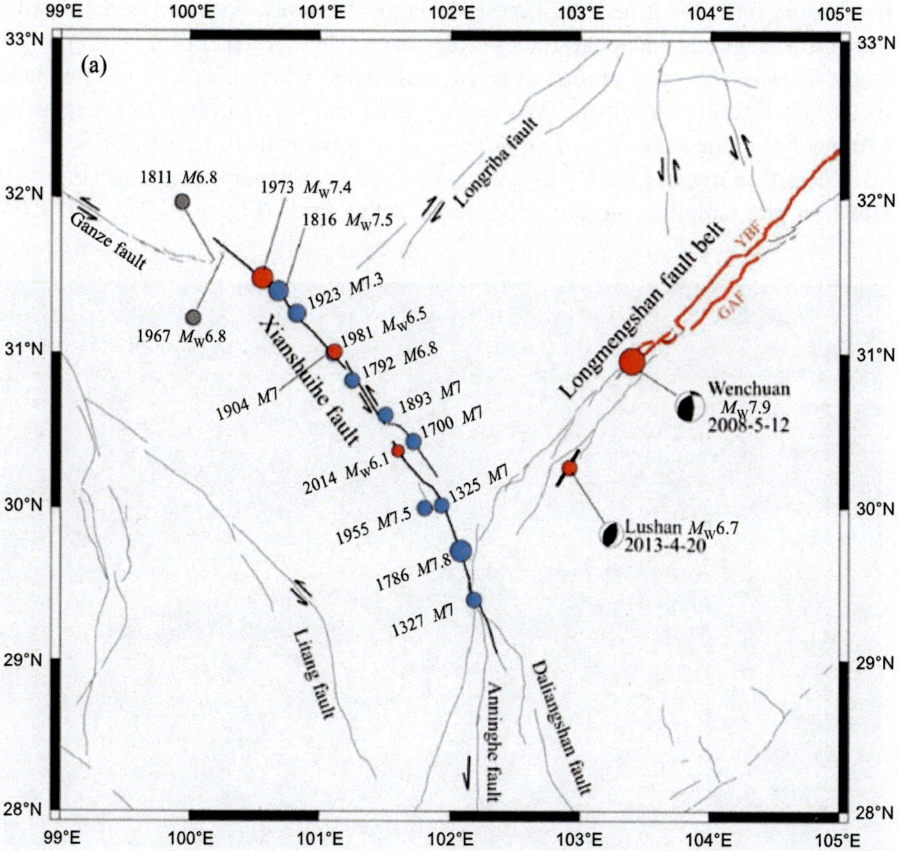

Fig. 5.3 Tectonic map and large historical earthquakes on the Xianshuihe fault and Longmenshan fault belt (modified from Cheng et al., 2020). Red solid lines show the surface ruptures of the 2008 Wenchuan M_w 7.9 earthquake

5.3.1 Xianshuihe Fault Model

The Xianshuihe fault that is a major strike-slip fault in the southeastern Tibetan Plateau, extends ~ 370 km in the entire length, trending NW–SE in its northwest portion and NNW-ESE in its southeast portion (Fig. 5.2). The Xianshuihe fault consists of five major fault segments: the Xianshuihe, Yalahe, Selaha, Zheduotang and Moxi faults from northwest to southeast, among which the Yalahe, Selaha and Zheduotang faults form a left-stepping en echelon pattern (Yan et al., 2017). At the southeastern end, the fault branches into two strands: the Anninghe fault (F3) in the west, and the Daliangshan fault (F4) in the east.

The Xianshuihe fault has a high strike-slip rate of ~ 10 mm/yr (Xu et al., 2003) and hosts frequent strong historical earthquakes. In the past > 100 year, a few dozens of large earthquakes rattled the Xianshuihe fault (Fig. 5.3), indicating one earthquake occurred on this fault every ~ 20–30 year in average (Cheng et al., 2020).

In the previous CFM model of the Xianshuihe fault, its surface traces were digitized based on maps from the Chinese Geological Survey, with additional detailed traces from regional and local studies (Allen et al., 1991; Riesner et al., 2017). In our study, we used the latest version of active fault traces which locations were updated in Google Earth environment (Wu et al., 2018) and the relocated hypocenters of earthquakes in the catalog, as well as the focal mechanisms from the CSES.

The surface trace of the Xianshuihe fault consists of several segments (Fig. 5.2). However, we simplified it along the major active fault (Fig. 5.4). The 3-D CFM

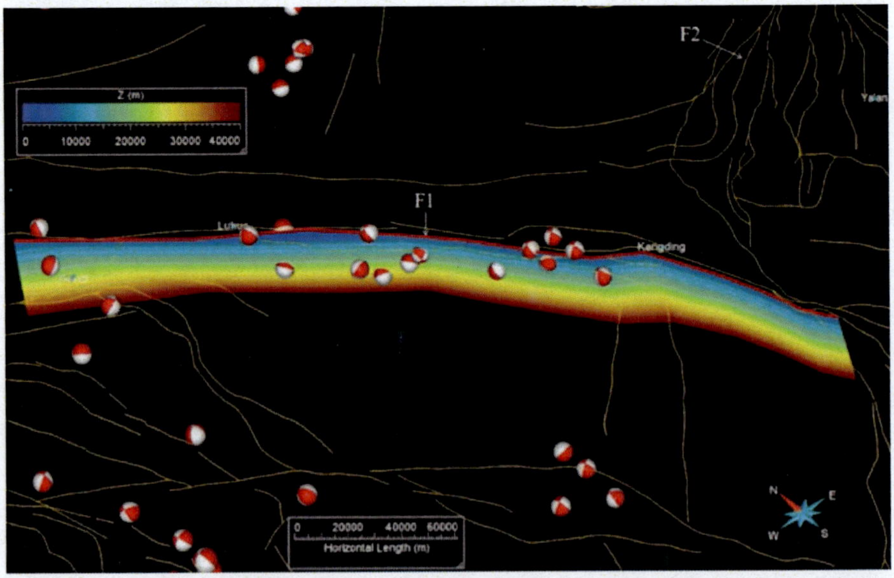

Fig. 5.4 3-D perspective image for the Xianshuihe fault. Red solid line is the fault surface trace of the simplified Xianshuihe fault. Yellow solid lines are the surface traces of other faults in the region. Beach balls show the focal mechanisms of major earthquakes (Fig. 5.3)

is also defined by the alignment of seismicity roughly beneath the fault surface trace, and by the focal mechanisms of 15 large earthquakes occurred on the fault. These mechanisms show left-lateral motion, with the principal nodal plane striking parallel to the fault trace and having a nearly vertical dip. Because there is lack of enough aftershocks recordings, and deep seismic reflection profiles and other effective constraints, the 3-D fault model of Xianshuihe fault is not yet well-defined in depth at present.

5.3.2 Longmenshan Fault Model

The 2008 Wenchuan earthquake ruptured two parallel faults within the Longmenshan (LMS) thrust belt (Fig. 5.3), consisting of a 240-km-long surface ruptures along the Yingxiu-Beichuan fault (YBF) with reverse and oblique-slip (Xu et al., 2009) and a 72-km-long surface rupture zone along the Guanxian-Anxian fault (GAF) with a pure thrusting (Liu-Zeng et al., 2009). Previous 2D fault models are useful to understand the seismotectonics of the 2008 Wenchuan M_w 7.9 earthquake (Hubbard et al., 2009; Zhang et al., 2010). However, they are difficult to reveal the relationships between the fault geometry and kinematics, as well as the complex ruptured process during the large earthquake (Liu-Zeng et al., 2009; Xu et al., 2009). Consensus has been reached that the fault surfaces are not simple planes but show a complex 3-D geometry (Hubbard et al., 2010; Li et al., 2010; Shen et al., 2009; Wang et al., 2011). The subsurface fault geometry is usually complicate in fold and-thrust belts, the LMS is no exception (Fig. 5.5). Thus, a detailed, 3-D fault geometry model could improve the assessments of the likelihood and potential size of multi-segment ruptures in this seismically active region.

Previously, a few studies relocated 2706 aftershocks with $M \geq 2.0$ using double-difference algorithm (e.g., Huang et al., 2008), and obtained relocations of 3775 events from May 12 to August 26, 2008 (Fig. 5.5). In order to reduce the influence of lateral variation in crustal and upper mantle velocity structure, the different velocity models for the east and west sides of Longmen Shan were adopted (Wu et al., 2009). In addition, they added seismic data from portable seismic stations close to the shocks to constrain focal depths. The uncertainty of these relocated aftershocks is small.

After the Wenchuan M_w 7.9 main shock, many strong aftershocks ($M \geq 5.0$) occurred and last several years. We chose these strong aftershocks as constraints on the 3-D fault model. Focal mechanism solutions and depths of 45 aftershocks ($M \geq 5.0$) with relatively high signal-to-noise ratio waveforms of the 2008 Wenchuan earthquake have been derived by CAP method (Yi et al., 2012). In addition, many petroleum seismic reflection profiles have combined with geological information, as well as some significant drilling wells (Jia et al., 2010; Li et al., 2010; Lu et al., 2014), providing constraints on the 3-D fault model.

The 3-D co-seismic fault models of GAF and YBF show their geometric relationship clearly (Fig. 5.5). The GAF is a thrust fault with gentle dip angle, and merges with the YBF above ~ 10 km at depth. Detailed 3-D model of YBF indicates this

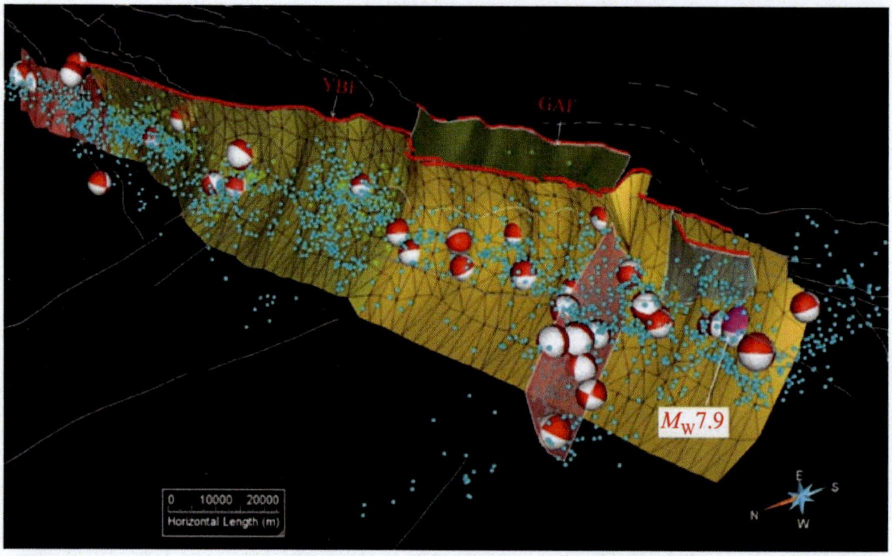

Fig. 5.5 3-D perspective image for the co-seismic fault models of the YBF and GAF in the Longmenshan fault belt. Red sloid lines are the surface ruptures of 2008 Wenchuan earthquake (Xu et al., 2009). The small dots show the relocated aftershocks (Huang et al., 2008), and beach balls show the focal mechanism of the main earthquakes (Yi et al., 2012). YBF: Yingxiu-Beichuan fault; GAF: Guanxian-Anxian fault

co-seismic fault has more complicate geometry. Although the first-order 3-D outline and shape are similar to previous studies (Shen et al., 2009; Wang et al., 2011), this 3-D model of YBF reveals the finer details in changes of fault plane along strike and depth. The YBF could divide into two segments along strike.

5.3.3 Lushan Blind Fault Model

Roughly five years after the devastating May 12, 2008, M_w 7.9 Wenchuan earthquake, the southern Longmen Shan was struck again by the M_w 6.7 Lushan earthquake on April 20, 2013 (Fig. 5.2). The epicenter of this earthquake was located ~ 80 km southwest of the Wenchuan earthquake (Fig. 5.3). There was no surface rupture of the Lushan earthquake found in field investigations, but the earthquake ruptured a deep portion of a blind thrust fault in the frontal Longmen Shan (Xu et al., 2013). Structural interpretation using high-resolution seismic reflection profiles suggested that the Lushan earthquake was occurred on a range front blind thrust (Lu et al., 2017; Wang et al., 2014).

In previous studies, aftershocks of the large earthquakes were used to model faults by fitting the fault surfaces to earthquake clusters in 3-D (Carena & Suppe, 2002). Approximate 2000 events ($M \geq 1.0$) of the Lushan earthquakes (Fig. 5.6a) were relocated using the double-difference method (Fang et al., 2015). Parallel to the strike

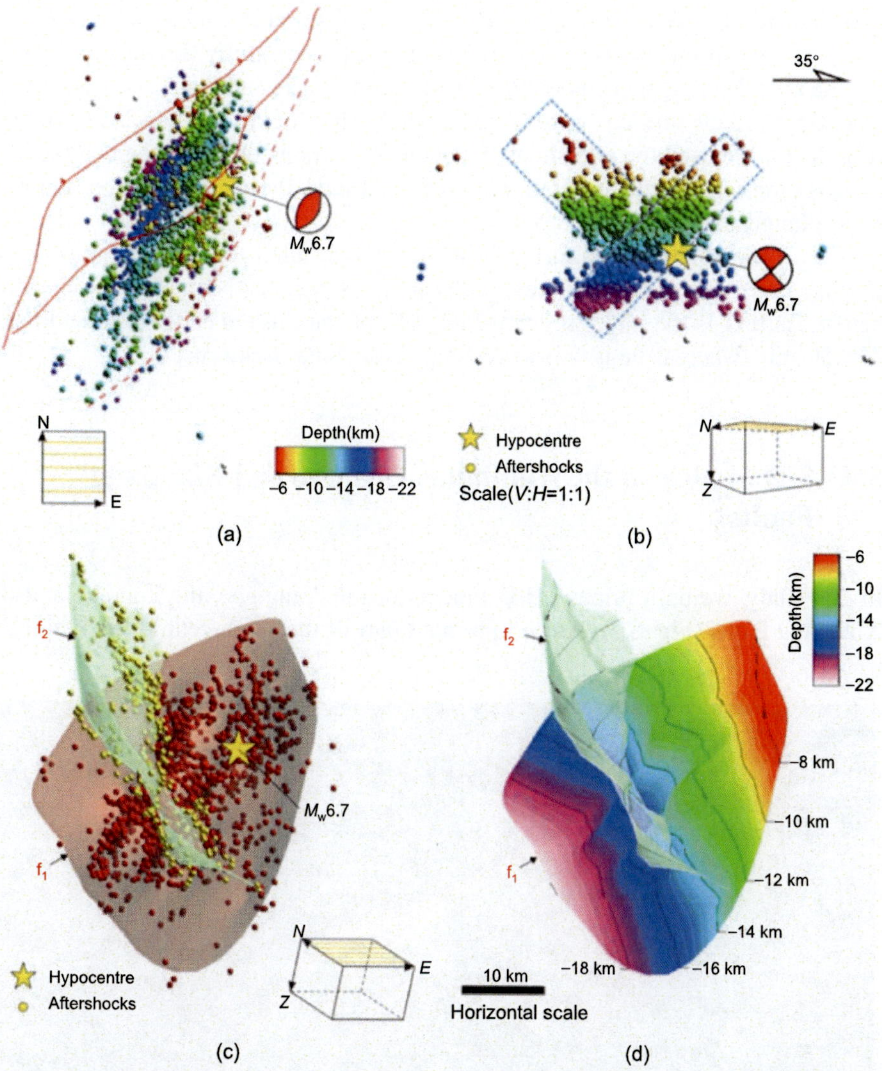

Fig. 5.6 3-D modeling of the co-seismic fault using relocated aftershocks (modified from Lu et al., 2017). **a** The map view of the ~ 2000 aftershock hypocenters. Red solid lines are the pre-existing fault traces on surface and dashed line is the surface projection of a blind fault that ruptured in the 2013 M_w 6.7 Lushan earthquake. **b** The depth-section viewing parallel to the fault strike; the strike direction of the aftershocks is ~ 55°. **c** 3-D images of the ruptured fault surfaces modeled based on the aftershock hypocenters. **d** Depth-contoured surfaces of the ruptured faults

of the LMS thrust belt, most of the relocated hypocenters were confined to a depth range between 7 and 19 km (Fig. 5.6b). In addition, the northwest-dipping earthquake cluster intersected the southeast-dipping earthquake cluster and together they formed a Y-shape configuration. The relocated aftershocks suggest the presence of at least two faults reactivated coseismically. Based on the distribution of the aftershocks, the 3-D geometry of the coseismic faults was mapped successfully (Lu et al., 2017).

The aftershock hypocenters of the Lushan earthquake and the 3-D images of the seismogenic fault surfaces were reconstructed using the SKUA-GOCAD software (Fig. 5.6c). We used the aftershock hypocenter locations of the Lushan earthquake to image the active faults in detail. Figure 5.6c, d show the same view direction with a 15° plunge angle and a 35° strike.

The 3-D depth-contoured surfaces illustrate the detailed geometry of faults f1 and f2 (Fig. 5.6d). The surfaces of these faults are non-planar and contain multiple wave bends. Fault f1 is the major seismogenic fault of the Lushan earthquake and has a 45°–50° dip, whereas fault f2 is a conjugate branching backthrust of f1.

5.4 3-D Models of the Anninghe, Zemuhe and Xiaojiang Faults

In this study, we built primary 3-D models of the Anninghe, the Zemuhe and the Xiaojiang faults (Fig. 5.7). As an important part of the south-central section of the

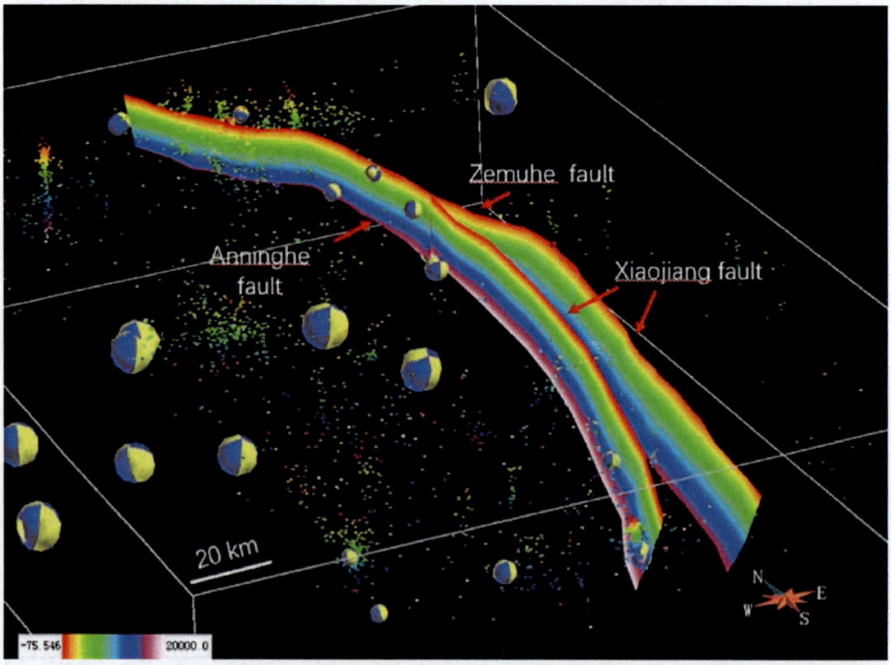

Fig. 5.7 3-D oblique view of the Aninghe, Zemuhe, and Xiaojiang faults. The color dots are the small earthquakes in the study area

North–South Earthquake Zone in China, these three faults form a significant seismically active and deeply rooted fault system in the southeastern Tibetan Plateau (He & Yasutakyr, 2007; Wen et al., 2008; Zhang, 2013). The construction of this 3-D model is based mainly on data such as fault surface traces, relocated earthquakes, and focal mechanism solutions. In the following, we briefly introduce the structural features of this fault system.

The length of the Anninghe fault is about 350 km, extending in a nearly north-to-south direction from Shimian County to Huili County in Sichuan Province. At the surface, the Anninghe fault is mainly developed in the pre-Sinian granite and the Sinian- to-Permian strata (Fig. 5.2). The surface outcrop of the Anning fault is mainly east-dipping and partially west-dipping, with a dip angle of 40°–60°. Based on the field investigation, the previous studies concluded that the Anninghe fault is a sinistral strike-slip fault with a certain component of thrust motion (He & Yasutakyr, 2007). In the 3-D initial model, the Anninhe fault is simplified as a single fault trace at the surface and characterized by a high angle (80°–90°) at depth, consistent with the feature of typical strike-slip fault. Unlike the Xianshuihe fault in the north, micro-seismic activity is sparse around the Anninghe fault.

The Zemuhe fault links to the Anninghe Fault in the north through a restraining bend near Xicang, and to the Xiaojiang Fault in the south, through a releasing bend near Qiaojia. The Zemuhe fault has a general strike of 330° (Fig. 5.8). According to the 3-D initial model, the Zemuhe fault is west-dipping strike-slip fault, with a dip angle between 70° and 80° at depth, which is smaller than the Anninhe fault.

The Xiaojiang fault generally extends in a nearly north–south direction and consists of two subparallel branching faults in the east and west, less than 20 km apart from each other. The average horizontal slip rate of the Xiaojiang fault is 9.4 ± 1.2 mm/year and shows the feature of sinistral strike-slip fault. The overall length of the Xiaojiang fault is about 300 km, and its eastern branch is connected to the Zemuhe fault. The initial 3-D model shows that both branch faults of the Xiaojiang fault dip to the west, and the dip angle of the faults is in the range of 50°–70°, generally smaller than that of the Anninghe and Zemuhe faults.

5.5 3-D Active Fault Model of Red River Fault

The active Red River fault (RRF) is a phenomenal structural discontinuity and stretches more than 1000-km-long, one of the largest right-lateral strike-slip faults of Asia, extending from eastern Tibetan Plateau, through the Hanoi Basin in Vietnam, and into the South China Sea (Fig. 5.9). This fault has been cited as a lithospheric-scale strike-slip fault (Leloup et al., 1995; Tapponnier et al., 1990). According to dissimilarities of structural framework, fault geometry, and seismicity, it can be divided into three segments, separated by segment boundaries near Midu and Yuanyang cities (Zhang, 2009).

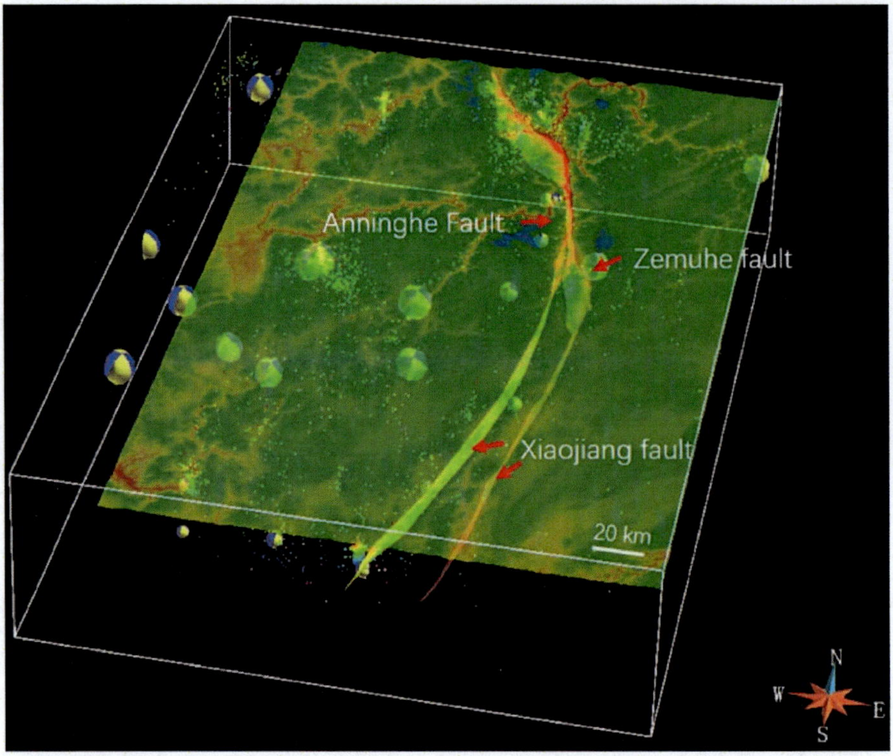

Fig. 5.8 3-D perspective view of the topography and the Aninghe, Zemuhe and Xiaojiang faults

The northern segment (called Eryuan-Midu segment) is transtensional in kinematics, and has experienced nine normal- and oblique-slip earthquakes of magnitude $M \geq 6$ (Guo et al., 2001). The central portion (called Midu-Yuanyang segment) consists of two parallel fault traces, including i.e. the Range Front fault and the Mid Valley fault. The Range Front fault grows along the baseline of the Ailao Shan mountain range and controls a series of triangular facets. The Mid Valley fault is situated 1–3 km northeast of the Range Front fault (Allen et al., 1984; Replumaz et al., 2001; Schoenbohm et al., 2006). The southern portion (Yuanyang-South China Sea) shows multiple strike-slip fault strands near its southeastern end in Vietnam and records some micro-seismicity (Hu et al., 2006; Zhang, 2009).

The RRF was considered as a slowly slipping fault with very low right-lateral strike-slip rate of 0.2–2 mm/year for all three segments, as indicated by Global Positioning System (GPS) (Shi et al., 2018, and references therein). Meanwhile, the cumulated geologic offsets and inferred time of onset of right-lateral slip suggest a long-term geological rate of 1–5 mm/year along the central segment of the RRF (Replumaz et al., 2001; Schoenbohm et al., 2006; Weldon et al., 1994). The existence for low modern-day right-lateral strike-slip rate and scarce large earthquake events are compatible with the lack of large historical earthquakes along the entire RRF.

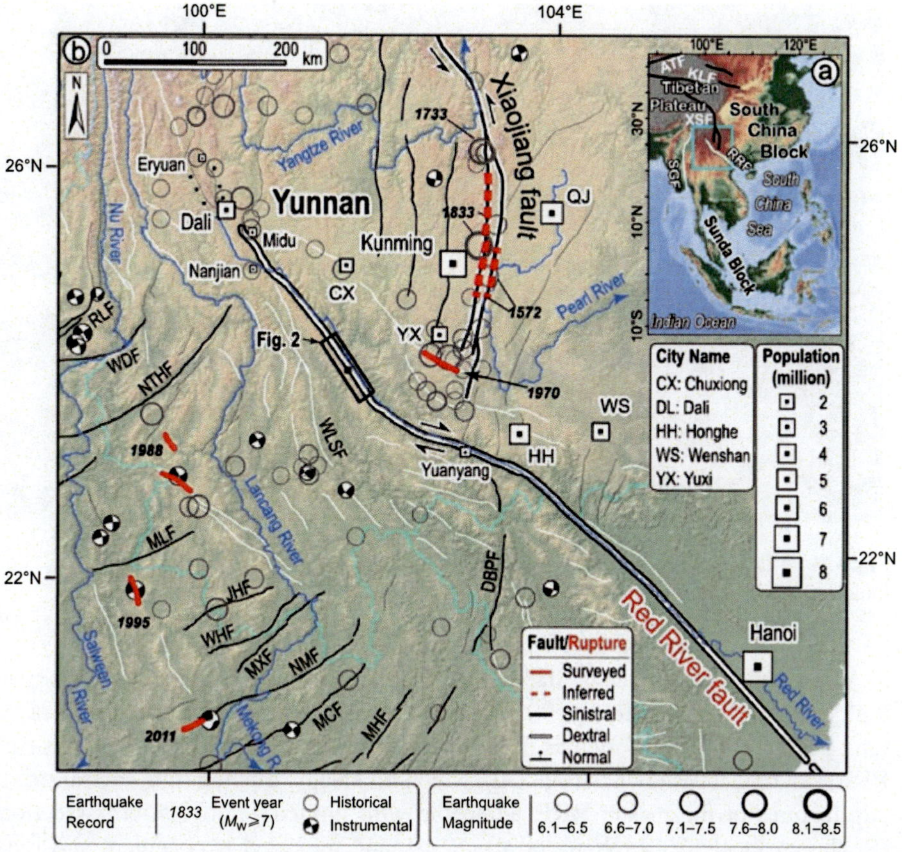

Fig. 5.9 Tectonic background and seismicity of the Red River fault and adjacent region [modified from Shi et al. (2018)]. ATF: Altyn Tagh fault; DBPF: Dien Bien Phu fault; JHF: Jinghong fault; KLF: Kunlun fault; MCF: Mae Chan fault; MHF: Muang Houn fault; MLF: Menglian fault; MXF: Mengxing fault; NMF: Nam Ma fault; NTHF: Nantinghe fault; RLF: Ruili fault; SGF: Sagaing fault; WDF: Wanding fault; WHF: Wan Ha fault; XSF: Xianshuihe-Xiaojiang fault

Building 3-D representations of the RRF is a crucial step to improve its seismic-hazard assessment. However, the fault geometry of the RRF is poorly defined at depth due to lack of subsurface data. We try to develop 3-D fault geometries of the RRF from surface to subsurface. The model of large-scale active fault mapping is based on field investigations and high-resolution remote-sensing image interpretations, integrated with the latest studies of the neotectonics and active tectonics of the entire RRF, and constrained by the precise fault traces at surface (Shi et al., 2018). Moreover, the fault geometry of the RRF at depth is mainly constrained by multi-source data, including topographic investigation of the upper crustal structure (Bai & Wang, 2003), geological cross sections (Replumaz et al., 2001; Wang et al., 2016), and relocated hypocenters and focal mechanism orientations of historic earthquakes (Li et al., 2018; Yin et al., 2018).

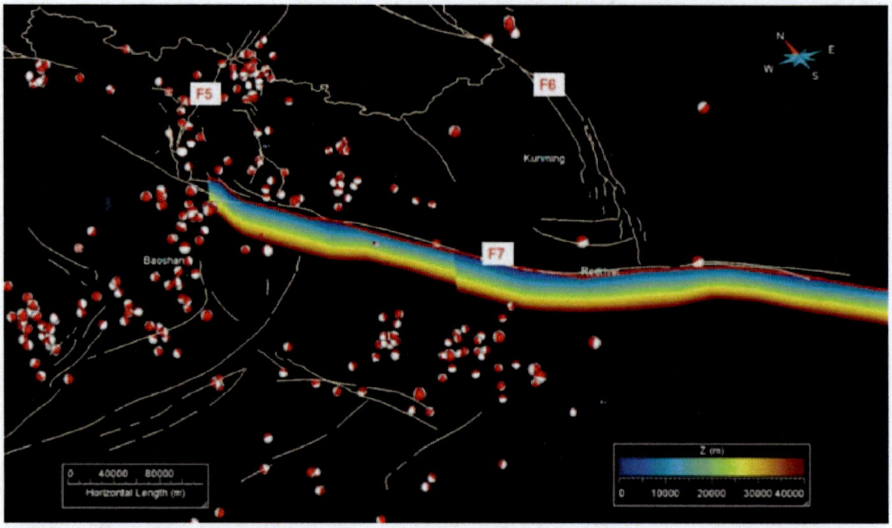

Fig. 5.10 3-D perspective images for the Red River fault and its splay faults. Red solid line is the fault trace of the simplified surface fault

We extended the surface fault traces into a 3-D fault model of the RRF and its splay fault using a constant fault dip (90°), because of lacking direct data constraints of the fault geometry at depth, such as the deep seismic reflection profiles. A preliminary RRF fault model is shown in Fig. 5.10, which fits steeply dipping right-lateral strike-slip features of the master RRF, agreeable with limited geological cross-sections (Replumaz et al., 2001; Wang et al., 2016), and relocated hypocenters and focal mechanisms of historic earthquakes (Li et al., 2018; Yin et al., 2018). However, more precise 3-D fault model is much needed to accurately define surface and subsurface fault geometry of the RRF, given its importance as a main active fault and a potential major source of seismic hazard in the southeast Tibet.

5.6 Results

The CFM (V2.1) database provides alternative representations for many of the faults in the CSES (Fig. 5.11). These alternative representations for a given fault may involve different levels of detail in the fault surfaces, different methods of interpolating or extrapolating fault surfaces between and within data constraints, or different viewpoints among contributing scientists about the geometry and position of the fault. However, there is still lack of sufficient data to constrain the fault geometry so that many fault models are preliminary at present. With the increase in active fault survey in the CSES, we will continue to add more constraints of data and materials to update the 3-D fault models, and effectively use them to earthquake studies.

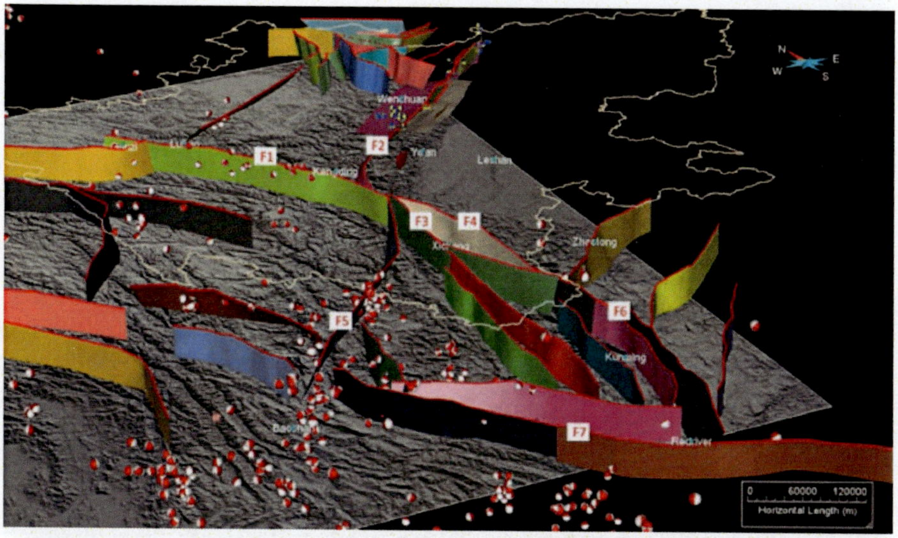

Fig. 5.11 3-D perspective image for the CFM (V2.1) of the CSES

Nowadays, the big data with advances in theory and software allow us to construct the 3-D models of more complex fault geometries in seismicity study. In addition, building CFM modeling is a long-term labor-intensive project, but provides the fundamental data for various studies in earthquake science. The CFM modeling involves the integration of multi-disciplinary and a huge variety of data, as well as continuous upgrades and updates; it also needs to meet the challenges of practical application for earthquake study in the future.

Acknowledgements This study received financial support from National Key Research and Development Program of China (2018YFC1504104), the China Earthquake Science Experiment (2017CESE0102), China National Science Foundation (41702202, 41872206). We are grateful to Prof. Yonggang Li for his constructive suggestions and helps that significantly improved this manuscript. We thank Prof. John Shaw, Andreas Plesch, John Suppe, Jonny Wu, Xiwei Xu, Dengfa He and Xiyan Wu for their long-term support and helpful discussions. The data of fault traces, relocated earthquakes and focal mechanisms provided by the CSES (http://www.cses.ac.cn/).

References

Allen, C. R., Gillespie, A. R., Yuan, H., Sieh, K. E., Buchun, Z., & Chengnan, Z. (1984). Red river and associated faults, Yunnan Province, China: Quaternary geology, slip rates, and seismic hazard. *Geological Society of America Bulletin, 95*(6), 686–700.

Allen, C. R., Luo, Z. L., Qian, H., Wen, X. Z., Zhou, H. W., & Huang, W. S. (1991). Field study of a highly active fault zone: The xianshuihe fault of southwestern China. *Geological Society of America Bulletin, 103*(9), 1178–1199.

Bai, Z. M., & Wang, C. Y. (2003). Tomographic investigation of the upper crustal structure and seismotectonic environments in Yunnan Province. *Acta Ssmologica Sinica, 16*(002), 127–139. (in Chinese with English abstract).

Carena, S., & Suppe, J. (2002). Three-dimensional imaging of active structures using earthquake aftershocks: The Northridge thrust, California. *Journal of Structural Geology, 24*, 887–904.

Carena, S., Suppe, J., Kao, H. (2004). Lack of continuity of the San Andreas fault in southern California: Three-dimensional fault models and earthquake scenarios. *Journal of Geophysical Research Solid Earth, 109*https://doi.org/10.1029/2003JB002643

Cheng, J., Chartier, T., & Xu, X. (2020). Multisegment rupture hazard modeling along the xianshuihe fault zone, southeastern Tibetan Plateau. *Seismological Research Letters, 91*, 1–14. https://doi.org/10.1785/0220200117

Deng, Q. D., Gao, X., Chen, G. H., & Yang, H. (2010). Recent tectonic activity of Bayankala fault block and the Kanlun-Wenchuan earthquake series of the Tibetan Plateau. *Earth Science Frontiers, 17*(5), 163–178. (in Chinese with English abstract).

Evans, W. S., Plesch, A., Shaw, J. H., Pillai, N. L., Yu, E., Meier, M., & Hauksson, E. (2019). A statistical method for associating earthquakes with their source faults in southern California. *Bulletin of the Seismological Society of America, 110*(1), 213–225. https://doi.org/10.1785/0120190115SCECContribution9057

Fang, L. H., Wu, J. P., Wang, W. L., Du, W. K., Su, J. R., Wang, C. Z., Yang, T., & Cai, Y. (2015). Aftershock observation and analysis of the 2013 Ms 7.0 Lushan Earthquake. *Seismological Research Letters, 86*, 1–8. https://doi.org/10.1785/0220140186

Field, E. H., Arrowsmith, R., Biasi, G. P., Bird, P., Dawson, T. E., Felzer, K. R., Jackson, D. D., Johnson, K., Jordan, T., Madden, C., Michael, A. J., Milner, K., Page, M. T., Parsons, T., Powers, P. M., Shaw, B. E., Thatcher, W. R., Weldon, R., & Zeng, Y. H. (2014). Uniform California Earthquake Rupture forecast, version 3 (UCERF3)—The time-independent model. *Bulletin of the Seismological Society of America, 104*, 1122–1180. https://doi.org/10.1785/0120130164

Guo, S., Ji, F., Xiang, H., Dong, X., Yan, F., Zhang, S., et al. (2001). *The Honghe active fault zone*. Ocean Press. (in Chinese with English abstract).

He, H. L., & Yasutakyr, I. (2007). Faulting on the Anninghe fault zone, southwest China in late Quaternary and its movement model. *Acta Seismologica Sinica, 29*(5), 537–548. (in Chinese with English abstract).

Hu, X. L., Zhang, J. G., Li, Z. C., & Lu, J. G. (2006). Relocation results of moderate-small earthquakes in the Red River fault zone of Yunnan and its relation with geologic structure. *Journal of Seismological Research, 29*(4), 349–354. (in Chinese with English abstract).

Huang, Y., Wu, J. P., Zhang, T. Z., & Zhang, D. N. (2008). Relocation of the M8.0 Wenchuan earthquake and its aftershock sequence. *Science in China (Series D), 51*(12), 1703–1712.

Hubbard, J., & Shaw, J. H. (2009). Uplift of the Longmen Shan and Tibetan Plateau, and the 2008 Wenchuan (M 7.9) earthquake. *Nature, 458*(1), 194–197.

Hubbard, J., Shaw, J. H., & Klinger, Y. (2010). Structural setting of the 2008 Mw 7.9 Wenchuan earthquake, China. *Bulletin of the Seismological Society of America, 100*(5B), 2713–2735.

Hubbard, J., Shaw, J. H., Plesch, A., Riesner, M., Smith, H., Wang, M., & Durand-Riard, P. (2012). A community fault model (CFM) for the Sichuan basin and Longmen Shan, presented at 2012 Fall Meeting, AGU, San Francisco, California, 3–7 December, Abstract T51G–08.

Jia, D., Li, Y. Q., Lin, A. M., Wang, M. M., Chen, W., Wu, X. J., Ren, Z. K., Zhao, Y., & Luo, L. (2010). Structural model of 2008 Mw 7.9 Wenchuan earthquake in the rejuvenated Longmen Shan thrust belt, China. *Tectonophysics, 491*(1–4), 174–184. https://doi.org/10.1016/j.tecto.2009.08.040

Li, C. J., Gan, W. J., Qin, S. L., Hao, M., & Song, S. W. (2018). Present-day activity of the main active faults in Central Yunnan area. *Journal of Seismological Research, 41*, 381–389. (in Chinese with English abstract).

Li, Y. Q., Jia, D., Shaw, J. H., Hubbard, J., Lin, A. M., Wang, M. M., Luo, L., Li, H. B., & Wu, L. (2010). Structural interpretation of the co-seismic faults of the Wenchuan earthquake: 3-D modeling of the Longmen Shan fold-and-thrust belt. *Journal of Geophysical Research Solid Earth, 115*, B04317. https://doi.org/10.1029/2009JB006824

Liu-Zeng, J., Zhang, Z., Wen, L., Tapponnier, P., Sun, J., Xing, X., Hu, G. Y., Qian, X., Zeng, L. S., Lin, D., Chen, J., & Hudnut, K. W. (2009). Co-seismic ruptures of the 12 May 2008, Ms 8.0 Wenchuan earthquake, Sichuan: East–west crustal shortening on oblique, parallel thrusts along the eastern edge of Tibet. *Earth and Planetary Science Letters, 286*, 355–370.

Leloup, P. H., Lacassin, R., Tapponnier, P., Schaerer, U., Zhong, D. L., Liu, X. H., Zhang, L. S., Ji, S. C., & Trinh, P. T. (1995). The Ailao Shan-Red River shear zone (Yunnan, China), tertiary transform boundary of Indochina. *Tectonophysics, 251*(1–4), 3–84.

Lu, R. Q., He, D. F., John, S., Wu, J. E., Liu, B., & Chen, Y. G. (2014). Structural model of the central Longmen Shan thrusts using seismic reflection profiles: Implications for the sediments and deformations since the Mesozoic. *Tectonophysics, 630*, 43–53.

Lu, R. Q., Xu, X. W., He, D. F., Suppe, J., Liu, B., Wang, F. Y., Tan, X. B., & Li, Y. Q. (2017). Seismotectonics of the 2013 Lushan Mw 6.7 earthquake: Inversion tectonics in the eastern margin of the Tibetan Plateau. *Geophysical Research Letters, 44*, 8236–8243.

Mallet, J.-L.L. (2002). *Geomodeling* (p. P624). Oxford University Press.

Marshall, S. T., Funning, G. J., Krueger, H. E., Qwen, S. E., & Loveless, J. P. (2017). Mechanical models favor a ramp geometry for the ventura-pitas point fault, California. *Geophysical Research Letters, 44*(3), 1311–1319.

Nicholson, C., Plesch, A., Sorlien, C. C., Shaw, J. H., Marshall, S. T., & Hauksson, E. (2019). Continued updates, expansion and improvements to the Community Fault Model (CFM version 5.3). Poster Presentation at the 2019 SCEC Annual Meeting, Palm Springs, California, 8–11 September 2019.

Plesch, A., Shaw, J. H., Benson, C., Bryant, W. A., Carena, S., Cooke, M. L., Dolan, J., Fuis, G., Gath, E., Ludwig, L. G., Hauksson, E., Jordan, T., Kamerling, M., Legg, M. R., Lindvall, S., Magistrale, H., Nicholson, C., Niemi, N. A., Oskin, M., … Yeats, R. S. (2007). Community fault model (CFM) for Southern California. *Bulletin of the Seismological Society of America, 97*, 1793–1820. https://doi.org/10.1785/0120050211

Plesch, A., Shaw, J. H., Ross, Z. E., & Hauksson, E. (2020). Detailed 3D fault representations for the 2019 Ridgecrest, California, earthquake sequence. *Bulletin of the Seismological Society of America, 110*(4), 1818–1831. https://doi.org/10.1785/0120200053

Replumaz, A., Lacassin, R., Tapponnier, P., & Leloup, P. H. (2001). Large river offsets and Plio-Quaternary dextral slip rate on the Red River fault (Yunnan, China). *Journal of Geophysical Research, 106*(B1), 819–836. https://doi.org/10.1029/2000JB900135

Riesner, M., Durand-Riard, P., Hubbard, J., Plesch, A., & Shaw, J. H. (2017). Building objective 3D fault representations in active tectonic settings. *Seismological Research Letters, 88*, 831–839. https://doi.org/10.1785/0220160192

Ross, Z. E., Cochran, E. S., Trugman, D. T., & Smith, J. D. (2020). 3D fault architecture controls the dynamism of earthquake swarms. *Science, 368*(6497), 1357–1361. https://doi.org/10.1126/science.abb0779

Schoenbohm, L. M., Burchfiel, B. C., Chen, L. Z., & Yin, J. Y. (2006). Miocene to present activity along the Red River fault, China, in the context of continental extrusion, upper-crustal rotation, and lower-crustal flow. *Geological Society of America Bulletin, 118*(5–6), 672–688. https://doi.org/10.1130/b25816.1

Shaw, J. H., Plesch, A., Tape, C., Suess, M. P., Jordan, T. H., Ely, G., Hauksson, E., Tromp, J., Tanimoto, T., Graves, R., Olsen, K., Nicholson, C., Maechling, P. J., Rivero, C., Lovely, P., Brankman, C. M., & Munster, J. (2015). Unified structural representation of the southern California crust and upper mantle. *Earth and Planetary Science Letters, 415*, 1–15.

Shen, Z. K., Sun, J. B., Zhang, P. Z., Wan, Y. G., Wang, M., Bürgmann, R., Zeng, Y. H., Gan, W. J., Liao, H., & Wang, Q. L. (2009). Slip maxima at fault junctions and rupturing of barriers during the 2008 Wenchuan earthquake. *Nature Geoscience, 2*(10), 718–724. https://doi.org/10.1038/ngeo0636

Shi, X. H., Sieh, K., Weldon, R., Zhu, C. N., Han, Y., Yang, J. W., & Robinson, S. W. (2018). Slip rate and rare large prehistoric earthquakes of the Red River fault, southwestern China. *Geochemistry, Geophysics, Geosystems, 19*, 2014–2031. https://doi.org/10.1029/2017GC007420

Stockmeyer, J. M., Shaw, J. H., & Guan, S. W. (2014). Seismic hazards of multi-segment thrust-fault ruptures: Insights from the 1906 Mw7.4-8.2 Manas, China. *Earthquake. Seismological Research Letters, 85*, 801–808.

Tapponnier, P., Lacassin, R., Leloup, P. H., Scharer, U., Zhong, D. L., Wu, H. W., Liu, X. H., Ji, S. C., Zhang, L. S., & Zhong, J. Y. (1990). The Ailao Shan/Red River metamorphic belt: Tertiary left-lateral shear between Indochina and South China. *Nature, 343*, 431–437.

Wang, M., Jia, D., Shaw, J. H., Hubbard, J., Plesch, A., Li, Y., & Liu, B. J. (2014). The 2013 Lushan earthquake: Implications for seismic hazards posed by the range front blind thrust in the Sichuan Basin, China. *Geology, 42*, 915–918.

Wang, Q., Qiao, X. J., Lan, Q. G., Freymueller, J., Yang, S. M., Xu, C. J., Yang, Y. L., You, X. Z., Tan, K., & Chen, G. (2011). Rupture of deep faults in the 2008 Wenchuan earthquake and uplift of the Longmen Shan. *Nature Geoscience, 4*, 634–640.

Wang, Y., Zhang, B., Schoenbohm, L. M., Zhang, J. J., Zhou, R. J., Hou, J. J., & Ai, S. (2016). Late Cenozoic tectonic evolution of the Ailao Shan-Red River fault (SE Tibet): Implications for kinematic change during plateau growth. *Tectonics, 35*, 1969–1988. https://doi.org/10.1002/2016TC004229

Wen, X. Z., Fan, J., Yi, G. X., Deng, Y. W., & Long, F. (2008). A seismic gap on the Anninghe fault in western Sichuan, China. *Science in China Series D: Earth Sciences, 51*(10), 1375–1387.

Weldon, R., Sieh, K., Zhu, C., Han, Y., Yang, J., & Robinson, S. (1994). Slip rate and recurrence interval of earthquake on the Hong He (Red River) Fault, Yunnan. International Workshop on Seismotectonics and Seismic Hazard in SE Asia, China, pp. 244–248.

Wu, J. P., Huang, Y., Zhang, T. Z., Ming, Y. H., & Fang, L. H. (2009). Aftershock distribution of the Ms8.0 Wenchuan earthquake and three dimensional P-wave velocity structure in and around source region. *Chinese Journal of Geophysics, 52*, 320–328. (in Chinese with English abstract).

Wu, X. Y., Xu, X. W., Yu, G. H., Cheng, J., Cheng, G. H., An, Y. F., & Wang, Q. X. (2018). Map preparation of earthquake surface ruptures in the national experimental field of earthquake Monitoring and perdiction in Sichuan and Yunnan Province. *Seismology and Geology, 40*(1), 27–41. (in Chinese with English abstract).

Xu, X. W., Wen, X. Z., Zheng, R. Z., Ma, W. T., Song, F. M., & Yu, G. H. (2003). Pattern of latest tectonic motion and its dynamics for active blocks in Sichuan-Yunnan region, China. *Science in China Series D: Earth Sciences, 46*, 210–226.

Xu, X. W., Wen, X. Z., Yu, G. H., Chen, G. H., Klinger, Y., Hubbard, J., & Shaw, J. (2009). Coseismic reverse- and oblique-slip surface faulting generated by the 2008 Mw 7.9 Wenchuan earthquake, China. *Geology, 37*(6), 515–518.

Xu, X. W., Wen., X. Z., Han, Z. J., Chen, G. H., Li, C. Y., Zheng, W. J., Zhang, S. M., Ren, Z. Q., Xu, C., Tan, X. B., Wei, Z. Y., Wang, M. M., Ren, J. J., He, Z. T., & Liang, M. J. (2013). Lushan Ms7.0 earthquake: A blind reserve-fault earthquake. *Chinese Science Bulletin, 58*(20), 1887–1893. (in Chinese with English abstract). https://doi.org/10.1007/s11434-013-5999-4

Yan, B., & Lin, A. M. (2017). Holocene activity and paleoseismicity of the Selaha Fault, southeastern segment of the strike-slip Xianshuihe Fault Zone, Tibetan Plateau. *Tectonophysics, 694*, 302–318.

Yi, G. X., Long, F., & Zhang, Z. W. (2012). Spatial and temporal variation of focal mechanisms for aftershocks of the 2008 Ms8.0 Wenchuan earthquake. *Chinese Journal of Geophysics, 55*, 1213–1227. (in Chinese with English abstract).

Yin, F. L., Jiang, C. S., Han, L. B., Zhang, H., & Zhang, B. (2018). Seismic hazard assessment for the Red River fault: Insight from Coulomb stress evolution. *Chinese Journal of Geophysics, 61*, 183–198. (in Chinese with English abstract).

Yue, L. F., Suppe, J., & Hung, J. H. (2005). Structural geology of a classic thrust belt earthquake: The 1999 Chi-Chi earthquake Taiwan (Mw = 7.6). *Journal of Structural Geology, 27*(11), 2058–2083. https://doi.org/10.1016/j.jsg.2005.05.020

Zhang, P. Z., Wen, X. Z., Shen, Z. K., & Chen, J. H. (2010). Oblique, high-angle, listric-reverse faulting and associated development of strain: The Wenchuan earthquake of May 12, 2008, Sichuan, China. *Annual Review of Earth and Planetary Sciences, 38*(1), 353–382. https://doi.org/10.1146/annurev-earth-040809-152602

Zhang, P. Z. (2013). A review on active tectonics and deep crustal processes of the Western Sichuan region, eastern margin of the Tibetan Plateau. *Tectonophysics, 584*, 7–22.

Chapter 6
CSES Community Deformation Models in Southwest China

Hua Wang, Jean-Philippe Avouac, Zhigang Shao, Xiaoxia Liu, Wenxin Wei, Songhui Zhan, Xiaolei Dai, and Yidong Lou

Abstract Geodetic measurements provide important constraints on crustal deformation and earthquake processes. In this study, we describe of such data can be used to produce community deformation models which quantify interseismic strain rates and fault slip rates, which can then be used to inform probabilistic earthquake forecast. We apply the framework to data from the China Seismic Experimental Site (CSES) in the Sichuan-Yunnan area. We followed a workflow, consisting of four steps. (1) We collected and compiled the most complete and up-to-date dataset of GNSS observations, then estimated the velocity at 618 sites after correction for seasonal and co- and post-seismic effects. (2) We determined a continuous 3D velocity field model using VELMAP from the joint inversion of these GNSS data with the addition of InSAR and precise leveling data. We estimated the uncertainties from cross-validation to 1.1 (1.5) and 0.9 (0.9) mm/yr w.r.t. the east and north components of the campaign (continuous) GNSS and 0.5 mm/yr for the leveling data (at the 68% confidence level). (3) We used this velocity field model to determine a continuous strain rate field, which shows the highest strain rates localized along the Xianshuihe-Xiaojiang fault system, where the second invariant of the strain rate tensor reaches 57 nanostrain/year. (4) We developed a block model for the CSES region which fits the east and north GNSS velocities to within 0.6 and 0.7 mm/year respectively (at the 68% confidence level). Our model implies uniform slip rates of 10–14 mm/year along the entire Xianshuihe-Xiaojiang fault system. In the central segment of this fault

The original version of this chapter was revised: Figure 6.5 has been corrected. The correction to this chapter is available at https://doi.org/10.1007/978-981-16-8607-8_13

H. Wang (✉) · S. Zhan
Department of Surveying Engineering, Guangdong University of Technology, Guangzhou, China
e-mail: ehwang@163.com

J.-P. Avouac
Department of Geological and Planetary Sciences, California Institute of Technology, Pasadena, CA, USA

Z. Shao · X. Liu · W. Wei
Institute of Earthquake Forecasting, China Earthquake Administration, Beijing, China

X. Dai · Y. Lou
GNSS Research Center, Wuhan University, Wuhan, China

© Higher Education Press 2022, corrected publication 2022

Y.-G. Li et al. (eds.), *China Seismic Experimental Site*,
https://doi.org/10.1007/978-981-16-8607-8_6

system, slip is partitioned between the Anninghe-Zemuhe and Daliangshan faults, with slip rates of ~ 6–7 and 4–5 mm/year respectively. The block model might underestimate faulting in the interior of blocks where minor faults exist and conversely overestimate the slip-rates on block boundaries, in particular for Xiaojiang where multiple NS-trending branches exist. Since the accuracy of the deformation models depends primarily on the geodetic accuracies and spatial resolution, future work will focus on increasing the geodetic observations using in particular InSAR for which large archive of acquisitions are available. While the approach presented here could certainly be improved, it provides an effective means to incorporate geodetic and remote sensing data in seismic hazard studies.

Keywords Community deformation models · Geodesy · Velocity field · Strain rates · Fault slip rates · CSES

6.1 Introduction

According to elastic rebound theory (Reid, 1910), an earthquake releases suddenly elastic strain that had accumulated gradually over the interseismic period (the period separating successive ruptures on a given fault segment). Nowadays we understand that the seismic cycle can be divided in different phases, commonly interseismic, preseismic, coseismic, and postseismic phases that had not been recognized by Reid (1910), but current mechanical models of the seismic cycle hold to the principle that, on average over the long term (meaning a period much longer than the return period of large earthquakes on a given fault), strain accumulation between earthquakes balances strain release during earthquakes (e.g., Avouac, 2015; Scholz, 2019). As a result, this conservation principle is widely used to incorporate geodetic measurements in seismic hazard studies.

Examples include the Uniform California Earthquake Rupture Forecast (UCERF3, Field et al., 2014), Seismic Hazard Inferred from Tectonics (SHIFT, Bird & Kreemer, 2015), Global Earthquake Activity Rate Model (GEAR1, Bird et al., 2015), National Seismic Hazard Maps (NSHM, Petersen et al., 2020), Global Earthquake Model (GEM) Seismic Hazard Map (Pagani et al., 2014). In most of these studies, the geodetic data are used to constrain long-term fault slip rates and the faults are assumed to slip, over the seismogenic depth range, only during large earthquakes. In fact, the geodetic measurements contain potentially more information as the interseismic strain field can provide tighter constraints on the distribution of fault locking and hence of the rate of accumulation of elastic strain to be released in earthquake. This is important as recent studies of continental settings have shown that faults can slip aseismically either gradually or episodically within the typical 0–15 km seismogenic depth range of continental faults (e.g., Avouac, 2015; Bürgmann et al., 2000; Cavalié et al., 2008; de Michel et al., 2011; Jolivet et al., 2012; Kaneko et al., 2013; Fattahi & Amelung, 2016). In this study, we aim at producing a geodesy-based deformation model of the China Seismic Experimental Site (CSES) (Fig. 6.1),

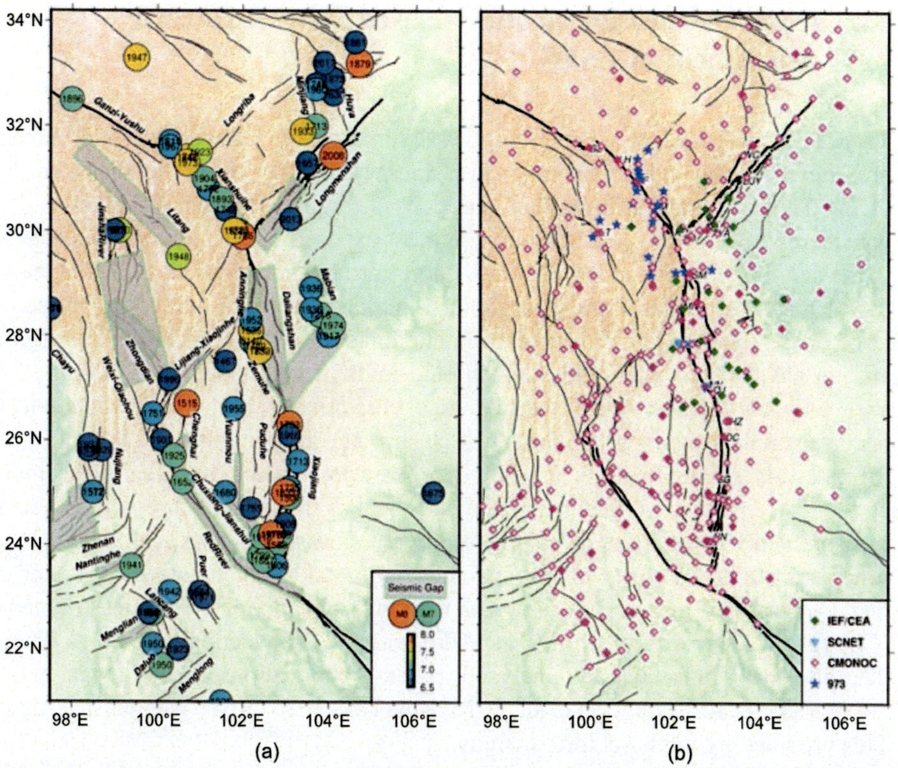

Fig. 6.1 **a** Topographic and tectonic map of Chuandian in Southwest China. Active faults are from the fifth edition of the national seismic zonation map of China (GB 18306, 2015). The color-coded circles with years of occurrence denote $M \geq 6.5$ historical earthquakes. The gray boxes indicate seismic gaps suggested by the Working Group of $M7$ (2012). **b** GNSS stations whose raw observations are available in our study. Magenta diamonds denote GNSS stations from the Crustal Movement Observation Network of China (CMONOC-I/II) project. Green diamonds denote stations from the Institute of Earthquake Forecasting, China Earthquake Administration (IEF/CEA). Cyan inverted triangles denote those along two profiles near Xichang and Ningnan from the Institute of Surveying and Mapping, Sichuan Earthquake Agency. Blue stars are those from the 973 Program. Solid and open symbols represent continuous and campaign stations, respectively

which can be used to estimate long term slip rates on the major faults in the area, as well as interseismic elastic strain build up on these faults. This effort requires the best possible density of data, and data curation to account for sources of deformation not related to interseismic strain.

Hereafter, we first describe the study area, the geodetic dataset and our proposed framework for seismic hazard studies. Next, we develop continuous velocity and strain rate field models by joint inversion of GNSS, InSAR and precise leveling data. Finally, we develop a block model to determine fault slip rates constrained by the GNSS velocities and compare them with the Late Quaternary geological slip rates.

6.2 A Framework for Geodesy-Based Seismic Hazard Assessment of CSES

The CSES covers a large area of Southwest China spanning the Sichuan and Yunnan provinces, abbreviated as Chuandian in Chinese (Fig. 6.1a). As the southeast border of the Tibetan Plateau, Chuandian experiences frequent large earthquakes and is considered to be of the area of China the most exposed to seismic hazards because of the dense population and steep relief. Large earthquakes occurred along the major faults, the Longmenshan and Xianshuihe-Xiaojiang faults in particular, but some large events did occur on rather minor faults. In addition, earthquakes in this region are not always associated with fast-slipping faults; for instance, the 2008 M_w 7.9 Wenchuan earthquake, one of the two largest historical earthquakes in Chuandian, occurred on the Longmenshan thrust which slips at a long-term rate of 1–2 mm/year (Shen et al., 2009; Zhang et al., 2013) or even less than 1 mm/year (Densmore et al., 2007). These two features of the earthquakes in Chuandian require geodetic measurements appropriate to resolve relatively slow strain rates, calling for long time series, a high density of stations and high geodetic accuracy. As a matter of fact, Chuandian is the area of China with the longest history of GNSS measurements (Chen et al., 2000; King et al., 1997) and the densest GNSS stations nowadays. These data have been presented and analyzed in many publications already (Shen et al., 2005; Meng et al., 2008; Hao et al., 2016; Rui et al., 2016, 2019; Wang et al., 2021). This provides a good basis for our study.

The flowchart of Fig. 6.2 shows how the GNSS data can be combined in principle with other types of data to build community deformation models for the CSES and inform probabilistic earthquake forecast. In this framework, GNSS and InSAR observations are combined with 50 years of precise leveling data and geological fault slip rates. These four sources of information are complementary.(1) GNSS can provide precise horizontal velocities at the sites which were observed in campaign mode or equipped with continuously operating stations. InSAR/leveling are more sensitive to vertical motion. InSAR has much higher spatial resolution than GNSS and leveling measurements. Combining these three kinds of geodetic data can help determine 3D deformation models. Geological fault slip rates provide long-term constraints on fault activities, some of which might appear low strain and underestimated seismic potential from geodetic data only. We integrate all these datasets using an improved version of VELMAP approach (Wang & Wright, 2012), to derive a continuous velocity field model. Based on the velocity field, we produce a continuous strain rate field model to describe interseismic strain and use a block model to determine fault slip rates. The strain rates and the fault slip rates are the standard products of our deformation models and can be used as inputs for the final probabilistic earthquake forecast models. In the following sections, we describe preliminary results from our effort to derive a community deformation model following this framework.

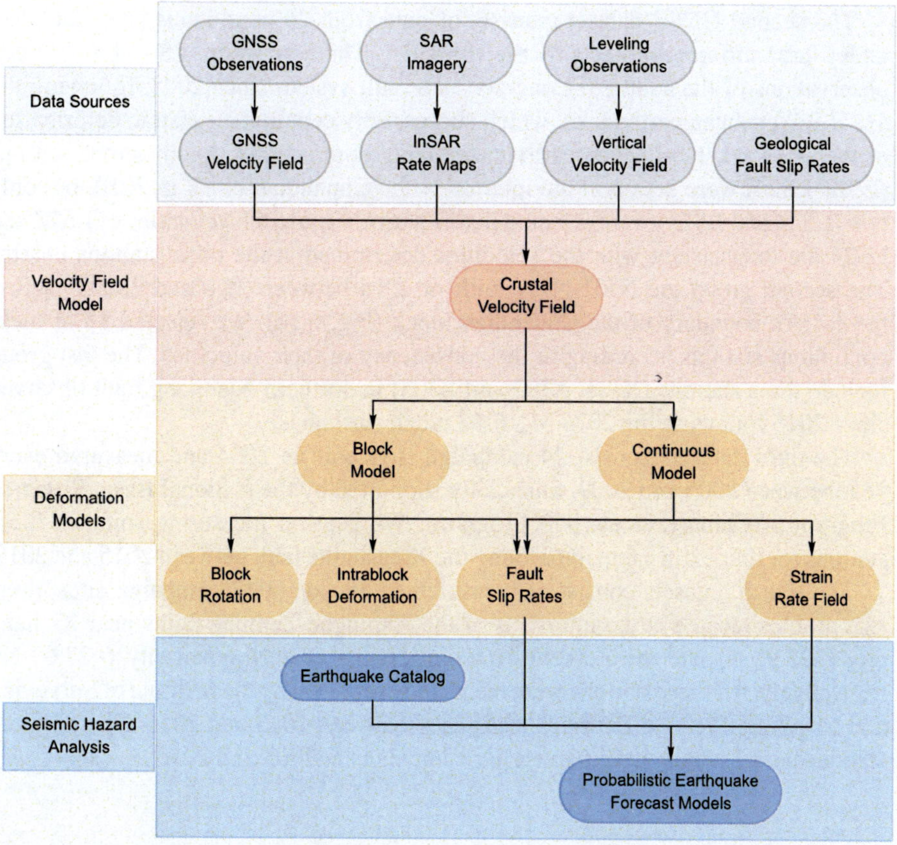

Fig. 6.2 Flowchart showing the dataset used to build CSES community deformation models and probabilistic earthquake forecast models

6.3 GNSS Velocity Field

6.3.1 GNSS Data Sources

In this study we used GNSS observations collected through four projects in southwest China. The majority of the data are from the CMONOC-I/II project, which includes data from 52 continuous and 343 campaign sites in the CSES region (Fig. 6.1b). The Phase I of this project was launched in 1998 and the Phase II has been ongoing since 2009. The campaign-mode GNSS stations were observed in 1999, 2001, 2004 and 2007 during CMONOC-I. Since CMONOC-II, they were measured every other year between 2009 and 2015, and every year in some key regions like Southwest China since 2017. As we used data until 2019, most of them have 11 campaigns for the Phase-I stations and 7 for the new Phase-II stations, with 4-day occupation in each survey.

The second GNSS dataset consists of data from 26 continuous sites deployed under three different projects by the IEF/CEA. The first group, LS01-LS10, made observations of the southern Longmenshan fault system since 2008, following the M_w 7.9 Wenchuan earthquake, which successfully captured coseismic deformation of the 2013 M_w 6.6 Lushan earthquake (Jiang et al., 2014). Another four stations (LS21-LS24) were set up in the interior of the Chuandian block in 2010, but only two (LS21 and LS23) were selected in this study because the velocities of LS22 and LS24 are inconsistent with the velocities determined at the other stations nearby. The second group are NJ01-NJ17 built in 2014 between 26.8°and 28.5°N across the eastern boundary of the Chuandian block (Fig. 6.1b). We selected 13 of these continuous stations according to the consistency of their velocities. The last group include three stations (XJ01, XJ02 and XJ03) in northern Xiaojiang fault observed since 2015 following the 2014 M_w 6.5 Ludian earthquake.

The third datasets consist 24 campaign sites built in 2004 and measured every year between 2005 and 2009, financially supported by the National Basic Research Program of China (abbrev. 973 Program). We kept on measuring some of these stations for four extra campaigns along the Xianshuihe fault between 2015 and 2019.

The fourth datasets consist of measurements made at 13 campaign sites along two profiles located at the intersects of the Anninghe-Zemuhe faults near Xichang city (~27.9° N) and the Zemuhe-Xiaojiang faults near Ningnan city (~27.0° N), respectively. All these stations were measured every year by the Institute of Surveying and Mapping, Sichuan Earthquake Agency between 2013 and 2016. The northern stations near Xichang have two earlier campaigns in 2005 and 2010.

6.3.2 GNSS Velocity Estimation

We used the PANDA software developed at Wuhan University (Shi et al., 2015) to process the GNSS data in single-day solutions with the Precise Point Positioning mode (Zumberge et al., 1997). We used precise ephemeris in IGS08 reference frame, which is the IGS realization of International Terrestrial Reference Frame 2008 (Rebischung et al., 2012). Detailed strategies for obtaining GNSS daily solution are listed in Table 1 in Zheng et al., (2017).

The geodetic position time series is composed of secular velocity, seasonal displacements in response to loading changes imposed upon the lithosphere, sudden offsets associated with coseismic deformation, equipment replacements, etc., and transient deformation such as postseismic, aseismic and slow-seismic creeping (Eq. 6.1). This procedure is standard and sometimes called trajectory modeling (Bevis & Brown, 2014). In tectonic geodesy, the four components are routinely described using a linear, sinusoidal, Heaviside, and logarithmic or exponential functions, respectively (Eq. 6.2).

$$x(t) = x_{\text{secular}} + x_{\text{seasonal}} + x_{\text{offsets}} + x_{\text{transient}} + \varepsilon \tag{6.1}$$

$$x(t) = x_0 + v(t - t_0) + \sum_{k=1}^{n_F} [s_k \sin(\omega_k t) + c_k \cos(\omega_k t)] + \sum_{j=1}^{n_J} b_j H\left(t - t_j\right) + \sum_{i=1}^{n_T} a_i \ln\left(1 + \frac{t - t_i}{\tau_i}\right) \quad (6.2)$$

where x_0 is initial coordinate at a time t_0, v is secular velocity, ω_k, s_k and c_k are frequency and amplitudes of the sine and cosine terms, respectively, b_j is instrumental or coseismic offset of the *j*-th earthquake at t_j, a_i and τ_i are amplitude and characteristic decay time of transient deformation following the *i*-th earthquake occurred at t_i. $H\left(t - t_j\right)$ is a Heaviside function. n_F is the total number of periodic terms, n_J and n_T are the total number of earthquakes showing coseismic and postseismic deformation, respectively.

In this study, our aim is to obtain secular velocities in southwest China. However, the GNSS time series were significantly affected by a few earthquakes in Chuandian, particularly the 2008 M_w 7.9 Wenchuan earthquake. The stations were even affected by coseismic deformation associated with the 2011 M_w 9.0 Tohoku-Oki earthquake and the 2004 M_w 9.1 Sumatra earthquake thousands of kilometers away from Chuandian. The velocity field looks messy especially around the 2008 M_w 7.9 Wenchuan earthquake if we don't correct all the factors carefully (Fig. 6.3). To improve the secular velocity reliability, we analyzed the GNSS time series according to the following steps:

(i) We preliminarily estimated secular velocity, annual and semi-annual periodic signal and offsets for continuous sites using an iterative weighted least-squares method. The offsets were furtherly adjusted by the difference of mean residuals from a few (typically 10) epochs before and after each jump.

(ii) We removed the offsets and detrended the time series with their secular velocities for the continuous sites. The detrended time series were then analyzed to extract more accurate seasonal signal using a variational Bayesian Independent Component Analysis (vbICA) approach (Choudrey & Roberts, 2003) as implement by Gualandi et al., (2016, 2017). Three independent components with the largest weight values $\boldsymbol{S}$ clearly represent a seasonal signal. They were thus subtracted from the original continuous time series. This procedure is refinement of the trajectory modeling approach as it allows to correct for non-harmonic seasonal signal with interannual variability. We recomposed the seasonal signal of campaign sites by remaining the same weight $\boldsymbol{S}$, doing interpolation for the spatial function $\boldsymbol{U}$ of the continuous sites, and stacking temporal function $\boldsymbol{V}$ for 365 days following Gualandi et al. (2017).

(iii) We applied again the first step to update the velocities and offsets for the continuous sites, which have no postseismic effect due to the 2008 Wenchuan earthquake.

(iv) Our study area is far away from the 2011 Tohoku-Oki earthquake. We found a uniform coseismic deformation field from the continuous sites. So, we

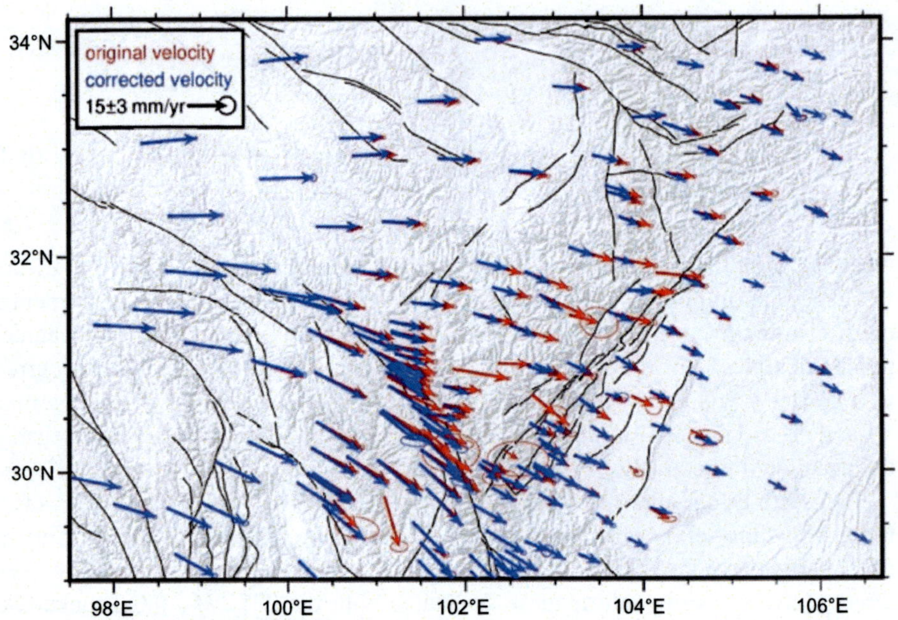

Fig. 6.3 Velocity field relative to Eurasia around the 2008 M_w 7.9 Wenchuan earthquake (red) without and (blue) with corrections. A few red arrows without corresponding blue ones indicate their corrected velocities are still much different from other stations nearby, so we removed them in our final results

conducted a bi-linear interpolation for all the campaign sites using the estimated offsets from continuous sites. For the other earthquakes, we estimated coseismic offsets directly in the trajectory model. Earthquakes were initially selected from the GCMT catalog according to the empirically relationship in Herring et al. (2016).

(v) In our datasets, no continuous station was built before the 2008 Wenchuan earthquake. We picked up campaign sites with the best preseismic and postseismic observations to estimate amplitudes a_i and characteristic decay time τ_i of the logarithmic function (Eq. 6.2). We then used VELMAP (Wang & Wright, 2012) to interpolate the amplitudes for all the other sites at 29.5° N northwards where we found clear postseismic deformation. The interpolated amplitudes and the decay time τ_i were then used to correct postseismic deformation of these sites.

(vi) After correcting postseismic deformation of the 2008 Wenchuan earthquake, we used the trajectory model again to estimate the resultant secular velocity for all sites. The velocity field of 458 sites is shown as blue arrows in Fig. 6.4a. In the northern part near the Wenchuan earthquake, the velocities are generally smooth for the sites with/without preseismic observations. This implies that postseismic deformation is well corrected in our results. As an example,

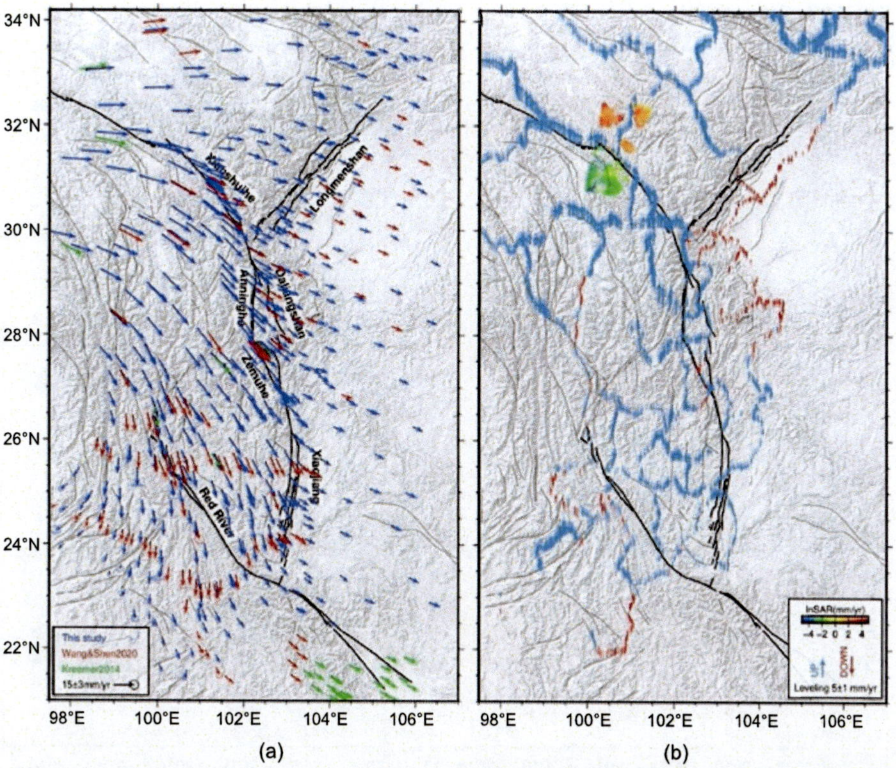

Fig. 6.4 Geodetic velocities used as input of the deformation models. **a** Merged GNSS velocity field relative to the Eurasia frame. The blue arrows indicate the GNSS stations which we processed from raw observations. The red and green ones are those incorporated from Wang and Shen (2020) and Kreemer et al. (2014). **b** Published leveling velocities (Hao et al., 2014) and InSAR interseismic rate map (Wang et al., 2009). The sky-blue arrows indicate uplift, and the red indicate subsidence

we showed the detrended time series of LS03 before and after all the above corrections in Fig. 6.5.

6.3.3 *Merged GNSS Velocity Field*

The 900-km-long right-slip Red River fault system extends southeastwards from Yunnan to northern Vietnam (Tapponnier et al., 1990). About 1/3 length of the Red River fault is outside China. The GNSS results in Vietnam are helpful to analyze regional crustal deformation in the CSES. In northern Vietnam, the published GNSS velocities are mainly from Nguyen et al. (2013) and Trần et al. (2013), which have been compiled by Kreemer et al. (2014). Recently, Wang and Shen (2020) published a comprehensive GPS velocity field in China. To increase the density of our GNSS data points, we incorporated the velocities reported in Wang and Shen (2020) and Kreemer et al. (2014) for the sites whose raw observations are not available for us. We

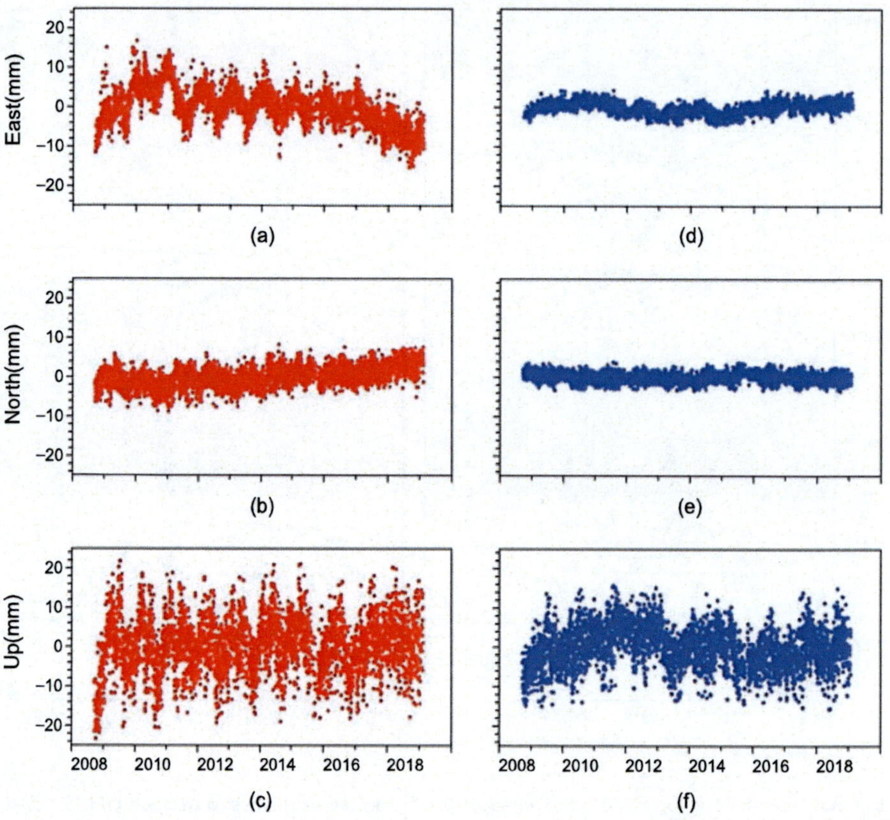

Fig. 6.5 Time series of LS03 detrended by a linear velocity **a–c** without and **d–f** with corrections. Note that we didn't correct postseismic deformation for vertical components as we only used horizontal GNSS velocities in this study

used the common sites to compute an Euler pole and converted all the complementary velocities from these two studies into our solutions. For the 242 common stations with Kreemer et al. (2014), the mean differences ($\pm$ 1 standard deviation) are 0.0 ± 1.4 and 0.0 ± 0.9 mm/yr for the east and north components, respectively. We have 367 common sites with Wang and Shen (2020) with mean differences of -0.2 ± 0.9 and -0.5 ± 0.7 mm/yr for the east and north components. Eventually we obtained 618 sites (Fig. 6.4a), of which 95, 112 and 218 stations are new w.r.t. Wang and Shen (2020), Zheng et al. (2017) and Kreemer et al. (2014), respectively.

6.4 Continuous Geodetic Velocity and Strain Rate Field Models

6.4.1 Velocity Field Model

In complement to GNSS, InSAR is able to provide high-resolution measurements in satellite's line-of-sight (LOS) direction, and repeated precise leveling data provide accurate measurements of vertical displacements. Here we used the VELMAP approach developed in Wang and Wright (2012) to make a consistent velocity field by combining all the geodetic deformation measurements in the region. For InSAR data, we used the published interseismic deformation along the northwestern Xianshuihe fault in Wang et al. (2009). For precise leveling data, we used the published vertical velocities from observations between 1970 and 2012 in Hao et al. (2014).

The idea of the VELMAP approach is to divide study area into a triangular mesh, and assume the velocity varies linearly with latitude and longitude within each triangle. Therefore, the geodetic observations within each triangle are related to the velocities of its vertices by a spherical shape function proposed in England and Molnar (2005). The previous studies used GNSS and InSAR data (e.g., Pagli et al., 2014; Wang & Wright, 2012; Wang et al., 2019; Weiss et al., 2020), here we improved the model by incorporating precise leveling data. The mathematical model can be written as:

$$\begin{bmatrix} \boldsymbol{G}_{sar} & \boldsymbol{G}_{orb} & \boldsymbol{G}_{atm} \\ \boldsymbol{G}_{gnss} & 0 & 0 \\ \boldsymbol{G}_{lev} & 0 & 0 \\ \kappa_s^2\nabla_s^2 & 0 & 0 \end{bmatrix} \begin{bmatrix} \boldsymbol{m}_{vel} \\ \boldsymbol{m}_{orb} \\ \boldsymbol{m}_{atm} \end{bmatrix} = \begin{bmatrix} \boldsymbol{d}_{sar} \\ \boldsymbol{d}_{gnss} \\ \boldsymbol{d}_{lev} \\ 0 \end{bmatrix} \tag{6.3}$$

where $\boldsymbol{G}_{sar}$ is the design matrix for InSAR rate maps composed of interpolation kernel and local LOS unit vectors; $\boldsymbol{G}_{orb}$ and $\boldsymbol{G}_{atm}$ are matrices for InSAR orbital ($\boldsymbol{m}_{orb}$) and topographically-correlated atmospheric delay correction models ($\boldsymbol{m}_{atm}$); $\boldsymbol{G}_{gnss}$ and $\boldsymbol{G}_{lev}$ are the design matrices composed of the interpolation kernel for GNSS and precise leveling velocities, respectively; $\boldsymbol{m}_{vel}$ is the unknown velocities of the triangular vertices; $\boldsymbol{d}_{sar}$, $\boldsymbol{d}_{gnss}$, and $\boldsymbol{d}_{lev}$ are LOS InSAR, GNSS and precise leveling velocities, respectively. We used smoothing constraints to remove short-wavelength features in InSAR rate maps that we assume have non-tectonic origin. ∇_s^2 is the Laplacian smoothing operator approximated by a scale-dependent umbrella operator (Desbrun, 1999), and the factor κ_s^2 determines the smoothing weight.

We solved the system of equations in (Eq. 6.3) to yield the velocity field under a least-squares criterion. The weight ratios among InSAR, GNSS and precise leveling data were determined using a variance component estimation approach (Grafarend, 1985). The resultant continuous 3-D velocity field and their one-sigma errors are shown in Figs. 6.6 and 6.7, respectively. The root-mean-square (RMS) errors of GNSS, InSAR and precise leveling data are 0.7, 0.8 and 0.4 mm/yr, respectively.

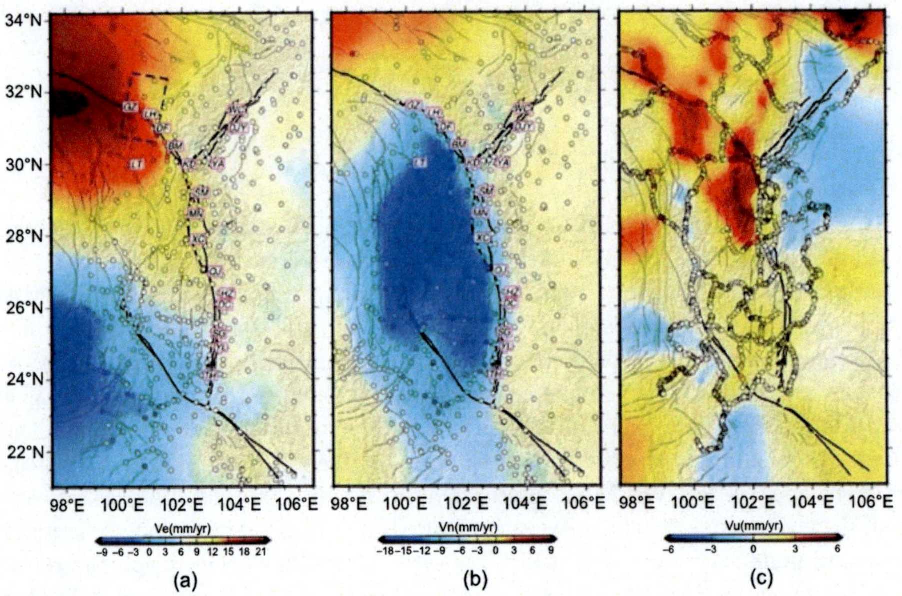

Fig. 6.6 Continuous 3-D velocity field model from GNSS, InSAR and precise leveling. **a** East components; **b** North components; **c** up components. The color-coded circles in (**a**–**b**) denote the corresponding observed east and north velocities from GNSS data, and the circles in (**c**) are the observed vertical velocities from precise leveling. The dashed box in (**a**) indicates InSAR coverage

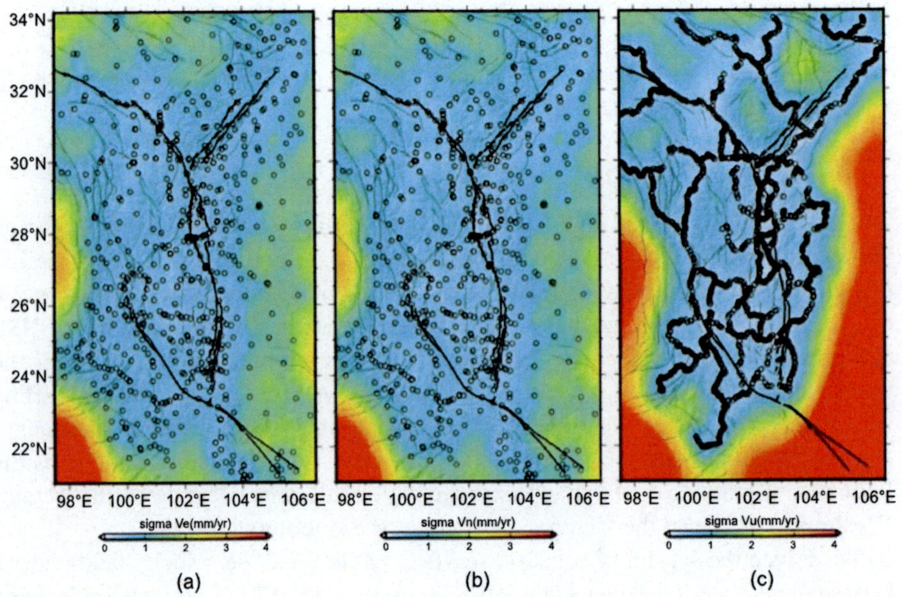

Fig. 6.7 One-sigma errors of the 3-D velocity field. **a** East components; **b** North components; **c** up components

Figure 6.6a, b shows clear clockwise rotation of the Chuandian block bounded by the Xianshuihe-Xiaojiang fault system. The vertical velocity field in Fig. 6.6c shows uplift in the majority of the region within the plateau interior and subsidence in the Sichuan Basin. The formal uncertainties of the velocity field depend on the spatial density of the observations. The mean one-sigma errors are 1.0, 0.9 and 1.5 mm/year for the east, north and up components, respectively. They are less than 1 mm/year along the major fault zones where there are fairly good data coverage. However, the uncertainties are larger than 4 mm/year where there are no observations, particularly for the vertical components. In these areas, the velocity field model is completely based on interpolation.

We used a bootstrapping strategy to estimate the accuracy of the velocity field. We excluded 10% of the observations, taking out a random selection of 10 continuous, 50 campaign GNSS and 200 leveling data points. Then we compared the observed and the predicted velocities of these sites (Fig. 6.8). The results show discrepancies

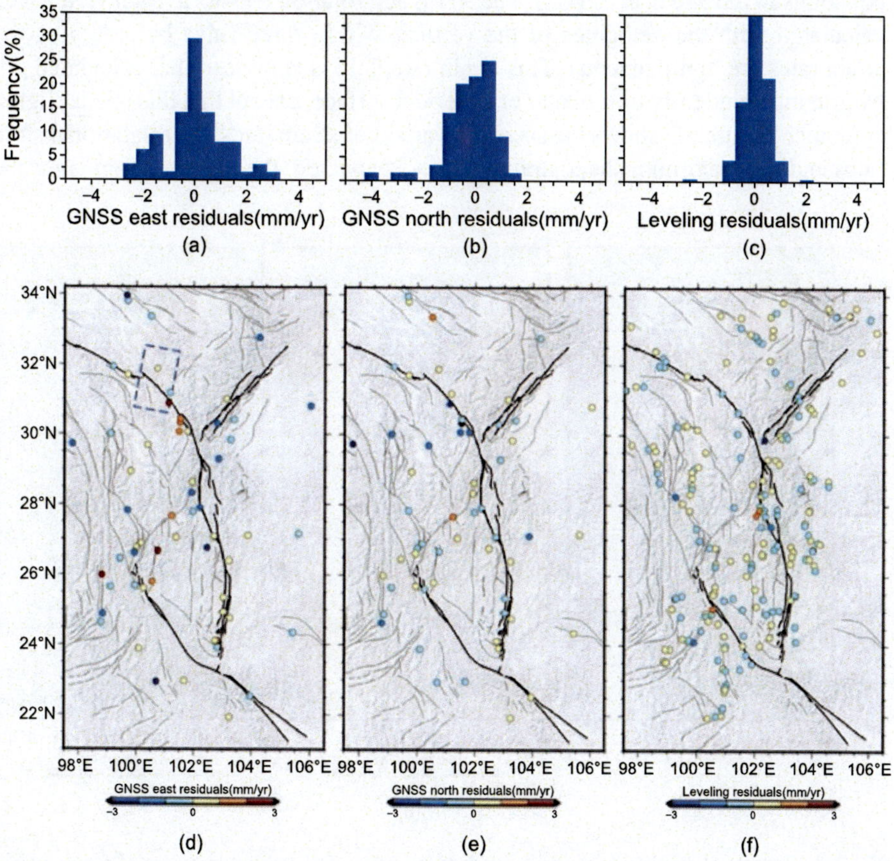

Fig. 6.8 Cross-validation of the GNSS and leveling data for the velocity field model. **a–b** Histograms and **d–e** spatial distribution of 60 GNSS residuals in east and north. **c** Histogram and **f** spatial distribution of 200 leveling residuals

of 0.0 ± 1.1 and −0.2 ± 0.9 mm/yr for the east and north components of campaign GNSS, −0.1 ± 1.5 and −0.5 ± 0.9 mm/yr for the continuous GNSS, and 0.0 ± 0.5 for the leveling data. The errors bars represent the standard deviation of the discrepancies and thus the uncertainty at the 68% confidence level.

6.4.2 Strain Rate Field Model

Several potential strategies exist for calculating a strain rate field if the data points are not uniformly distributed. One option is to calculate strain rate tensor in each triangle directly using the spherical shape function determined in the VELMAP inversion, similar to England and Molnar (2015). However, we find that such forward calculation is very sensitive to the noise of local velocities for dense meshes. Here we inverted the strain and rotation tensors on the surface of spherical Earth using equations in Savage et al. (2001). The strain and rotation tensors at each vertex were calculated with the velocities of the vertices within three rings by assuming that strain rates are homogeneous. This strain rate field is similar to that calculated grid by grid independently on a planar or spherical surface, except that this one is slightly smoother. Figure 6.9 shows the second invariant of strain rates, the dilatational strain rates and the maximum shear strain rates. As expected, the highest strain is located

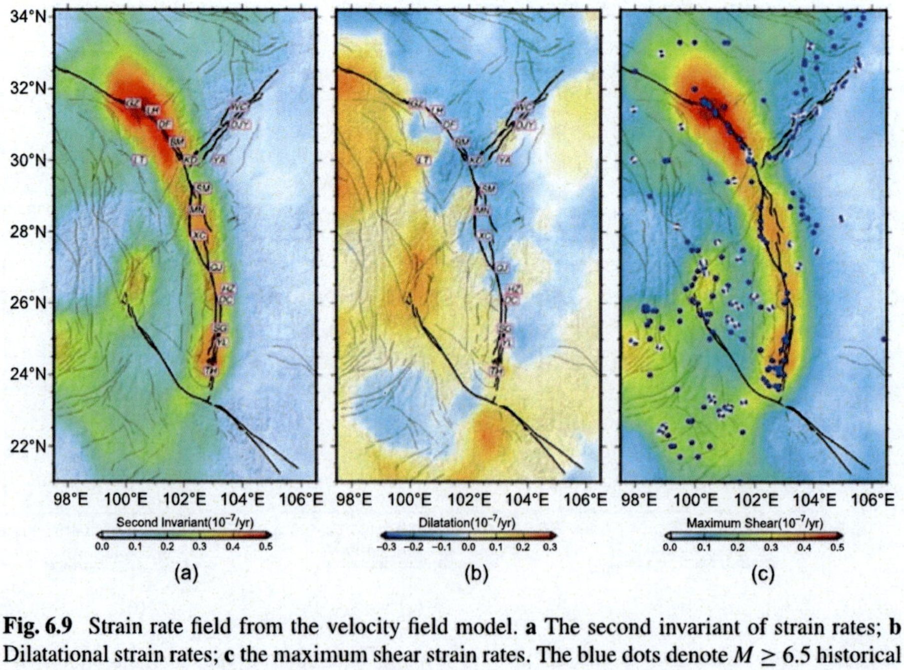

Fig. 6.9 Strain rate field from the velocity field model. **a** The second invariant of strain rates; **b** Dilatational strain rates; **c** the maximum shear strain rates. The blue dots denote $M \geq 6.5$ historical earthquakes before 1990. The "beach-balls" show focal mechanisms of the $M \geq 6.5$ earthquakes occurred since 1991

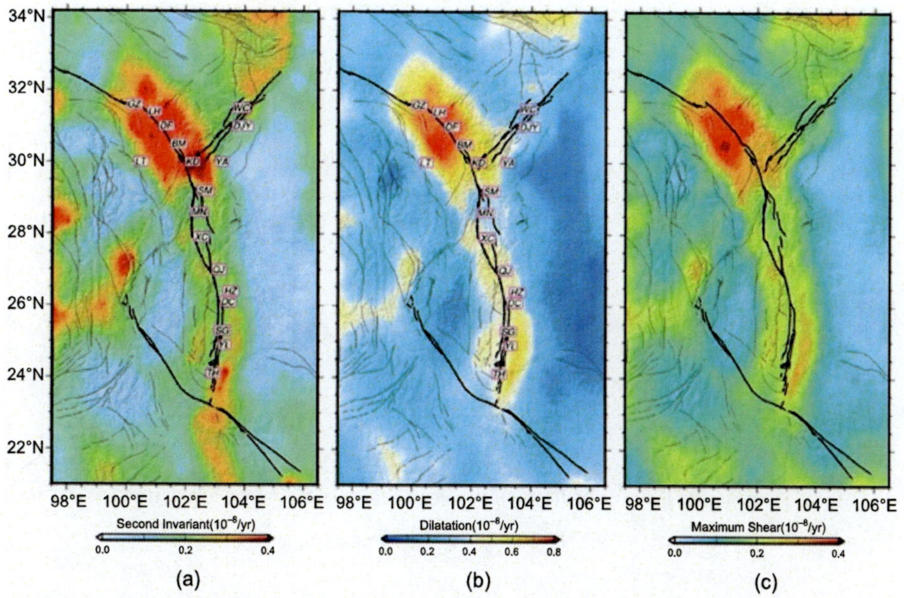

Fig. 6.10 One-sigma errors of **a** the second invariant of strain rates, **b** dilatational strain rates and **c** the maximum shear strain rates

along the Xianshuihe-Xiaojiang fault system with largest value of 57 nanostrain/yr between Ganze and Luhuo on the Xianshuihe fault. Figure 6.10 shows their one-sigma uncertainties, with mean values of 1.9, 3.2 and 1.6 nanostrain/yr. The mean relative error of the second invariant of strain rates is about 10%.

6.5 Block Model

Block models have been widely used to describe kinematics of regional or global deformation and fault activity for complex faulting network (e.g., Avouac & Tapponnier, 1993; Floyd et al., 2010; Graham et al., 2018; Hammond et al., 2011; McCaffrey, 2009; Meade, 2007; Meade & Hager, 2005; Thatcher, 2007). Meade and Loveless (2009) developed a toolbox *Blocks* which allows modeling long term deformation in term of rigid blocks moving on spherical Earth (so each block is associated to Euler pole describing its rotation as done in plate tectonics). Interseismic elastic strain accumulation on block-bounding faults is modeled with kinematically consistent slip rates and spatially variable coupling. Fault slip rates are then calculated by projecting velocities of the rigid blocks onto the block-bounding faults. Here we used this toolbox to model regional deformation in Southwest China. Considering very limited coverage of InSAR rate maps at present, we only used the GNSS velocity field in the modeling presented here.

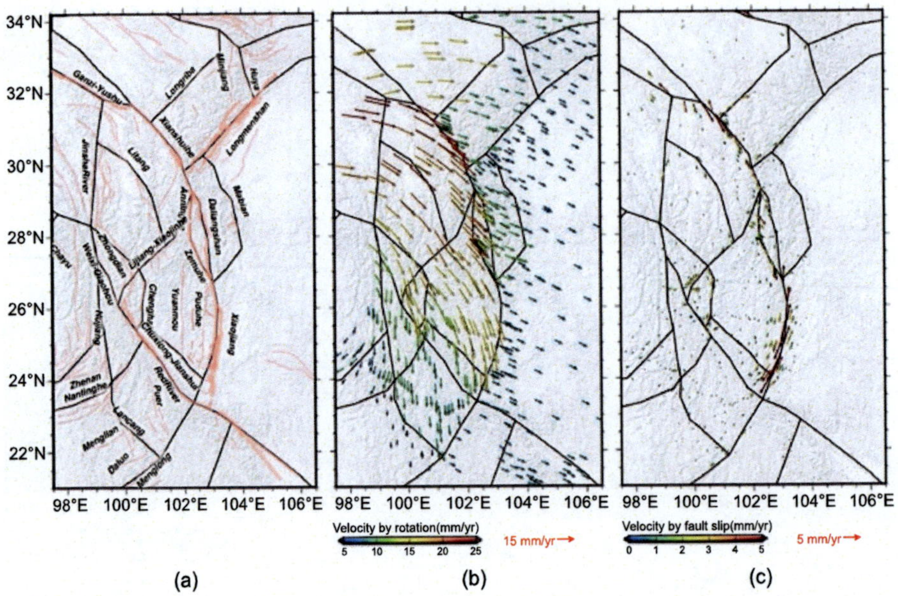

Fig. 6.11 **a** Comparison of block boundaries (black lines) with active faults (red lines in the background). **b–c** Modeled GNSS velocities from rigid block rotation and strain accumulation on bounding faults

In Southwest China, we subdivided the region into 18 blocks as shown in Fig. 6.11 bounded by major faults including Xianshuihe-Anninghe-Zemuhe-Xiaojiang fault system, Litang fault, Longmenshan fault, Longriba fault, Lijiang-Xiaojinhe fault, Daliangshan fault, Mabian fault, Red River fault, Zhongdian fault, Jinsha River fault, Chenghai fault, Jianchuan fault, Nujiang fault, Nantinghe fault, Lancang River fault. Outside the region, we used the blocks and fault geometries from Loveless and Meade (2011) directly. In Southwest China, we used the GNSS velocities obtained in this study. In the other regions, we updated the GNSS velocity field by compiling the published results in Loveless and Meade (2011), Kreemer et al., (2014) and Zheng et al., (2017). We allow for intrablock strain in one block in west Xiaojiang where multiple parallel faults exist, and the data are insufficient to refine the model locally. Figure 6.11b, c show predicted velocities from the long-term slip rate and the interseismic strain accumulation along the block boundaries, respectively. The RMSs are 0.6 and 0.7 mm/yr for the east and north velocities. The residuals in Fig. 6.12b show no systematic misfits, implying our block model fit the GNSS velocities very well. Figure 6.13 shows the rotation rates of the blocks and the strike- and dip-slip rates of the bounding faults. The results show highest slip rates of 12–14 mm/yr along the Xianshuihe and Xiaojiang faults. Between these two faults, deformation is partitioned by the Anninghe-Zemuhe and Daliangshan faults, with slip rates of 5–7 mm/yr and 4–5 mm/yr, respectively. In addition to the east boundary, we also find moderate slip rates of about 5–6 mm/yr along Lijiang-Xiaojinhe, Zhongdian, Litang and Longriba faults.

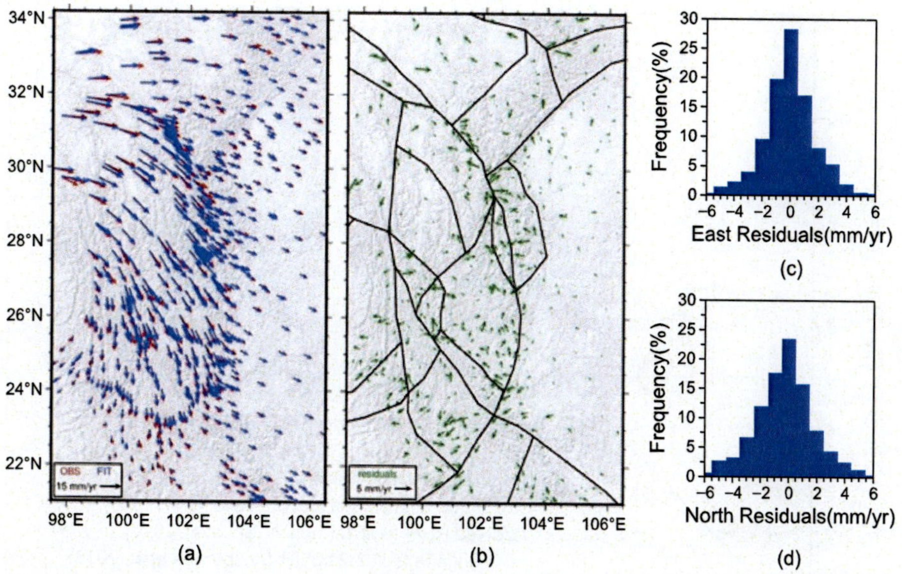

Fig. 6.12 **a** Observed and fitted velocity field. **b** Spatial distribution of the residuals. **c–d** Histograms of the residuals

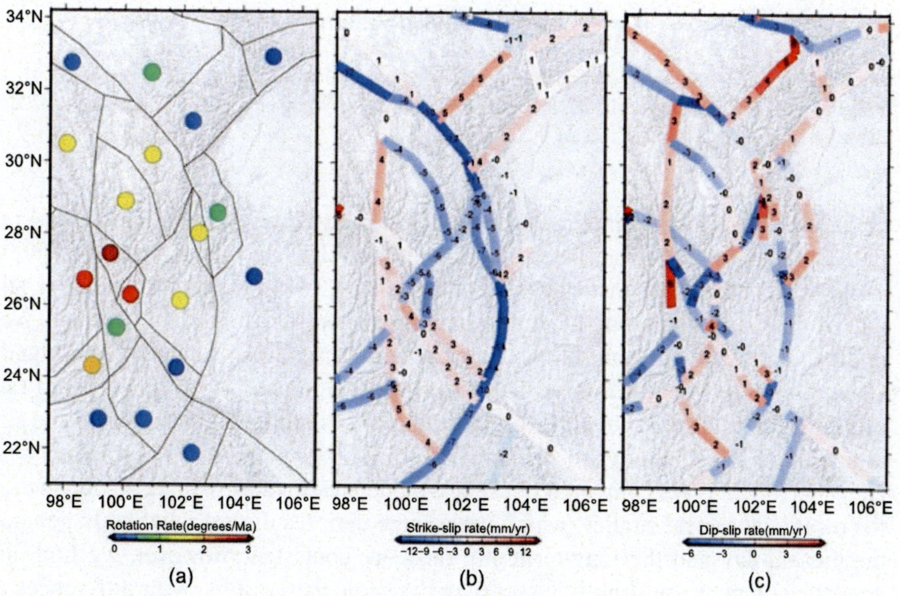

Fig. 6.13 **a** Block rotation rates in degrees/Ma, clockwise as positive. **b–c** Interseismic fault strike-slip and dip-slip rates. Right-lateral and reverse/closing sense slip are given as positive. Each number on the faults denotes slip rate of the fault segment

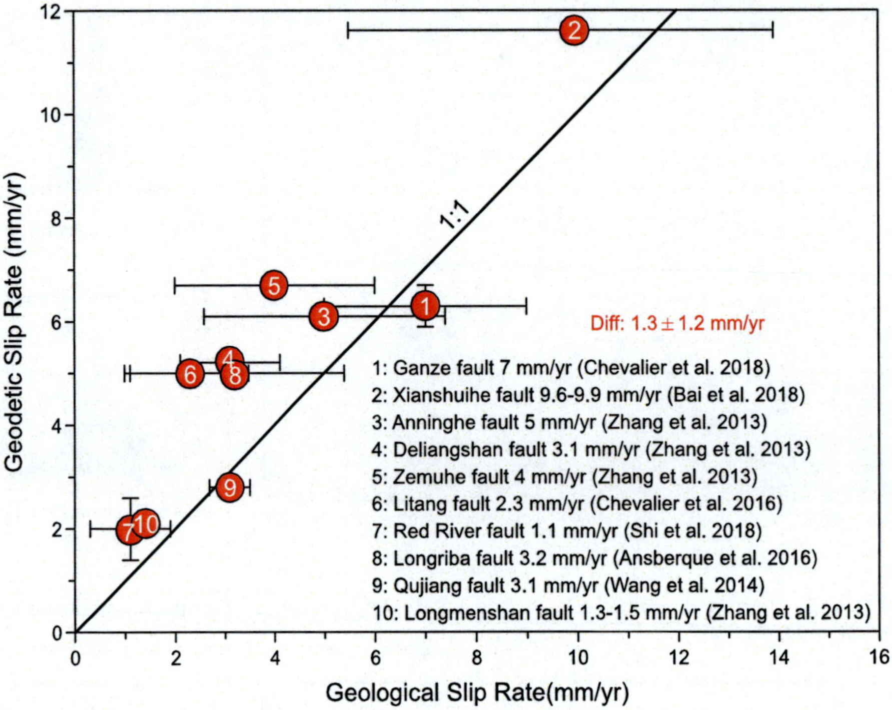

Fig. 6.14 Comparison between geodetic and geological fault slip rates. (Ansberque et al., 2016; Bai et al., 2018; Chevalier et al., 2016,2018; Shi et al., 2018; Wang et al., 2014; Zhang et al., 2013) Error bars indicate 2-sigma errors of the slip rates

Figure 6.14 shows comparison of fault slip rates between block model and the Late Quaternary geological investigations. In fact, slip rates are diverse in different geological studies. For instance, Chevalier et al. (2018) obtained 6–8 mm/yr slip rate of the Ganzi fault, but 14 mm/yr was suggested by Xu et al. (2003) and Wen et al. (2003). The slip rate of the northwestern Xianshuihe fault is 17 mm/yr in Chen et al. (2016) or 14 mm/yr in Xu et al. (2003), much larger than the slip rate of 9.6–9.9 mm/yr in Bai et al. (2018). For the NS-trending Xiaojiang fault, the slip rate is up to 14–22 mm/yr in He and Oguchi (2008), while it is only 7 mm/yr in Han et al. (2017). We compared our block-modeled slip rates with those obtained in ten recent geological studies, which we assume their results are reliable. In general, the present-day and the long-term slip rates are consistent, however, we find that geodetic slip rates are slightly larger than the geological results, with differences of 1.3 ± 1.2 mm/yr. We argue that geodetic slip rates might be overestimated because the block model assumes a single fault trace acting as block boundary even if there are a few important splays like the Xiaojiang fault. This requires us to improve the block models in future work as described in the next section.

6.6 Summary and Future Work

We developed a framework to incorporate GNSS with InSAR and leveling data in crustal deformation models usable for seismic hazard assessment and provided preliminary results to determine a community deformation model for the CSES area. Our major accomplishments have been to compile and curate the most complete and up-to-date GNSS dataset, yielding velocities at 618 sites, continuous velocity and strain rate field models by combining of GNSS, InSAR and leveling data, and a block model of the CSES area. Outputs of our workflow, the interseismic strain rates and fault slip rates, can in principle be used as inputs for probabilistic earthquake forecasting.

The quality of the community deformation model significantly depends on the spatial resolution and the accuracy of data used as inputs, i.e., GNSS, InSAR, precise leveling and geological fault slip rates. As our current work focuses on analyzing GNSS observations, we only used the published InSAR and precise leveling data to test the feasibility of the whole workflow in producing deformation models. Over the past six years, European Space Agency's Sentinel-1A/B twin satellites have already acquired over 160 images on the ascending orbits and 130 images on the descending with about 12-day revisit time covering the entire CSES. The fairly good temporal/perpendicular baselines and the large archive of datasets are promising to make a continuous coverage of InSAR interseismic deformation field like we obtained in south-central Tibet (Wang et al., 2019). The high-resolution InSAR results are particularly important for the CSES where there is very complex fault network. Therefore, generating InSAR rate maps will be our primary future work for improving the velocity field model. In addition, our framework can also incorporate long-term constraints from late Quaternary fault slip rates to improve further the deformation models.

Future work could also aim at improving the block modeling approach. Due to complex tectonic setting in southwest China, it is challenging to fit crustal deformation here with end-member continuous or block models. The continuous models tend to underestimate strain accumulation on fault zones, as discussed in Weiss et al. (2020). The block models ignore fault activity inside blocks. Nevertheless, there are plenty of $M \geq 6.5$ historical earthquakes inside blocks even though the block sizes are fairly small in this study. Probabilistic earthquake forecast based on block models undoubtedly will underestimate earthquake risk in blocks. In the future, we are going to develop a continuous model embedded with discrete mapped faults. This model will enable us to estimate continuous strain rates and fault slip rates, which will act as inputs for the probabilistic earthquake forecast models.

Acknowledgements This work is funded by the National Key Research and Development Program of China (2017YFC1500501), the China Seismic Experimental Site (2019CESE0110) and the National Natural Science Foundation of China (41672205, 41774036). We thank all Chinese colleagues who devoted in the extremely hard GNSS field work in Chuandian over twenty years. The first author thanks Adriano Gualandi for sharing ICAIM toolbox, Pond Sirorattanakul for help on using ICAIM, Jake Loveless for sharing block geometries and help on using Blocks, Kaihua Ding and Yang Liu for discussion on GNSS data processing.

References

Ansberque, C., Bellier, O., Godard, V., Lasserre, C., Wang, M. M., Braucher, R., et al. (2016). The Longriqu fault zone, eastern Tibetan Plateau: Segmentation and Holocene behavior. *Tectonics, 35*, 565–585.

Avouac, J.-P. (2015). From geodetic imaging of seismic and aseismic fault slip to dynamic modeling of the seismic cycle. *Annual Review of Earth and Planetary Sciences, 43*, 233–271.

Avouac, J. P., & Tapponnier, P. (1993). Kinematic model of active deformation in Central-Asia. *Geophysical Research Letters, 20*, 895–898.

Bai, M., Chevalier, M. L., Pan, J., Replumaz, A., Leloup, P. H., Métois, M., et al. (2018). Southeastward increase of the late Quaternary slip-rate of the Xianshuihe fault, eastern Tibet. Geodynamic and seismic hazard implications. *Earth and Planetary Science Letters, 485*, 19–31.

Bevis, M., & Brown, A. (2014). Trajectory models and reference frames for crustal motion geodesy. *Journal of Geodesy, 88*(3), 283–311.

Bird, P., & Kreemer, C. (2015). Revised tectonic forecast of global shallow seismicity based on version 2.1 of the Global Strain Rate Map. *Bulletin of the Seismological Society of America, 105*(1), 152–166.

Bird, P., Jackson, D. D., Kagan, Y. Y., Kreemer, C., & Stein, R. S. (2015). GEAR1: A global earthquake activity rate model constructed from geodetic strain rates and smoothed seismicity. *Bulletin of the Seismological Society of America, 105*(5), 2538–2554.

Bürgmann, R., Schmidt, D., Nadeau, R. M., d'Alessio, M., Fielding, E., Manaker, D., et al. (2000). Earthquake potential along the northern Hayward fault. *Science, 289*, 1178–1182.

Cavalié, O., Lasserre, C., Doin, M. P., Peltzer, G., Sun, J., Xu, X., & Shen, Z. K. (2008). Measurement of interseismic strain across the Haiyuan fault (Gansu, China), by InSAR. *Earth and Planetary Science Letters, 275*(3–4), 246–257.

Chen, G., Chen, X., Wen, X., & Chen, Y. G. (2016). Late Quaternary slip-rates and slip partitioning on the Southeastern Xianshuihe fault system, eastern Tibetan Plateau. *Acta Geologica Sinica (english Edition), 90*(2), 537–554.

Chen, Z., Burchfiel, B. C., Liu, Y., King, R. W., Royden, L. H., Tang, W., Wang, E., Zhao, J., & Zhang, X. (2000). Global positioning system measurements from eastern tibet and their implications for India/Eurasia intercontinental deformation. *Journal of Geophysical Research: Solid Earth, 105*(B7), 16215–16227.

Chevalier, M. L., Leloup, P. H., Replumaz, A., Pan, J., Liu, D., Li, H., et al. (2016). Tectonic-geomorphology of the Litang fault system, SE Tibetan Plateau, and implication for regional seismic hazard. *Tectonophysics, 682*, 278–292.

Chevalier, M. L., Leloup, P. H., Replumaz, A., Pan, J., Métois, M., & Li, H. (2018). Temporally constant slip rate along the Ganzi fault, NW Xianshuihe fault system, eastern Tibet. *GSA Bulletin, 130*(3–4), 396–410.

Choudrey, R. A., & Roberts, S. J. (2003). Variational mixture of Bayesian independent component analyzers. *Neural Computation, 150*(1), 213–252.

de Michele, M., Raucoules, D., Rolandone, F., Briole, P., Salichon, J., Lemoine, A., & Aochi, H. (2011). Spatiotemporal evolution of surface creep in the Parkfield region of the San Andreas fault (1993–2004) from synthetic aperture radar. *Earth and Planetary Science Letters, 308*, 141–150.

Densmore, A., Ellis, M. A., Li, Y., Zhou, R., Hancock, G. S., & Richardson, N. (2007). Active tectonics of the Beichuan and Pengguan faults at the eastern margin of the Tibetan Plateau. *Tectonics, 26*, TC4005. https://doi.org/10.1029/2006TC001987

Desbrun, M., Meyer, M., Schroder, P., & Barr, A. H. (1999). Implicit fairing of irregular meshes using diffusion and curvature flow. *SIGGRAPH, 99*, 317–324.

England, P., & Molnar, P. (2005). Late Quaternary to decadal velocity fields in Asia. *Journal of Geophysical Research: Solid Earth, 110*, B12401. https://doi.org/10.1029/2004JB003541

Fattahi, H., & Amelung, F. (2016). InSAR observations of strain accumulation and fault creep along the Chaman Fault system, Pakistan and Afghanistan. *Geophysical Research Letters, 43*, 8399–8406.

Field, E. H., Arrowsmith, R. J., Biasi, G. P., Bird, P., Dawson, T. E., Felzer, K. R., Jackson, D. D., Johnson, K. M., Jordan, T. H., Madden, C., Michael, A. J., Milner, K. R., Page, M. T., Parsons, T., Powers, P. M., Shaw, B. E., Thatcher, W. R., Weldon, R. J., II., & Zeng, Y. (2014). Uniform California earthquake rupture forecast, version 3 (UCERF3)—The time-independent model. *Bulletin of the Seismological Society of America, 104*(3), 1122–1180.

Floyd, M. A., Billiris, H., Paradissis, D., Veis, G., Avallone, A., Briole, P., McClusky, S., Nocquet, J. M., Palamartchouk, K., Parsons, B., & England, P. C. (2010). A new velocity field for Greece: Implications for the kinematics and dynamics of the Aegean. *Journal of Geophysical Research: Solid Earth, 115*, B10403. https://doi.org/10.1029/2009JB007040

GB18306 (2015). Seismic ground motion parameter zonation map of China. Standard Press of China (in Chinese).

Grafarend, E. W. (1985). Variance-covariance component estimation: Theoretical results and geodetic applications. *Statistics and Decision, Supplement, 2*, 407–441.

Graham, S. E., Loveless, J. P., & Meade, B. J. (2018). Global plate motions and earthquake cycle effects. *Geochemistry, Geophysics, Geosystems, 19*, 2032–2048.

Gualandi, A., Serpelloni, E., & Belardinelli, M. E. (2016). Blind source separation problem in GPS time series. *Journal of Geodesy, 90*(4), 323–341.

Gualandi, A., Avouac, J. P., Galetzka, J., Genrich, J. F., Blewitt, G., Adhikari, L. B., Koirala, B. P., Gupta, R., Upreti, B. N., Pratt-Sitaula, B., & Liu-Zeng, J. (2017). Pre- and post-seismic deformation related to the 2015, M_w7.8 Gorkha earthquake, Nepal. *Tectonophysics, 714*, 90–106.

Hammond, W. C., Blewitt, G., & Kreemer, C. (2011). Block modeling of crustal deformation of the northern Walker Lane and Basin and Range from GPS velocities. *Journal of Geophysical Research: Solid Earth, 116*, B04402. https://doi.org/10.1029/2010JB007817

Han, Z. J., Dong, S. P., Mao, Z. B., Hu, N., Tan, X.-B., Yuan, R., & Guo, P. (2017). The Holocene activity and strike-slip rate of the southern segment of Xiaojiang fault in the Southeastern Yunnan region, China. *Seismology and Geology, 39*(1), 1–19. (in Chinese).

Hao, M., Wang, Q., Shen, Z., Cui, D., Ji, L., Li, Y., & Qin, S. (2014). Present day crustal vertical movement inferred from precise leveling data in eastern margin of Tibetan Plateau. *Tectonophysics, 632*, 281–292.

Hao, M., Freymueller, J. T., Wang, Q., Cui, D., & Qin, S. (2016). Vertical crustal movement around the southeastern Tibetan Plateau constrained by GPS and GRACE data. *Earth and Planetary Science Letters, 437*, 1–8.

He, H. L., & Oguchi, T. (2008). Late Quaternary activity of the Zemuhe and Xiaojiang faults in southwest China from geomorphological mapping. *Geomorphology, 96*(1–2), 62–85.

Herring, T. A., Melbourne, T. I., Murray, M. H., Floyd, M. A., Szeliga, W. M., King, R. W., Phillips, D. A., Puskas, C. M., Santillan, M., & Wang, L. (2016). Plate boundary observatory and related networks: GPS data analysis methods and geodetic products. *Reviews of Geophysics, 54*, 759–808.

Jiang, Z., Wang, M., Wang, Y., Wu, Y., Che, S., Shen, Z. K., Burgmann, R., Sun, J., Yang, Y., Liao, H., & Li, Q. (2014). GPS constrained coseismic source and slip distribution of the 2013 Mw6.6 Lushan, China, earthquake and its tectonic implications. *Geophysical Research Letters, 41*(2), 407–413.

Jolivet, R., Lasserre, C., Doin, M.-P., Guillaso, S., Peltzer, G., Dailu, R., Sun, J., Shen, Z.-K., & Xu, X. (2012). Shallow creep on the Haiyuan Fault (Gansu, China) revealed by SAR Interferometry. *Journal of Geophysical Research: Solid Earth, 117*, B06401. https://doi.org/10.1029/2011JB008732

Kaneko, Y., Fialko, Y., Sandwell, D. T., Tong, X., & Furuya, M. (2013). Interseismic deformation and creep along the central section of the North Anatolian Fault (Turkey): InSAR observations and implications for rate-and-state friction properties. *Journal of Geophysical Research: Solid Earth, 118*, 316–331.

King, R. W., Shen, F., Clark Burchfiel, B., Royden, L. H., Wang, E., Chen, Z. L., Liu, Y. P., Zhang, X. Y., Zhao, J. X., & Li, Y. L. (1997). Geodetic measurement of Crustal motion in southwest China. *Geology, 25*, 179–182.

Kreemer, C., Blewitt, G., & Klein, E. C. (2014). A geodetic plate motion and global strain rate model. *Geochemistry, Geophysics, Geosystems, 15*, 3849–3889.

Loveless, J. P., & Meade, B. J. (2011). Partitioning of localized and diffuse deformation in the Tibetan Plateau from joint inversions of geo-logic and geodetic observations. *Earth and Planetary Science Letters, 303*(1–2), 11–24.

McCaffrey, R. (2009). Time-dependent inversion of three-component continuous GPS for steady and transient sources in northern Cascadia. *Geophysical Research Letters, 36*, L07304. https://doi.org/10.1029/2008GL036784

Meng, G. J., Ren, J. W., Wang, M., Gan, W. J., Wang, Q., Qiao, X. J., & Yang, Y. L. (2008). Crustal deformation in western Sichuan region and implications for 12 May 2008 Ms 8.0 earthquake. *Geochemistry, Geophysics, Geosystems, 9*, Q11007. https://doi.org/10.1029/2008GC002144

Meade, B. J., & Hager, B. H. (2005). Block models of crustal motion in southern California constrained by GPS measurements. *Journal of Geophysical Research: Solid Earth, 110*, B03403. https://doi.org/10.1029/2004JB003209

Meade, B. J. (2007). Present-day kinematics at the India-Asia collision zone. *Geology, 35*(1), 81–84.

Meade, B. J., & Loveless, J. P. (2009). Block modeling with connected fault-network geometries and a linear elastic coupling estimator in spherical coordinates. *Bulletin of the Seismological Society of America, 99*(6), 3124–3139.

Nguyen, A. D., Sagiya, T., Kimata, F., To, T. D., Hai, V. Q., Cong, D. C., Binh, N. X., & Xuyen, N. D. (2013). Contemporary horizontal crustal movement estimation for northwestern Vietnam inferred from repeated GPS measurements. *Earth, Planets and Space, 65*(12), 1399–1410.

Pagani, M., Monelli, D., Weatherill, G., Danciu, L., Crowley, H., Silva, V., et al. (2014). OpenQuake engine: An open hazard (and risk) software for the global earthquake model. *Seismological Research Letters, 85*(3), 692–702.

Pagli, C., Wang, H., Wright, T. J., Calais, E., & Lewi, E. (2014). Current plate boundary deformation of the Afar rift from a 3-D velocity field inversion of InSAR and GPS. *Journal of Geophysical Research: Solid Earth, 119*, 8562–8575.

Petersen, M. D., Shumway, A. M., Powers, P. M., Mueller, C. S., Moschetti, M. P., Frankel, A. D., et al. (2020). The 2018 update of the US National Seismic Hazard Model: Overview of model and implications. *Earthquake Spectra, 36*, 5–41.

Rebischung, P., Griffiths, J., Ray, J., Schmid, R., Collilieux, X., & Garayt, B. (2012). IGS08: The IGS realization of ITRF2008. *GPS Solutions, 16*(4), 483–494.

Reid H. F. (1910). The mechanics of the earthquake. The California Earthquake of April 18, 1906, Report of the State Earthquake Investigation Commission (Vol. 2, pp. 1–192). Carnegie Institution.

Rui, X., & Stamps, D. S. (2016). Present-day kinematics of the eastern Tibetan Plateau and Sichuan Basin: Implications for lower crustal rheology. *Journal of Geophysical Research: Solid Earth, 121*, 3846–3866.

Rui, X., & Stamps, D. S. (2019). Strain accommodation in the Daliangshan Mountain area, southeastern margin of the Tibetan Plateau. *Journal of Geophysical Research: Solid Earth, 124*, 9816–9832.

Savage, J. C., Gan, W., & Svarc, J. L. (2001). Strain accumulation and rotation in the Eastern California Shear Zone. *Journal of Geophysical Research: Solid Earth, 106*(10), 21995–22007.

Scholz, C. H. (2019). *The mechanics of earthquakes and faulting* (3rd ed.). Cambridge University Press.

Shen, Z.-K., Lü, J., Wang, M., & Bürgmann, R. (2005). Contemporary crustal deformation around the southeast borderland of the Tibetan Plateau. *Journal of Geophysical Research: Solid Earth, 110*, B11409. https://doi.org/10.1029/2004JB003421

Shen, Z. K., Sun, J., Zhang, P., Wan, Y., Wang, M., Bürgmann, R., Zeng, Y., Gan, W., Liao, H., & Wang, Q. (2009). Slip maxima at fault junctions and rupturing of barriers during the 2008 Wenchuan earthquake. *Nature Geoscience, 2*, 718–724.

Shi, C., Li, M., Zhao, Q., Wang, C., & Zhang, H. (2015). WHU analysis center technical report 2014. *IGS Technical Report, 2014*, 81–88.

Shi, X., Wang, Y., Sieh, K., Weldon, R., Feng, L., Chan, C. H., & Liu-Zeng, J. (2018). Fault slip and GPS velocities across the Shan Plateau define a curved southwestward crustal motion around the eastern Himalayan syntaxis. *Journal of Geophysical Research: Solid Earth, 123*(3), 2502–2518.

Tapponnier, P., Lacassin, R., Leloup, P. H., Schärer, U., Zhong, D. L., Wu, H. W., Xiaohan, L., Shaocheng, J., Lianshang, Z., & Jiayou, Z.(1990). The Ailao Shan/Red River metamorphic belt: Tertiary left-lateral shear between Indochina and South China. *Nature, 343*(6257), 431–437.

Thatcher, W. (2007). Microplate model for the present-day deformation of Tibet. *Journal of Geophysical Research: Solid Earth, 112*, B01401. https://doi.org/10.1029/2005JB004244

Trần, ĐT., Nguyễn, T. Y., Dương, C. C., Vy, Q. H., Zuchiewicz, W., & Nguyễn, V. N. (2013). Recent crustal movements of northern Vietnam from GPS data. *Journal of Geodynamics, 69*, 5–10.

Wang, H., Wright, T. J., & Biggs, J. (2009). Interseismic slip rate of the northwestern Xianshuihe Fault from InSAR data. *Geophysical Research Letters, 36*, L03302. https://doi.org/10.1029/2008GL036560

Wang, H., & Wright, T. J. (2012). Satellite geodetic imaging reveals internal deformation of western Tibet. *Geophysical Research Letters, 39*, L07303. https://doi.org/10.1029/2012GL051222

Wang, H., Wright, T. J., Liu-Zeng, J., & Peng, L. (2019). Strain rate distribution in south-central Tibet from two decades of InSAR and GPS. *Geophysical Research Letters, 46*, 5170–5179.

Wang, M., & Shen, Z.-K. (2020). Present-day crustal deformation of continental China derived from GPS and its tectonic implications. *Journal of Geophysical Research: Solid Earth, 125*, e2019JB018774.

Wang, W., Qiao, X., & Ding, K. (2021). Present-day kinematics in southeastern Tibet inferred from GPS measurements. *Journal of Geophysical Research: Solid Earth, 126*, e2020JB021305.

Wang, Y., Zhang, B., Hou, J. J., & Xu, X. W. (2014). Structure and tectonic geomorphology of the Qujiang fault at the intersection of the Ailao Shan-Red River fault and the Xianshuihe-Xiaojiang fault system, China. *Tectonophysics, 634*, 156–170.

Wen, X., Xu, X., Zheng, R., Xie, Y., & Wan, C. (2003). Average slip-rate and recent large earthquake ruptures along the Ganzi-Yushu fault. *Science in China Series d: Earth Sciences, 46*, 276–288.

Weiss, J. R., Walters, R. J., Morishita, Y., Wright, T. J., Lazecky, M., Wang, H., et al. (2020). High-resolution surface velocities and strain for Anatolia from Sentinel-1 InSAR and GNSS data. *Geophysical Research Letters, 47*, e2020GL087376.

Working Group of M7 (2012), Study on the mid- to long-term potential of large earthquakes on the Chinese Continent (336 p.). Seismological Press.

Xu, X., Wen, X., Zheng, R., Ma, W., Song, F., & Yu, G. (2003). Pattern of latest tectonic motion and its dynamics for active blocks in Sichuan-Yunnan region, China. *Science in China Series d: Earth Sciences, 46*(2), 210–226.

Zumberge, J. F., Heflin, M. B., Jefferson, D. C., Watkins, M. M., & Webb, F. H. (1997). Precise point positioning for the efficient and robust analysis of GPS data from large networks. *Journal of Geophysical Research: Solid Earth, 102*(B3), 5005–5017.

Zhang, P. Z. (2013). A review on active tectonics and deep crustal processes of the Western Sichuan region, eastern margin of the Tibetan Plateau. *Tectonophysics, 584*, 7–22.

Zheng, G., Wang, H., Wright, T. J., Lou, Y., Zhang, R., Zhang, W., Shi, C., Huang, J., & Wei, N. (2017). Crustal deformation in the India-Eurasia collision zone from 25 years of GPS measurements. *Journal of Geophysical Research: Solid Earth, 122*(11), 9290–9312.

Chapter 7
Interseismic Coupling Along the Xianshuihe-Xiaojiang Fault System

Yanchuan Li and Xinjian Shan

Abstract We use Global Positioning System (GPS) data to quantify the interseismic coupling along the Xianshuihe-Xiaojiang fault system (XXFS). Our results confirm 7–11 mm/yr left-lateral strike-slip motion along the XXFS. South of ~ 29.5° N, high interseismic coupling is found down to 20 km-depth along the Anninghe-Zemuhe fault. A second 130 km-long section, highly coupled down to 10–15 km-depth, is identified along the northern Xiaojiang fault. North of 30.2° N, the Xianshuihe fault appears to be predominantly creeping at depth. We quantitatively compared our results from GPS with the seismic moment release and accumulation on XXFS. Results highlight that along the Anninghe, Zemuhe and Xiaojiang faults, moment deficits for $M_w > 7$ earthquakes have now accumulated.

Keywords Eastern Tibet · Xianshuihe-Xiaojiang fault system · Interseismic fault coupling · Geodesy · Earthquake potential

7.1 Introduction

The left-lateral strike-slip Xianshuihe fault is one of the large strike-slip faults cutting through eastern Tibet and contributing its eastward motion (e.g., Allen et al., 1991; Tapponnier & Molnar, 1977; Tapponnier et al., 2001). The Xianshuihe fault extends over several hundreds of kilometers farther east outside Tibet along the Anninghe, Zemuhe and Xiaojiang faults in the Sichuan and Yunnan provinces, where the fault finally merges with the Red River fault (Fig. 7.1). The 1400 km-long Xianshuihe-Xiaojiang fault system (XXFS) has been one of the most seismically active faults in the mainland of China, as attested by a detailed history of past earthquakes for the last three centuries (e.g., Allen et al., 1991; Wen et al., 2008a). During that period, macro-seismic intensities based on historical records indicate more than 35 $M \geq$

Y. Li (✉) · X. Shan
Institute of Geology, China Earthquake Administration, Beijing 100029, China
e-mail: yanchuan@ies.ac.cn

X. Shan
e-mail: xjshan@ies.ac.cn

© Higher Education Press 2022
Y.-G. Li et al. (eds.), *China Seismic Experimental Site*,
https://doi.org/10.1007/978-981-16-8607-8_7

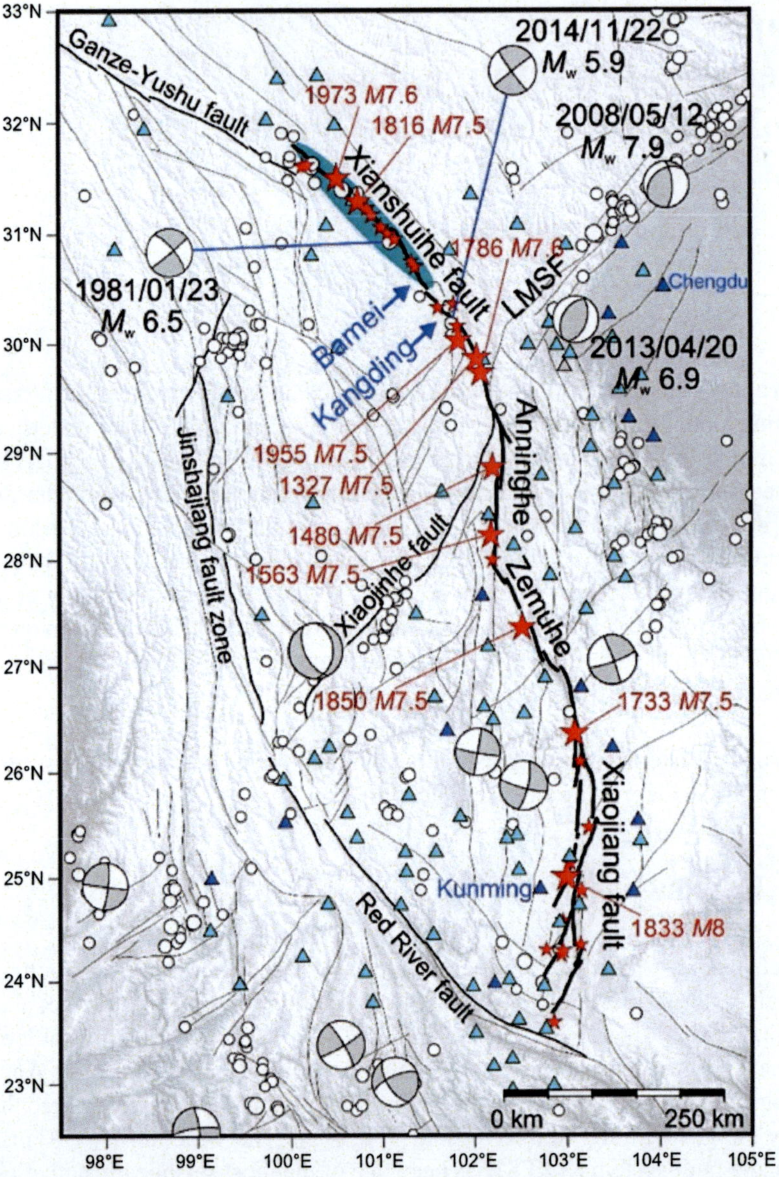

Fig. 7.1 Regional tectonic map of the eastern Tibetan Plateau. Bold black lines correspond to the Xianshuihe-Anninghe-Zemuhe-Xiaojiang fault system (XXFS). Thin black lines indicate active faults (Deng et al., 2003). Beach-balls represent focal mechanism solutions ($M_W > 6$, 1976–2015) come from the Global Centroid Moment Tensor (GCMT). Thin white circles represent earthquakes ($M_W > 5$, 1960–2015) from the United States Geological Survey (USGS). Red stars mark the epicenters of historical earthquakes ($M > 6$; Wen et al., 2008a). Triangles show the locations and populations of cities (cyan: < 0.5 million; blue: > 0.5 million). LMSF denotes the Longmenshan fault

6.0 earthquakes on this fault system. Additional historical records dating back to the XIVth century indicate that each of the four faults has hosted at least one large $M >$ 7.5 earthquake, and the largest event (M8.0) occurred along the Xiaojiang fault in 1833. According to the earthquake history summarized by Wen et al. (2008a), the Xianshuihe fault hosted large events with $M > 7.5$ earthquakes regularly in 1327, 1786, 1816, 1955 and 1973 (Fig. 7.1).

Since the 1973 M7.6 Luhuo earthquake, the XXFS has been relatively silent with only two earthquakes (1981 M6.9, 2014 M_w5.9) along the northwestern sections. Several studies assessed the short-term seismic potential along XXFS. Ran et al. (2008) and Wen et al. (2008a) show that for the Xianshuihe fault, from the Kangding County to Bamei town (hereafter refer to the Kangding-Bamei section; Fig. 7.1) and for the Anninghe fault, the time elapsed since the last large earthquake (232 and 528 years respectively) is close to the average earthquake recurrence time (230–350 and 520–660 years respectively) proposed from paleoseismology studies (Allen et al., 1991). Interseismic fault coupling inversions of GPS velocity solution for the 1999–2013 period also find that the Kangding-Bamei section and the Anninghe fault have accumulated slip deficit for characteristic earthquakes (M_w7.0 and M_w7.5 respectively; Jiang et al., 2015). Since the region crossed by the XXFS hosts a dense population of more than 50 million, assessing the seismic potential of the XXFS is a major issue in a rapidly developing area in China.

We first derive GPS velocity solution for eastern Tibet in 1999–2014. Then, using elastic block modeling, we estimate the spatial distribution of interseismic fault coupling (ISC) along the XXFS. Using the available earthquake history, we derive balances between moment accumulation and release, and discuss the implications in terms of seismic potential.

7.2 Data and Model

7.2.1 GPS Data Processing

The 1999–2014 GPS velocity solution is used as constraints for block model. It consists of two parts:

The Crustal Movement Observation Network of China (CMONOC) GPS data were acquired between 1999 and 2014, including a total of 300 sites in the study area (244 campaign and 56 continuous stations); these data were processed by Wang et al. (2017).

The GNSS/GPS Network of the Sichuan Earthquake Administration and the Sichuan Bureau of Surveying, Mapping and Geoinformation, including 177 sites that were occupied between 2004 and 2014. These GPS observations were processed by Rui and Stamps (2016).

The two GPS solutions represent interseismic velocities for eastern Tibet in 1999–2014. The two GPS velocity solutions were originally referenced to different Eurasia plate reference frames. We combine them into a single consistent solution by solving for an Euler vector, which minimize the velocity residuals between co-located sites. We chose the solution of Wang et al. (2017) as the master solution and rotated the solution of Rui and Stamps (2016) onto it. To highlight the relative motion between Tibetan Plateau and South China Block, Fig. 7.2 shows our final combined velocity field being taken as a South China reference frame.

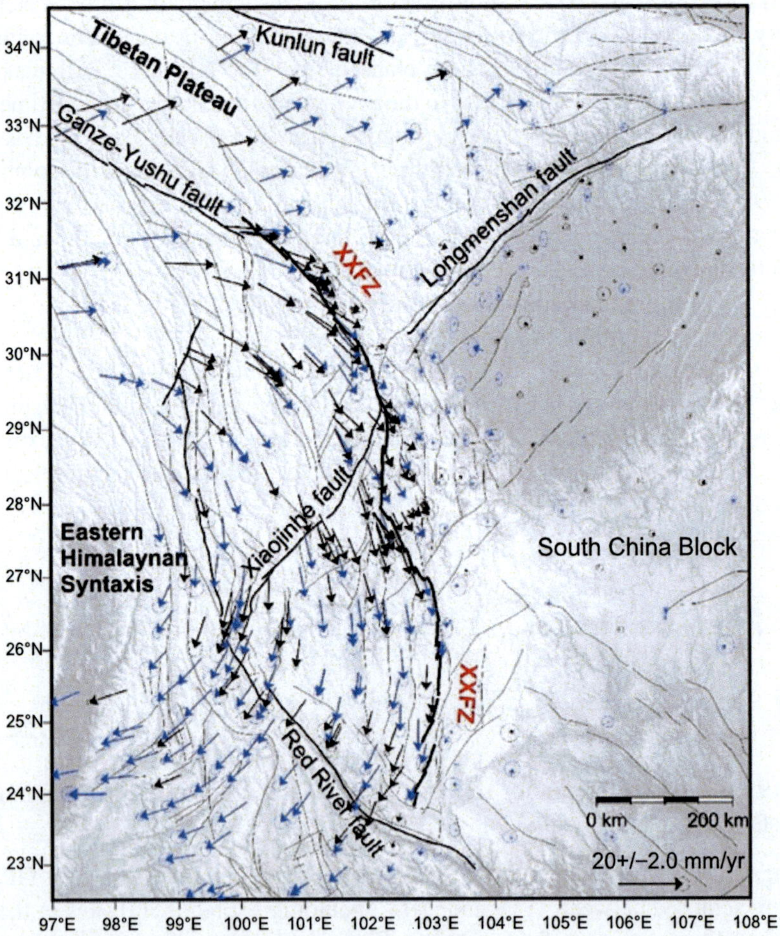

Fig. 7.2 Global Positioning System (GPS) velocity fields (1999–2014) with respect to the South China block. Arrows in blue represent the velocity solution from Wang et al. (2017); arrows in black represent the velocity solution from Rui and Stamps (2016). Error ellipses show 68% confidence levels. Thin black lines correspond to active faults. XXFZ donates the Xianshuihe-Xiaojiang fault zone

7.2.2 GPS Data Modeling

We use the elastic block modelling implemented in DEFNODE package (McCaffrey et al., 2007) to obtain a regional kinematic model. We simultaneously solve for block rotation rates, strain rates within blocks, and spatial distribution interseismic coupling along block-bounding faults. In the model, faults are represented by a set of 3D nodes (average distance of 30 km along-strike, 5 km along-dip) defining the fault surface and the ISC at each node is estimated during the inversion. We did not include tight constraints in our inversion, that is, no smoothing factor was applied for ISC variations between different nodes. The fault surface between the fault nodes is divided into patches of 10 km length along strike and 1.7 km along dip. A bilinear interpolation is applied to obtain the ISC for each rectangle. We minimized data misfit, defined by the reduced chi-squared statistic (χ_n^2):

$$\chi_n^2 = \sum_1^n \left(\frac{\gamma_i}{\sigma_i}\right)^2 /(n - m) \tag{7.1}$$

where n is the number of observations, m is the number of free parameters, γ_i is the residuals (observed minus calculated velocities), and σ_i is the data uncertainty for observation i.

The eastern Tibetan Plateau is divided into five tectonic blocks, i.e., the South China, Longmenshan, Bayanha, North Yunnan and South Yunnan blocks, in accordance previous studies with (e.g., Jiang et al., 2015; Loveless & Meade, 2011; Wang et al., 2017). Block boundaries follow the surface traces of major active faults (Fig. 7.1). The South China block is assumed to be rigid and we solve internal strains for the Longmenshan block, the Bayanha block, the North Yunnan block and the South Yunnan block respectively.

In the model, the Longmenshan fault was modeled to be listric in shape, consistent with the geometry proposed in previous studies (Thompson et al., 2015). With reference to previous studies (e.g., Allen et al., 1991; Deng et al., 2003; Jiang et al., 2015; Wen et al., 2008a), we model the XXFS as an 80° southwestward dipping fault while other block bounding faults are taken vertical.

7.3 Results

7.3.1 Overall Characteristics

The XXFS is dominated by left-lateral strike-slip motion with the fastest rates of 10–11 mm/yr found along the northwesternmost section of the Xianshuihe fault, progressively decreasing to 7 mm/a along the Anninghe, Zemuhe and Xiaojiang faults (Fig. 7.3). These estimates are consistent with both geological (~ 9–10 mm/a;

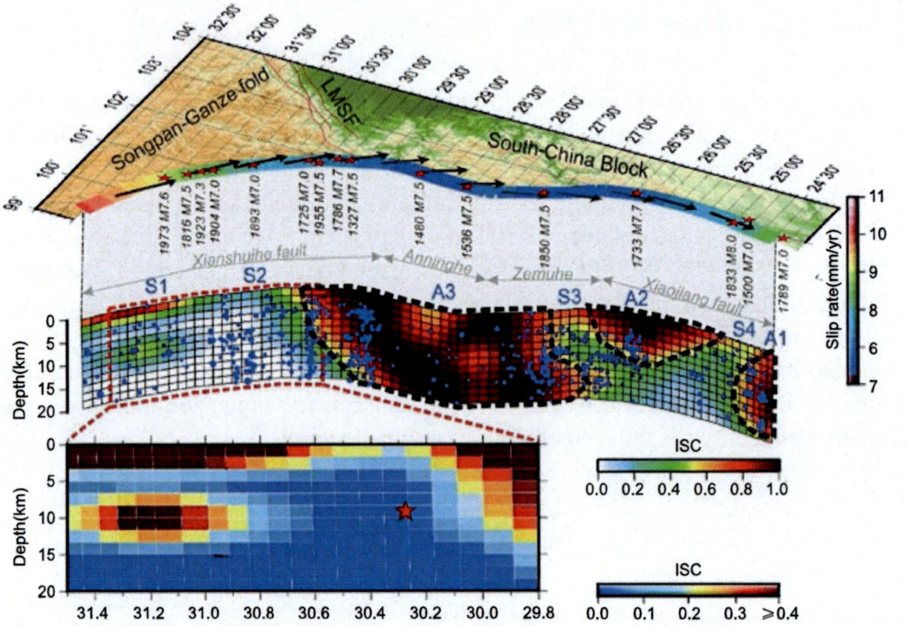

Fig. 7.3 Fault slip and interseismic fault coupling (ISC) along the Xianshuihe-Anninghe-Zemuhe-Xiaojiang fault system (XXFS). Top panel: Estimated total magnitude of fault slip rate along the XXFS is shown with color, with black arrows indicate fault slip direction. Red stars denote locations of historical earthquakes. Middle panel: ISC distribution on the XXFS plane. Blue dots indicate hypocenters of microseismicity (1970–2015), for which data come from China Earthquake Data Center. Thick black dotted lines outline the asperities, which are named as A1, A2 and A3, respectively. Bottom panel: Enlarged view of the ISC distribution on the Xianshuihe fault. The red star shows the hypocenter of the 2014 M_{W}5.9 Kangding earthquake

e.g., Yan & Lin, 2017; Bai et al., 2018) and previous geodetic (~7–13 mm/yr; e.g., Shen et al., 2005; Loveless & Meade, 2011; Wang et al., 2017) results. Small fault-perpendicular rates of 0.2–2.1 mm/yr (shortening) and 0–1.3 mm/yr (extension) are found along the Xianshuihe-Anninghe fault and Zemuhe-Xiaojiang fault respectively.

Our ISC model highlights two opposite behaviors. From latitude 32°N to 30°N, the Xianshuihe fault appears to be predominantly creeping, while the Anninghe, Zemuhe and Xiaojiang faults show significant coupling. In the detail, high coupling occurs along three locked regions hereafter referred as A1 (~ 24.3°–24.6° N), A2 (~ 24.9°–26.6° N), and A3 (~ 26.8°–29.7° N; A is an abbreviation for Asperity), separated by narrow sections of shallower (~ 5 km depth) or weaker coupling (Fig. 7.3). A1 located along the southernmost part of the Xiaojiang fault appears to be locked down to ~ 20 km depth. However, since this location corresponds to the edge of the model, we cannot rule out the possibility of boundary effects so that we do not interpret this pattern in the following analysis. A2 along the Xiaojiang fault is confined to ~ 10 km depth but spans ~ 130 km along strike. The largest A3 is found to be locked down to ~ 20 km and extends laterally over ~ 240 km along the Anninghe and Zemuhe faults.

In addition, other remarkable features of the fault coupling are the shallow locking depth found along several sections (hereafter refer to S1, S2, S3 and S4; S is an abbreviation for Section in Fig. 7.3). S1 is located along the northwestern section of the Xianshuihe fault between latitude 30.4°–32° N, with a very shallow locking depth of 2–3 km. S1 overlaps the section experiencing surface creep found by Allen et al. (1991) and Zhang et al. (2018). Adjacent to S1, the S2 section extends ~ 30 km along the Xianshuihe fault between latitude 30.2°–30.4° N and overlaps the Kangding-Bamei section. ISCs along S2 are estimated to be smaller than 20%, possibly reflecting creep reaching the surface. From our block model, the surface creep rate is estimated to be ~ 7.5 mm/yr, comparable with the estimated long-term slip rate (7.9 ± 1.6 mm/yr). S3 (~ 26.6°–6.8° N) and S4 (~ 24.6°–24.9° N) are short sections separating the locked asperities A2–A3 and A1–A2 respectively, with fault locking depths in the range of ~ 2–5 km.

7.3.2 *Fault Behavior and Moment Budget Along the XXFS*

7.3.2.1 Northern Xianshuihe Fault

We divide the Xianshuihe fault into two sub-sections Sa1 and Sa2 (Fig. 7.4). Sa1 is located north of the Bamei town and encompasses shallowly locked section S1 (Fig. 7.3). Sa2 from Bamei town to Shimian County, encompasses the creeping section S2 and part of Asperity A3 (Fig. 7.3).

Along Sa1, we compare the rate of moment accumulation predicted by our model to the moment released in large earthquakes. Taking an elastic shear modulus of 30 GPa, our model indicates an accumulation rate of moment deficit rate at 4.68×10^{17} Nm/yr. The amount of moment released by earthquakes since 1700 is 7.26×10^{20} Nm (Fig. 7.4), corresponding to ~ 1500 years of moment accumulated at the present-day rate with no earthquakes. Even considering that historical earthquake magnitudes were overestimated by 0.2, the resulting time span would be ~ 780 years. There is clearly a discrepancy here, which originates from either an abnormally high frequency of earthquakes during the last three centuries or a too weak amount of present-day accumulation moment rate that would then not be representative of the longer term average fault coupling. As a point of comparison, if we assume the 205 km-long fault to be entirely locked down to 15–20 km depth, a moment accumulation rate of $9.23–12.3 \times 10^{17}$ Nm/yr (10 mm/yr × 15–20 km × 205 km × 30 GPa) leads to a moment deficit of $6.03–8.04 \times 10^{20}$ Nm over the 1327–1981 period, a value fully consistent with the moment released by earthquakes over the same period. This calculation supports the view that at the scale of several centuries, the northern Xianshuihe fault is locked and that the current behavior represents a transient behavior, progressively moving towards a relocking of the fault and perhaps initiating a new cycle.

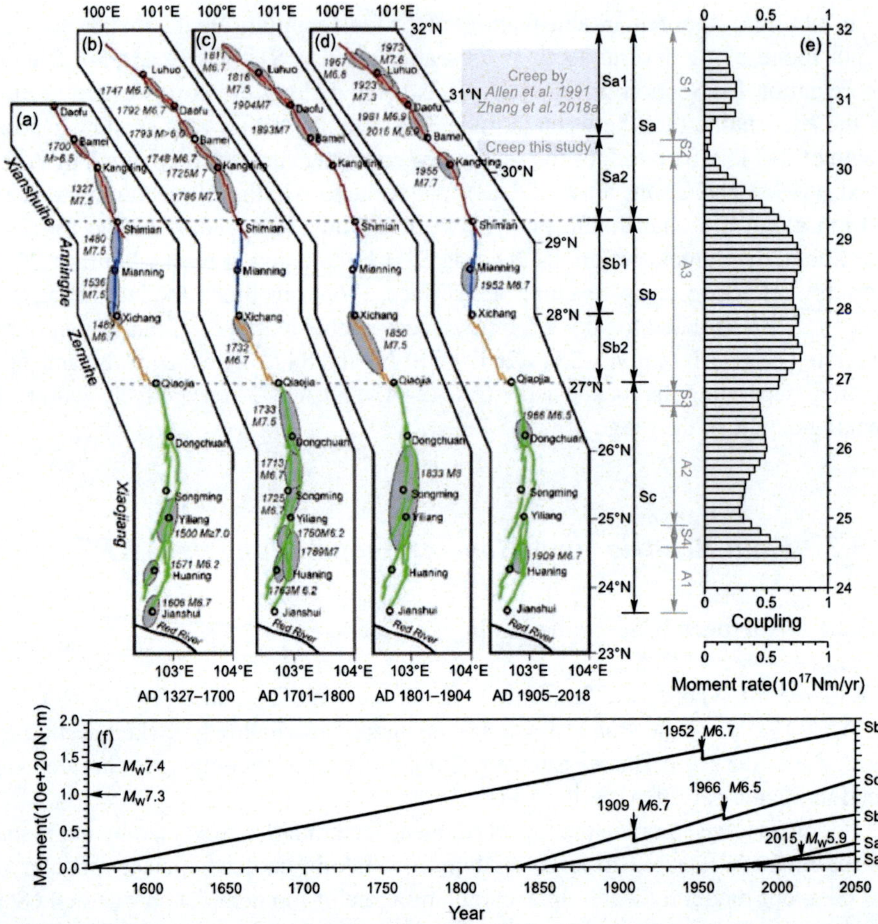

Fig. 7.4 **a–d** Maps for four historical periods of relative severely damaged areas of 36 M ≥ 6 earthquakes along the XXFS (modified from Wen et al., 2008a). Gray ellipses relative severely damaged areas of earthquakes with date and magnitude indicated. The red, blue, yellow and green lines correspond to the Xianshuihe fault, the Anninghe fault, the Zemuhe fault and the Xiaojiang fault respectively. **e** Red line corresponds to the vertical averaged coupling coefficients above 20 km depth. Grey histograms represent cumulative seismic moments every 0.2° bin. **f** Cumulative moment deficit corrected from large earthquakes moment released since 1536

7.3.2.2 Southern Xianshuihe Fault

Along Sa2, our ISC result shows two opposite behaviors, with almost full decoupling (locking coefficient of < 20%) along the Bamei-Kangding section (i.e., S2 in Fig. 7.3), and a transition towards full coupling down to ~ 10–20 km depth along the Kangding-Shimian section (Figs. 7.3 and 7.4e).

Our model provides an accumulation rate of moment deficit rate at 1.87 × 10^{17} Nm/yr, equivalent to a total moment of 1.17 × 10^{20} Nm between 1327 and 1955. This amount is less than 15% of the released moment through earthquakes

(8.41×10^{20} Nm) during the same period. Now considering that this fault section is 100% coupled down to 15–20 km, the accumulation rate of moment deficit is $(5.57–7.43) \times 10^{17}$ Nm/a, resulting in a moment of $(3.50–4.66) \times 10^{20}$ Nm for the 1327–1955 period. Such values are 70–90% of the moment release in earthquakes and can be considered as being consistent to a first order, given the uncertainty for magnitudes of historical earthquakes.

Combining the results for both Sa1 and Sa2, the Xianshuihe fault appears to be in a transient state following a rich sequence of earthquakes for the last three centuries, and then initiating a new cycle of progressive increasing locking. For the locked area, i.e., from Kangding to Shimian, if we assume the 1955 *M*7.7 to have released all the previously accumulated elastic strain along the section, the current moment available for an earthquake is M_w6.7.

7.3.2.3 Anninghe-Zemuhe Fault

The Anninghe-Zemuhe fault involves the main coupled part of the XXFS system (A3 in Fig. 7.3), where full locking down to depth ~ 20 km-depth along a 240 km-long section is obtained from our block model (Figs. 7.3 and 7.4e). The history of earthquakes shows that the northern section from Shimian to Xichang towns did not experienced any large earthquake for more than 4 centuries (last $M > 7$ earthquake occurred in 1563), while the southern section last ruptured in 1850. A slip deficit of 2–4 m is thus expected along the Anninghe-Zemuhe fault, corresponding to high potential of a large earthquake, a result consistent with the seismic gap proposed by Wen et al. (2008b).

The northern section (Sb1) has a length 150-km able to host an M_w7.6 earthquake according to fault-length magnitude scaling laws (Wells & Coppersmith, 1994). For the moment balance calculation here, we assume that the 1536 M7.5 earthquake released all the accumulated moment for this section. Under this assumption, we find a geodetic total moment deficit of 1.96×10^{20} Nm for the 1536–2050 period at an accumulation rate of 4.06×10^{17} Nm/yr from our block model. This amount is equivalent to an M_w7.5 earthquake if all the accumulated moment be released in a single event (Fig. 7.4f).

Along the southern section Sb2, the 130 km-long section is capable of rupture in an M_w7.5 earthquake. We thus consider that the 1850 *M*7.5 earthquake as our starting date for the moment balance assessment. Our model predicts a moment deficit accumulation rate of 3.54×10^{17} Nm/yr. By 2050, this amount leads to a moment of 7.08×10^{19} Nm which can be released during an M_w7.2 earthquake (Fig. 7.4f).

7.3.2.4 Xiaojiang Fault

The southern part of the XXFS fault shows along strike varying ISC distribution, with a ~ 180 km-long highly locked section (A2 in Fig. 7.3) which is surrounded by relatively narrow shallowly locked sections (S3 and S4 in Fig. 7.3). High locking close to the southern tip of the XXFS is also imaged but with a lower resolution due to possible side effects at the merging of the Xiaojiang fault with the Red River fault (A1 in Fig. 7.3).

From the past history of earthquakes dating back to the XVIth century, we do not observe a clear segmentation of earthquakes with the one suggested by our ISC models. Indeed, S3 and S4 were at least partially ruptured in the 1733 $M7.5$ and 1833 $M8$ earthquakes respectively. Given the time elapsed since these earthquakes, shallow locking here probably reflects lateral frictional variations along the Xiaojiang fault rather than transient creep following large earthquakes. Here, for the moment budget calculation, we assume that the 1833 $M8$ earthquake released all elastic strain along the Xiaojiang fault. At the estimated moment deficit rate from our ISC model, we find an equivalent single event of $M_w7.4$ to be required in 2050 to balance the accumulated moment since 1833 (Fig. 7.4f).

7.3.3 *Seismic Potential Along XXFS*

According to earthquake rupture length-magnitude scaling relationship (Wells & Coppersmith, 1994), the maximum magnitudes of earthquakes are $M_w8.1$, $M_w7.6$ and $M_w7.5$ along the Xiaojiang fault (Sc, ~ 360 km-long), the Anninghe fault (Sb1, 150 km-long) and the Zemuhe fault (Sb2, 130 km-long), respectively. The accumulated moment (Fig. 7.4f) and the above calculation suggest that the Anninghe and Zemuhe faults are ready for large earthquakes in the next decades, as already pointed out by several studies (e.g., Ran et al., 2008; Wen et al., 2008a; Jiang et al., 2015). Besides, we find that the Xiaojiang fault has a relatively high seismic potential with an accumulated moment deficit allowing for an $M > 7$ earthquake to happen.

Along the Kangding-Bamei section (~30.0°–30.5° N), paleoseismological studies (e.g., Ran et al., 2008; Wen et al., 2008a) indicate that the time elapsed since the last large earthquake (232 years) is close to the average earthquake recurrence time (230–350 years). Geodetic inversions (e.g., Jiang et al., 2015) stress the fault have accumulated slip deficit for characteristic earthquakes ($M_w7.0$). However, our block model results indicate that this section has very shallow fault locking depths (0–5 km), and includes a creeping section (S2 in Fig. 7.3) between latitude ~ 30.2°–30.4° N. Low moment accumulation rate and magnitude of accumulated slip deficit suggest that large earthquakes are unlikely, in the next decades, along this part of the fault system.

7.4 Conclusions

In this study, we present descriptions of the crustal deformation along the XXFS, which is one of the most seismically active and potentially most damaging one in

China. Firstly, using the GPS velocity solution (1999–2014) for eastern Tibet and the elastic block model, we obtained left-lateral slip rates of 7–11 mm/a along the fault system. In the northern part of the XXFS from 32° N to 30° N, the Xianshuihe fault appears to be predominantly creeping, whereas the Anninghe, Zemuhe and Xiaojiang faults show significant coupling.

Quantitative calculation of moment budget highlights relatively high seismic potential on the Anninghe, Zemuhe and Xiaojiang faults, where the elastic strain deficit could be balanced by $M_w7.5$, $M_w7.2$ and $M_w7.4$ earthquakes respectively.

Acknowledgements We are grateful to all colleagues who constructed the GPS network and collected GPS measurements. Robert McCaffrey shared the block model program (https://robmccaffrey.github.io/TDEFNODE/TDEFNODE.html). This work was supported by the National Key Research and Development Program of China (Grant No. 2019YFC1509205) and the National Natural Science Foundation of China (grant number 41631073). Majority of figures were generated using the Generic Mapping Tools (GMT) software (Wessel et al., 2013).

References

Allen, C. R., Zhuoli, L., Hong, Q., Xueze, W., Huawei, Z., & Weishi, H. (1991). Field study of a highly active fault zone: The Xianshuihe fault of southwestern China. *Geological Society of America Bulletin, 103*(9), 1178–1199.

Bai, M., Chevalier, M. L., Pan, J., Replumaz, A., Leloup, P. H., Métois, M., & Li, H. (2018). Southeastward increase of the late Quaternary slip-rate of the Xianshuihe fault, eastern Tibet: Geodynamic and seismic hazard implications. *Earth and Planetary Science Letters, 485*, 19–31.

Deng, Q., Zhang, P., Ran, Y., Yang, X., Min, W., & Chu, Q. (2003). Basic characteristics of active tectonics of China. *Science of China. Series D, 46*, 356–372.

Jiang, G., Xu, X., Chen, G., Liu, Y., Fukahata, Y., Wang, H., Yu, G., Tan, X., & Xu, C. (2015). Geodetic imaging of potential seismogenic asperities on the Xianshuihe-Anninghe-Zemuhe fault system, southwest China, with a new 3-D viscoelastic interseismic coupling model. *Journal of Geophysics Research Solid Earth, 120*(3), 1855–1873.

Loveless, J. P., & Meade, B. J. (2011). Partitioning of localized and diffuse deformation in the Tibetan Plateau from joint inversions of geologic and geodetic observations. *Earth and Planetary Science Letters, 303*(1–2), 11–24.

McCaffrey, R., Qamar, A. I., King, R. W., Wells, R., Khazaradze, G., Williams, C. A., Stevens, C. W., Vollick, J. J., & Zwick, P. C. (2007). Fault locking, block rotation and crustal deformation in the Pacific Northwest. *Geophysical Journal International, 169*(3), 1315–1340.

Ran, Y., Chen, L., Cheng, J., & Gong, H. (2008). Late Quaternary surface deformation and rupture behavior of strong earthquake on the segment north of Mianning of the Anninghe fault. *Science of China. Series D, 51*(9), 1224–1237.

Rui, X., & Stamps, D. S. (2016). Present-day kinematics of the eastern Tibetan Plateau and Sichuan Basin: Implications for lower crustal rheology. *Journal of Geophysics Research Solid Earth, 121*(5), 3846–3866.

Shen, Z. K., Lü, J., Wang, M., & Bürgmann, R. (2005). Contemporary crustal deformation around the southeast borderland of the Tibetan Plateau. *Journal of Geophysics Research Solid Earth, 110*, B11409.

Tapponnier, P., & Molnar, P. (19770. Active faulting and tectonics in China. *Journal of Geophysics Research Solid Earth, 82*(20), 2905–2930.

Tapponnier, P., Zhiqin, X., Roger, F., Meyer, B., Arnaud, N., Wittlinger, G., & Yang, J. (2001). Oblique stepwise rise and growth of the Tibet Plateau. *Science, 294*(5547), 1671–1677.

Thompson, T. B., Plesch, A., Shaw, J. H., & Meade, B. J. (2015). Rapid slip-deficit rates at the eastern margin of the Tibetan Plateau prior to the 2008 Mw 7.9 Wenchuan earthquake. *Geophysics Research Letter, 42*(6), 1677–1684.

Wang, W., Qiao, X., Yang, S., & Wang, D. (2017). Present-day velocity field and block kinematics of Tibetan Plateau from GPS measurements. *Geophysical Journal International, 208*(2), 1088–1102.

Wells, D. L., & Coppersmith, K. J. (1994). New empirical relationships among magnitude, rupture length, rupture width, rupture area, and surface displacement. *Bulletin of the Seismological Society of America, 84*(4), 974–1002.

Wen, X., Ma, S., Xu, X., & He, Y. (2008a). Historical pattern and behavior of earthquake ruptures along the eastern boundary of the Sichuan-Yunnan faulted-block, southwestern China. *Physics of the Earth and Planetary Interiors, 168*(1–2), 16–36.

Wen, X., Fan, J., Yi, G., Deng, Y., & Long, F. (2008b). A seismic gap on the Anninghe fault in western Sichuan, China. *Science of China. Series D, 51*(10), 1375–1387.

Wessel, P., Smith, W. H., Scharroo, R., Luis, J., & Wobbe, F. (2013). Generic mapping tools: Improved version released. *EOS, Transactions of the American Geophysical Union, 94*(45), 409–410.

Yan, B., & Lin, A. (2017). Holocene activity and paleoseismicity of the Selaha Fault, southeastern segment of the strike-slip Xianshuihe Fault Zone, Tibetan Plateau. *Tectonophysics, 694*, 302–318.

Zhang, J., Wen, X. Z., Cao, J. L., Yan, W., Yang, Y. L., & Su, Q. (2018). Surface creep and slip-behavior segmentation along the northwestern Xianshuihe fault zone of southwestern China determined from decades of fault-crossing short-baseline and short-level surveys. *Tectonophysics, 722*, 356–372.

Chapter 8
Three-Dimensional Crustal Structure and Seismicity Characteristics of Changning-Xingwen Area in the Southwestern Sichuan Basin, China

Kezhen Zuo, Cuiping Zhao, Haijiang Zhang, and Xiaodong Zhang

Abstract The Changning $M_S6.0$ earthquake, occurred on June 17, 2019, is located in the southwestern margin of the Sichuan Basin. Using seismic data recorded on permanent and temporary stations around Changning area, the high-resolution three-dimensional crustal V_P, V_S, V_P/V_S models and earthquake locations in Changning-Xingwen area are obtained by using the V_P/V_S model consistency-constrained double-difference seismic tomography method. The results show that crustal structures in the source area of the 2019 Changning $M_S6.0$ earthquake have significant variations, especially in the depth of 0–7 km. Seismic activity in Shuanghe and Yutan anticline area before the Changning $M_S6.0$ earthquake outlined several NE trending stripes, implying pre-existing small-scale faults which are perpendicular to major NW-striking faults in Changning-Shuanghe anticline system. We found that the Changning $M_S6.0$ earthquake broke through these pre-existing small-scale faults and extended from the Shuanghe to Yutan anticline. Both the rupture process and aftershock activity were influenced by the pre-existing small-scale faults. Most earthquakes within the Changning area are located in a slant zone that gradually deepen from Shuanghe anticline on the east to Yutan anticline on the west with the maximum depth from 5 to 10 km, which are associated with obvious high V_S and low V_P/V_S features. The relocated seismic clusters in Luochang-Jianwu syncline area have different strikes and dips, which are mainly located at the edge of low velocity anomaly bodies, and correspond to the low V_P/V_S area.

Supplementary Information The online version contains supplementary material available at (https://doi.org/10.1007/978-981-16-8607-8_8).

K. Zuo (✉) · C. Zhao · X. Zhang
Institute of Earthquake Forecasting, China Earthquake Administration, Beijing, China

C. Zhao
e-mail: zhaocp@ief.ac.cn

H. Zhang
School of Earth and Space Sciences, Laboratory of Seismology and Physics of Earth's Interior, University of Science and Technology of China, Hefei, China

© Higher Education Press 2022
Y.-G. Li et al. (eds.), *China Seismic Experimental Site*,
https://doi.org/10.1007/978-981-16-8607-8_8

Keywords Changning area · Induced earthquake · 3D velocity structure · Seismogenic mechanism

8.1 Introduction

On June 17, 2019, Changning $M_S6.0$ earthquake occurred in Sichuan Province, China, and caused casualties and severe economic losses. As of August 31, 2019, more than 6000 aftershocks have been recorded, including four $M_S > 5$ aftershocks. The Changning $M_S6.0$ earthquake is located in the southwestern margin of the Sichuan Basin (Fig. 8.1), and seismic activity in this area has significantly increased since 2006 (Ruan et al., 2008; Sun et al., 2017). After 2015, the frequency and strength of seismicity in this area dramatically increased again (Meng et al., 2019), and the Xingwen $M_S5.7$ on December 16, 2018 and Gongxian $M_S5.3$ on January

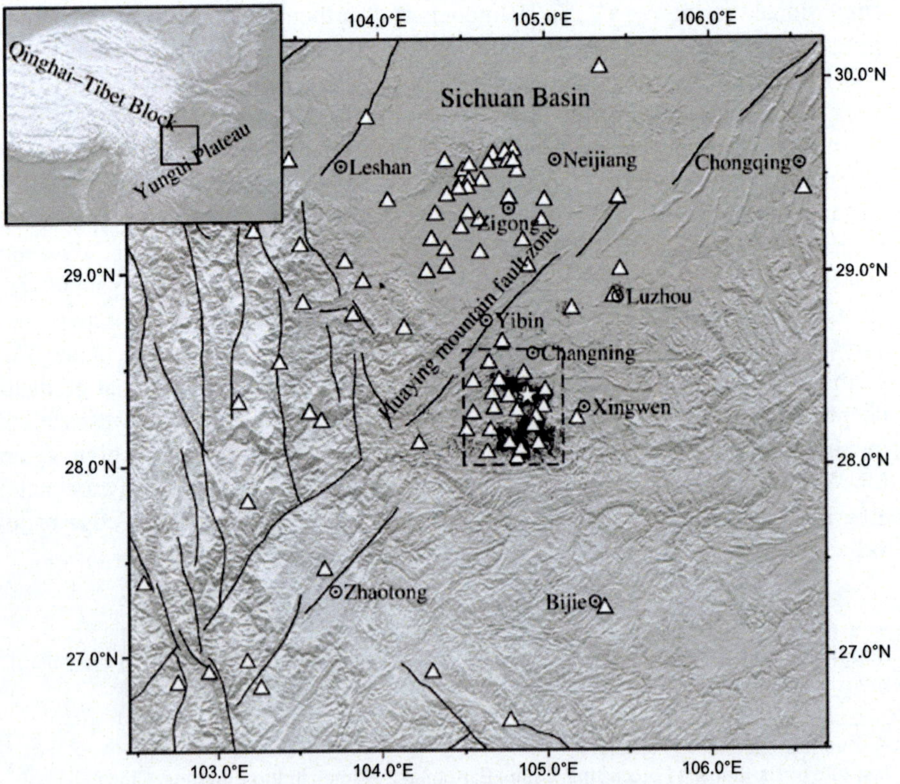

Fig. 8.1 Distribution of epicenters and seismic stations used in the study. The box with black dashed line is the study area, white triangles represent seismic stations, black dots denote the initial earthquake locations, the white star denote the 2019 Changning $M_S6.0$ earthquake, and black lines are main active faults. The box in the inset map shows the position of the whole region in Southwest China

3, 2019 occurred successively, only a dozen kilometers away from the Changning M_S6.0 earthquake (Fig. 8.2).

From north to south, the main industry activities in Changning and Xinwen are salt and shale gas mining, respectively. Correspondingly, the seismic activities in the two regions are basically separated (Lei et al., 2019a). The Changning salt mine is located in the Changning-Shuanghe anticline system in the north of the study area, with the buried depth of salt deposit greater than 2.5 km. The depths of injection wells are about 2.8–3.0 km, and the distance between the wells is hundreds of meters (Ruan et al., 2008). Salt mining in the area began in 1990 (Sun et al., 2017), and long-term water injection in the Changning salt mine have induced very frequent seismicity (Ruan et al., 2008; Sun et al., 2017; Lei et al. 2019a). The shale gas block is located in the Jianwu syncline in the south of the study area. The main buried depths of shale gas resources in the study area are 2.0– 3.5 km (Ma et al., 2020). The systematic hydraulic fracturing operation started in 2014 (Lei et al., 2017), and the

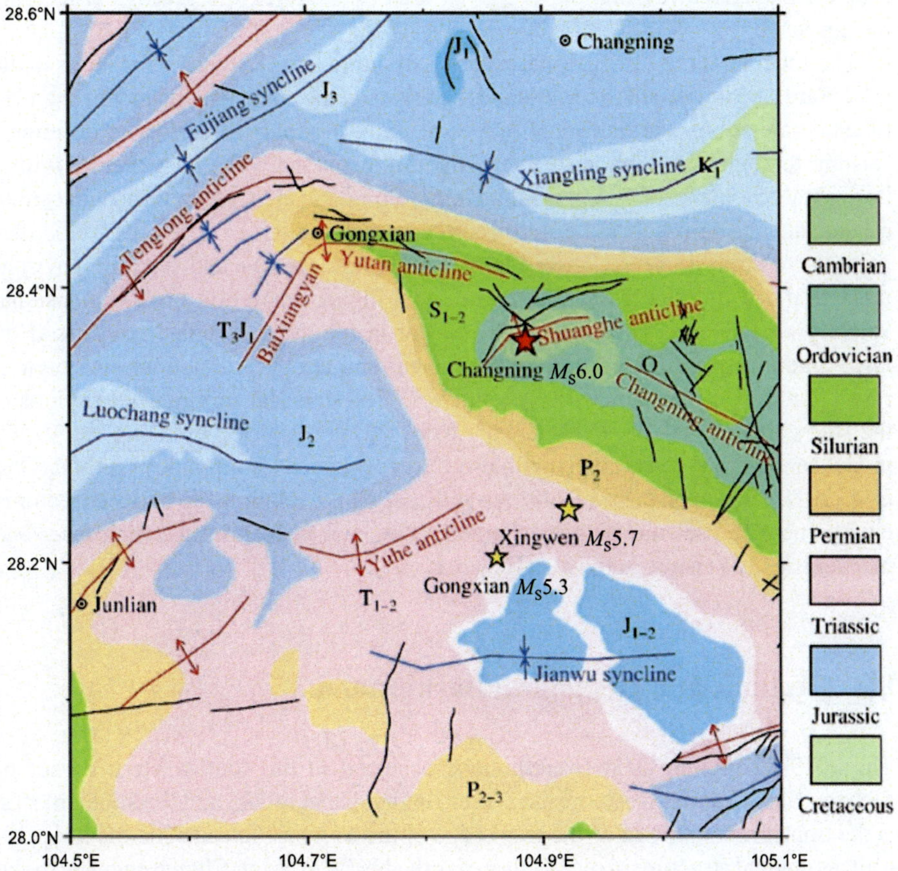

Fig. 8.2 Tectonic and geological settings of the study region. Red lines denote anticlines, blue lines denote synclines, and black lines are small-scale faults. The background geological map is from the 1:1 000 000 geological map of Pang et al. (2017)

number of well pads continued to increase. For hydraulic fracturing operations in this area, the average fracturing pressure is about 40 –70 MPa, average injection rate is 10 –2 m^3/min, and flow back rate is 15%, respectively (Meng et al., 2019). The water required for a single well pad with six horizontal wells is more than 150,000 m^3 (Lei et al., 2017). Previous studies show that there is a strong spatio-temporal correlation between seismic activity and hydraulic fracturing in the shale gas block (Lei et al., 2017, 2019a; Meng et al., 2019; Tan et al., 2020). Lei et al. (2019b) suggested that Xingwen M_S5.7 earthquake and Gongxian M_S5.3 earthquake were caused by hydraulic fracturing, and Xingwen M_S5.7 earthquake may have promoted the occurrence of Changning M_S6.0 earthquake (Lei et al., 2019a). However, other studies argue the occurrence of frequent seismic activities and moderate-strong earthquakes in Changning-Xingwen area is due to natural mechanism. Yi et al. (2013) speculated that the frequent moderate-strong earthquakes in Changning area reflected the higher stress level in the eastern part of the Sichuan-Yunnan junction. He et al. (2019) thought that the injection of over-pressured fluids is not the main cause of moderate magnitude earthquakes in Changning-Xingwen area by studying the geological tectonic background.

The accurate three-dimensional velocity structure and earthquake locations in the source area are important to understand the seismogenic environment and mechanism of moderate-strong earthquake. Long et al. (2020) adopted the double-difference seismic tomography method to image three-dimensional velocity structure with a horizontal resolution of 10 km in Changning-Gongxian area, and revealed that small earthquake clusters occurred primarily in regions with high P- or S-wave velocities. In order to study the relationship between seismic activity and tectonic environment in detail, it is necessary to determine a higher-resolution three-dimensional velocity model, especially the V_P/V_S model in the area. In this study, based on large amounts of data recorded by permanent and temporary stations, the higher-resolution three-dimensional V_P, V_S and V_P/V_S models and earthquake locations in the study area (28.0°–28.6° N, 104.6°–105.1° E) are inverted by using the V_P/V_S model consistency-constrained double-difference seismic tomography method (Guo et al., 2018). Based on these results, we will analyze the relationship between tectonic environment and seismic activity characteristics, and further discuss the causes and mechanisms of seismic activity in the area.

8.2 Tectonic and Geological Background

The 2019 Changning M_S6.0 earthquake occurred in the southwestern corner of Sichuan Basin, which western side is the Qinghai-Tibet block, and the southern side is the transition mountain of the front edge of the Yungui Plateau. The most significant geological structure in the study area is the NWW striking Changning-Shuanghe anticline system, which is a compound anticline, mainly composed of Tenglong anticline, Baixiangyan-Yutan anticline, Shuanghe anticline and Changning anticline. The north limb of Changning-Shuanghe anticline system is steeper than the south limb, and Cambrian strata are exposed in the core. The direction of the maximum horizontal principal stress axis in the Changning-Shuanghe anticline system is about 77° (Lei et al., 2019a). The aftershocks of Changning M_S6.0 earthquake are generally distributed along the axis of the anticline. The main active fault around the source area is Huayingshan fault zone with a strike of northeast-southwest, and its main movement characteristics are thrusting and dextral strike-slip (Deng et al., 2003). In addition, a series of small-scale faults have developed in the core and limbs of the Changning-Shuanghe anticline system, most of which are high-angle thrust faults. The focal mechanism solutions reveal that the direction of the maximum horizontal principal stress axis in the shale gas block in the south of the Changning-Shuanghe anticline system is about 117° (Lei et al., 2017), which is consistent with the maximum horizontal principal stress direction of 100°–115° obtained from the measurement of in-situ rock stress (Wang et al., 2016). Xingwen M_S5.7 earthquake and Gongxian M_S5.3 earthquake are located in the north of Jianwu syncline. Both Jianwu syncline and Luochang syncline are broad and gentle synclines composed of Jurassic strata on the surface. The tectonic fractures are generally developed, and the strikes of large-scale fractures are mainly near EW, NE, and NNW directions, most of which are high-angle fractures (Shi et al., 2019).

8.3 Data and Method

8.3.1 Method

Zhang and Thurber (2003, 2006) developed a double-difference seismic tomography method, which uses absolute and relative arrival times to invert the three-dimensional velocity structure and the seismic location. Compared with standard tomography method, this method has obvious advantages for inversion of the accurate velocity structure in the source area. Double-difference seismic tomography method was extended by Zhang et al. (2009) to directly invert V_P/V_S using S-P data (Ts-Tp). The V_P/V_S model obtained from S-P data is reliable, but its resolution is relatively low due to the limited amount of available S-P data (Guo et al., 2018). In contrast, V_P/V_S model can also be obtained by dividing V_P model by V_S model. However, due to the differences in seismic phase data quality and ray paths between P-wave and S-wave, the V_P/V_S model obtained in this way has higher resolution but lower reliability, which makes the results often difficult to interpret (Eberhart-Phillips, 1990; Huang et al., 2014). Guo et al. (2018) developed a new V_P/V_S model consistency-constrained double-difference seismic tomography method (TomoDDMC) that can produce a higher-resolution and reliable V_P/V_S models using P-wave data, S-wave data, S-P data, while V_P and V_S models and earthquake relocations can also be obtained.

8.3.2 Data and Processing

We collected the seismic phase arrivals of 34,296 earthquakes in the study area from January 1, 2015 to August 31, 2019 recorded by 86 permanent and temporary stations. The distributions of epicenters and stations are shown in Fig. 8.1. First, we selected phase data with epicenter distances less than 150 km and removed the data with larger deviations according to the time-distance curves (Fig. 8.3a). Then, each earthquake is required to have at least 8 phase arrivals. For the differential times, interevent distance of event pairs is less than 10 km, and the maximum number of neighbors is 20. In total, 24,640 earthquakes and their phase data are used for inversion, including 161,276 P-wave absolute arrival data, 1,575,384 P-wave relative arrival data, 161,515 S-wave absolute arrival data, 1,569,036 S-wave relative arrival data, 148,818 S-P absolute arrival data, 1,447,388 S-P relative arrival data, respectively. The average interevent distance of seismic pairs is 1.48 km.

The initial one-dimensional velocity model used in this study refers to the results of Lei et al. (2017), and the average V_P/V_S is 1.73. According to the distributions of earthquakes and stations, the center of the study area (104.8° E, 28.3° N) is used as the coordinate origin, with the *X* axis oriented east and the Y axis oriented north. The grid nodes in the *X* axis and *Y* axis are located at −300, −60, −45, −30, −20, −15, −10, −5, 0, 5, 10, 15, 20, 30, 45, 60, 300 km. And the grid nodes in the vertical

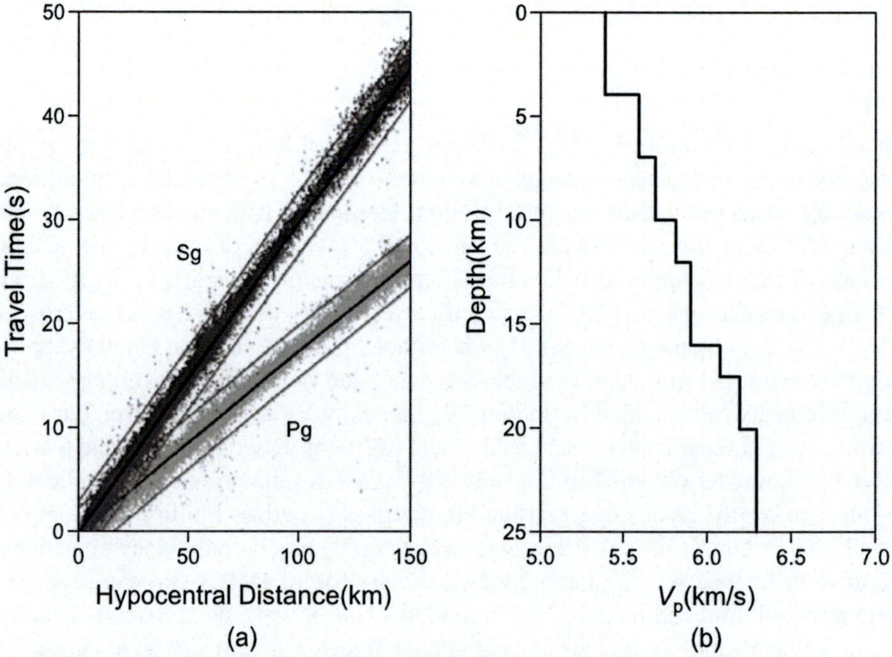

Fig. 8.3 **a** Time-distance curves of P- and S-wave data. Black lines are the linear fitting of travel time versus hypocentral distance, and data outside the gray lines are removed due to large deviations. **b** Initial 1D P-wave velocity model

direction are located at the depths of −100, 0, 4, 7, 10, 13, 16, 20, 30, 200 km. Figure 8.4 shows the distributions of seismic ray paths and inversion grids in the study area. The optimal damping and smoothing parameters used in the inversion are determined by constructing trade-off curves of solution variance and data variance (Eberhart-Phillips, 1986). The damping parameter is used to constrain the model's slowness and seismic location during the inversion, and its optimal value is set to be 600; the smoothing parameter is only used to constrain the model's slowness, and its optimal value is set to be 40 (Fig. 8.5).

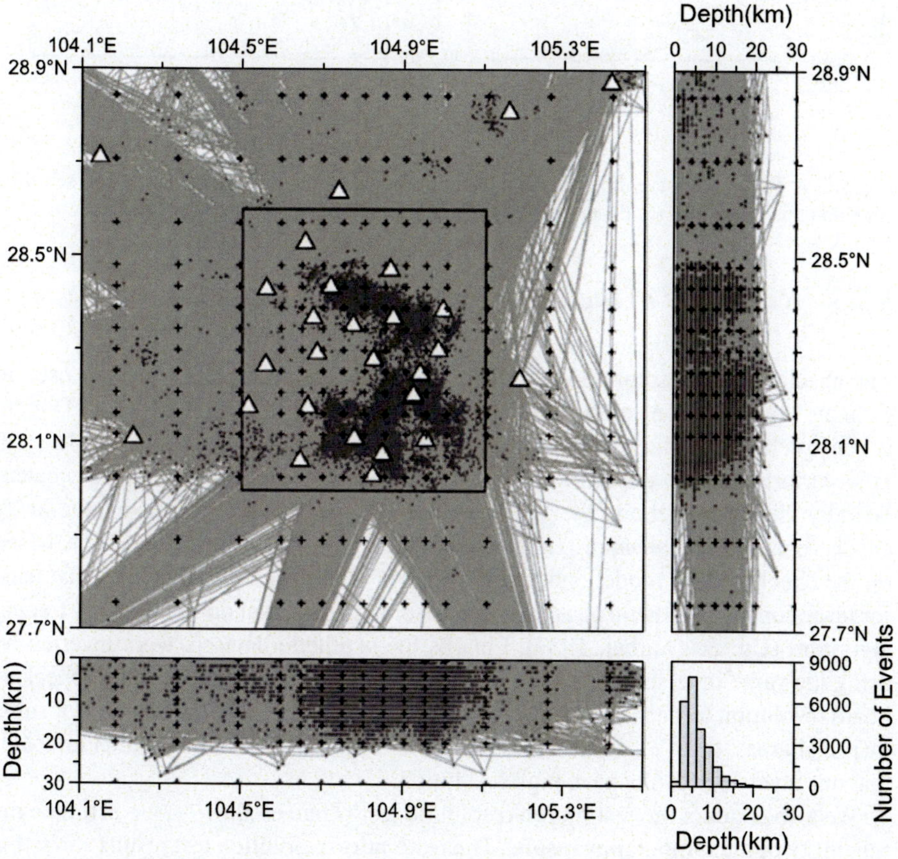

Fig. 8.4 Distribution of seismic ray paths and inversion grids. The box is the study area, black dots denote the initial locations of earthquakes, white triangles represent stations, black crosses denote inversion grids, and gray lines denote seismic ray paths

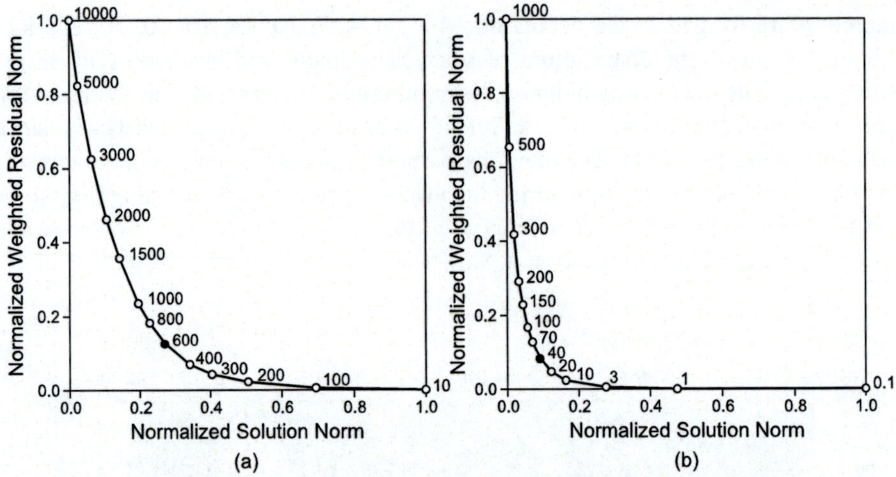

Fig. 8.5 **a** Trade-off curve for selecting damping parameter. **b** Trade-off curve for selecting smoothing parameter. Black dots represent the optimal parameters selected

8.3.3 Reliability of the Inversion Results

The checkerboard resolution test (Humphreys & Clayton, 1988) is widely used to estimate the resolution and reliability of the model. A synthetic checkerboard model is constructed by adding ± 5% perturbations to the initial V_P model, and adding ± 10% perturbations to the initial V_P/V_S model. Since the initial V_S model is calculated by the V_P model and the V_P/V_S model, the perturbations of the synthetic V_S checkerboard model are approximately + 5.5% and − 4.5% alternately. Then, based on the checkerboard model and distributions of stations and earthquakes that used for inversion, the synthetic phase arrivals are obtained by pseudo-bending ray tracing algorithm (Um & Thurber, 1987). Finally, the synthetic datasets were inverted by using the same inversion strategy as used in real inversion. The results of checkerboard resolution test are shown in Fig. 8.6. At the depth of 0–16 km, except for some marginal areas, the V_P, V_S and V_P/V_S models can be well restored, which indicates that our inversion results have high resolution.

We also conduct the restoration resolution test (Zhao et al., 1992) to estimate the reliability of the tomography results. The restoration resolution test results show that the velocity anomalies in the study area can be well reconstructed (Figs. S1–S4).

The uncertainties of earthquake relocations are analyzed by using the bootstrap resampling method (Efron & Gong, 1983; Efron & Tibshirani, 1991). First, the phase arrival data used in the study are randomly resampled, and then the new resampled datasets were inverted with the same inversion strategy as the real inversion to obtain the earthquake relocations. By repeating the above process 100 times, the standard deviations of earthquake relocations can be estimated, which can be treated as the uncertainties for earthquake relocations. The statistical results of relocation uncertainty are shown in Fig. 8.7.

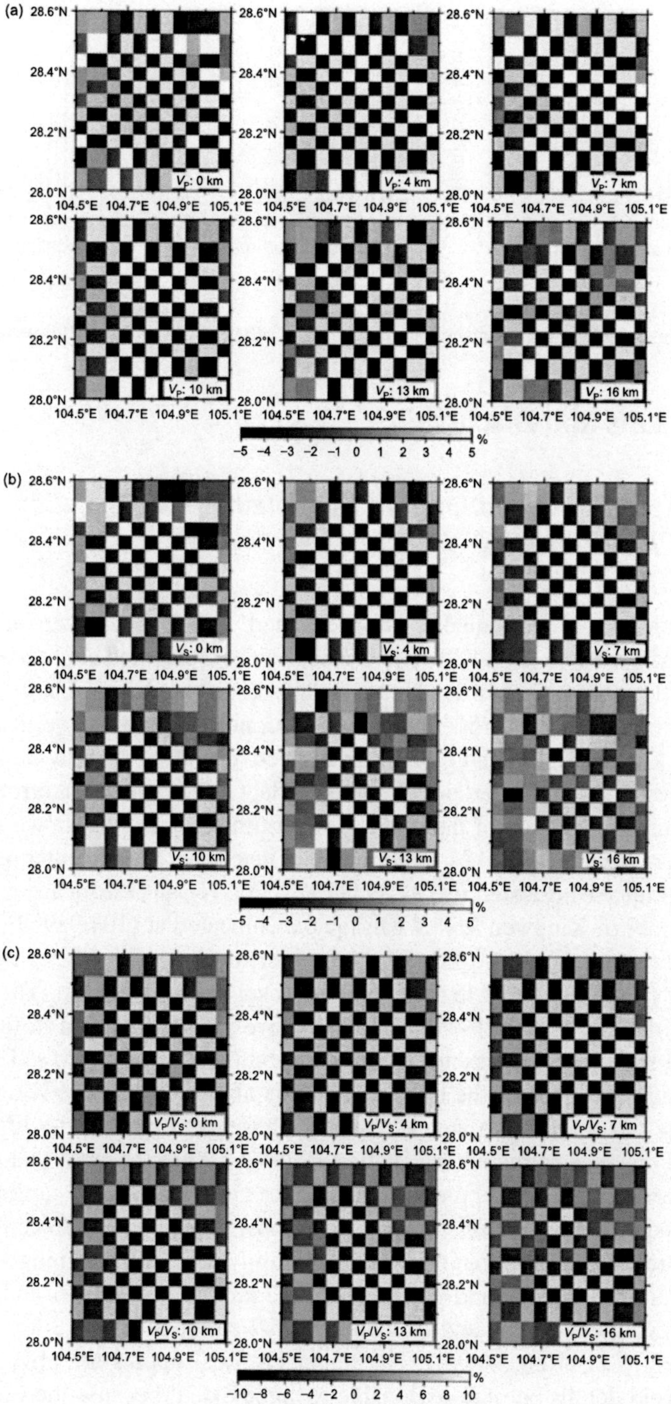

Fig. 8.6 Recovered checkerboard anomalies for **a** V_P, **b** V_S, and **c** V_P/V_S models

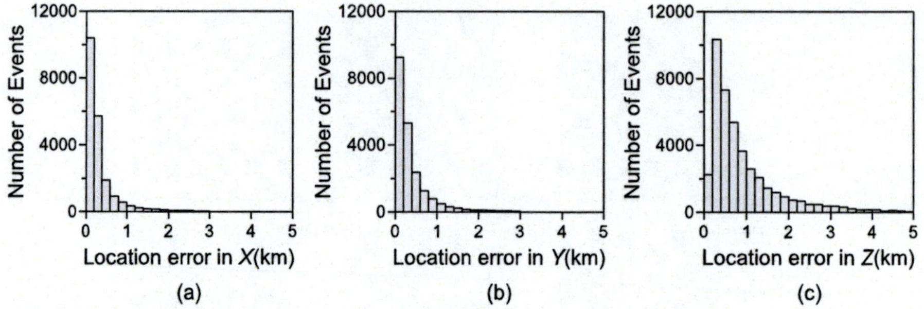

Fig. 8.7 The uncertainties of earthquake relocations in the **a** *X*, **b** *Y*, and **c** *Z* directions

8.4 Results and Discussion

8.4.1 Seismic Relocation and Spatiotemporal Migration Characteristics

A total of 21,711 earthquakes are relocated after joint inversion using the TomoDDMC method (Fig. 8.8). The distributions of relocated earthquakes are clustered tightly, and the focal depths are mainly distributed at 2–7 km. The median values of the relocation errors in the east–west, north–south and vertical directions are 0.201 km, 0.232 km and 0.633 km, respectively. As shown in Guo and Zhang (2017), these location uncertainties mostly reflect relative location errors. The root-mean-square distribution of the seismic travel time residuals is shown in Fig. 8.8c, and its average value is 0.012 s. After the joint inversion, the epicenter of Changning M_S6.0 earthquake is located at (104.894° E, 28.363° N), and the focal depth is 7.8 km; The epicenter of Xingwen M_S5.7 earthquake is located at (104.929° E, 28.223° N) and the focal depth is 12.0 km; The epicenter of Gongxian M_S5.3 earthquake is located at (104.859° E, 28.191° N) and the focal depth is 8.1 km. The Changning M_S6.0 earthquake has 22 P-wave and 4 S-wave phases involved in the inversion. Because a station (CNI) was not working during the Changning M_S6.0 earthquake, the epicentral distance of the nearest station is about 12 km. The Xingwen M_S5.7 earthquake has 11 P-wave and 10 S-wave phases involved in the inversion. The epicentral distance of the nearest station is about 17 km, because many temporary stations were deployed after the earthquake. The Gongxian M_S5.3 earthquake has 19 P-wave and 9 S-wave phases involved in the inversion, and the epicentral distance of the nearest station is about 6 km. The centroid depth of Changning M_S6.0 earthquake obtained by the cut-and-paste method is 3 km (Yi et al., 2019), and the centroid depths of Xingwen M_S5.7 and Gongxian M_S5.7 earthquakes are 3 km and 1.8 km, respectively (Lei et al., 2019b). The relocated focal depths are obviously deeper than centroid depths but it is reasonable to some extent because the centroid depth of earthquake represents the mean depth of earthquake rupture area rather than the initial point. Zhang et al. (2019) studied the spatial distribution of earthquakes in this

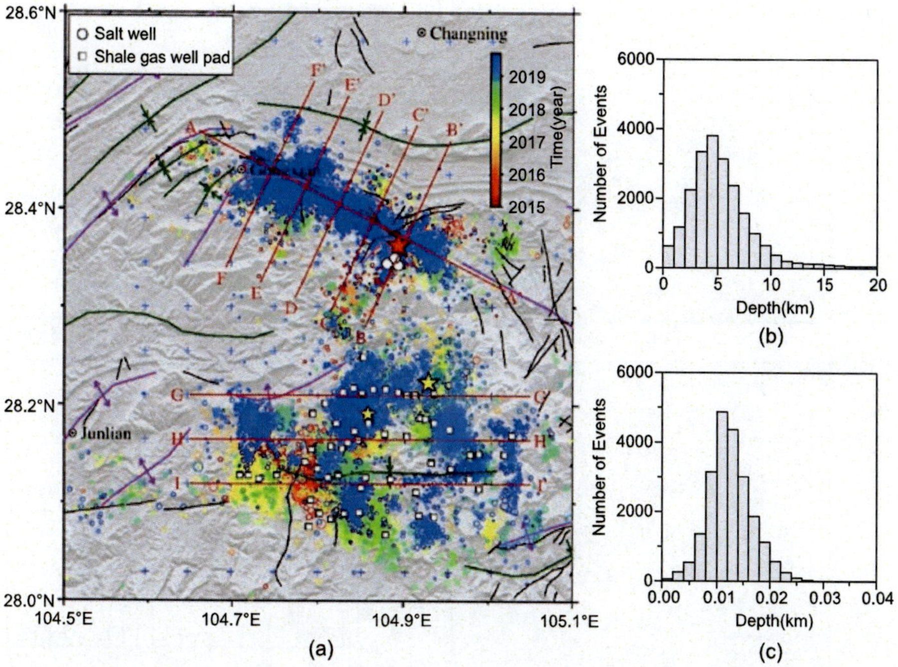

Fig. 8.8 **a** Distribution of epicenters colored in time after relocation. The locations of shale gas well pads (white squares) are obtained from the study of Lei et al. (2017). Blue lines represent the location of profiles, purple lines denote anticlines, dark green lines denote synclines, and black lines are small-scale faults. **b** Distribution of focal depth of earthquakes after inversion. **c** Distribution of RMS travel time residuals of earthquakes after inversion

area from January 2008 to May 2015, and also found that earthquakes with larger magnitude have relatively deeper focal depths.

The spatiotemporal migration process of seismicity before and after the major earthquake can provide important information on the source rupture process and fault structures (Liu et al., 2019; Tang et al., 2014). Figures 8.9 and 8.10 show the spatial distribution pattern of earthquakes at different time periods in the source region of the Changning M_S6.0 earthquake. According to Fig. 8.9a, we find that there are lots of events around the Changning salt mine before the Changning M_S6.0 earthquake. Because there are many stone pits in the west of Shuanghe anticline (within the area outlined by red line in Fig. 8.9a), the earthquake catalogue may contain a large number of blasting events (Lei et al., 2019a). Figure 8.9d displays the statistical results of stacked hourly distribution, indicating that the blasting events are mainly distributed between 11:00 and 20:00 local time. But for the earthquakes with $M \geq 2$ in the earthquake catalogue of this area, the stacked hourly distribution is relatively even and the spatial distribution is not affected by the location of stone pits (Fig. 8.9b, e). Therefore, we speculate that most of the events with $M \geq 2$ are earthquakes. According to Fig. 8.9b, we can find that the earthquakes are generally distributed along the NWW striking Yutan anticline, and there are NE-extending

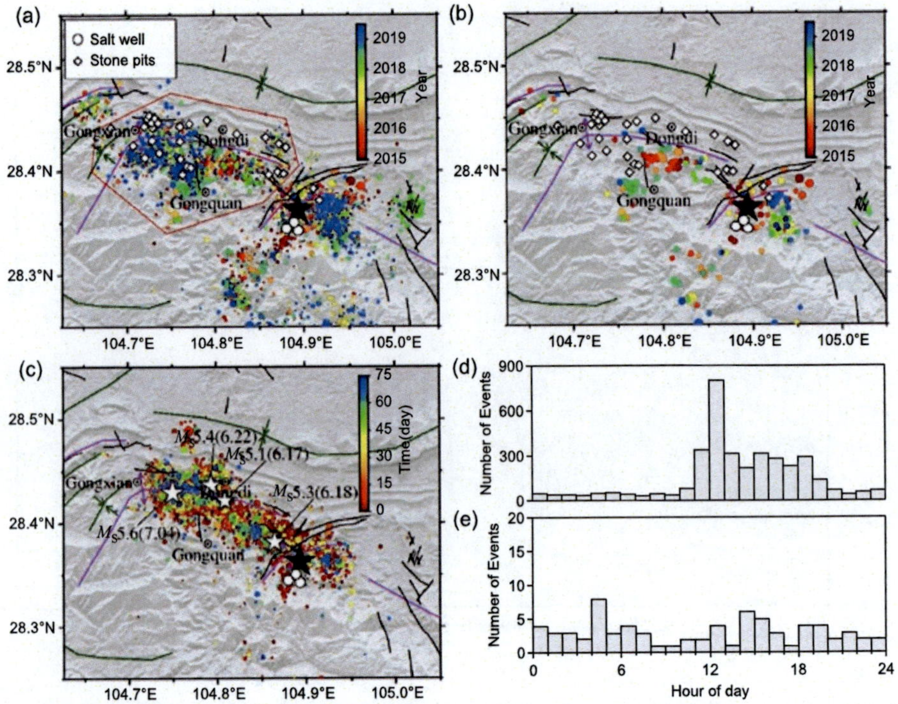

Fig. 8.9 Epicenter distribution in different time periods. **a** Distributions of events before Changning M_S6.0 earthquake. **b** Distributions of events with $M \geq 2$ before Changning M_S6.0 earthquake. **c** Distributions of events after Changning M_S6.0 earthquake. **d** Stacked hourly distribution of events in the red polygonal area of (**a**). **e** Stacked hourly distribution of events with $M \geq 2$ in the red polygonal area of (**a**). The black star represents the Changning M_S6.0 earthquake, and white stars represent aftershocks with $M_S > 5$. The locations of stone pits were identified via satellite map

seismic clusters on the NE striking Shuanghe anticline and near Gongquan town. The seismic activity forms several NE-extending seismic clusters and is confined in Yutan-Shuanghe anticline, indicating that there are many pre-existing faults or fractures striking along the NE direction.

From Fig. 8.9c, we can see that after the Changning M_S6.0 earthquake, events clustered mainly along the north boundary of Changning-shuanghe anticline system, and NE-extending seismic clusters still appear on the western end of Changning-shuanghe anticline system and near the M_S6.0 epicenter, respectively. The focal mechanism solutions of the aftershocks calculated by Yi et al. (2019) also show that the overall nodal strike distribution of the aftershocks is scattered. The above pattern indicates that both the rupture process of Changning M_S6.0 and aftershock activity are strongly affected by the complex tectonic background and related to the pre-existing faults.

Figure 8.10b–f shows the focal depth distribution along the AA' profile in different time periods separated by the four $M_S > 5$ aftershocks. From Fig. 8.10b, c, we find that the aftershocks are evenly distributed during the period from the occurrence of

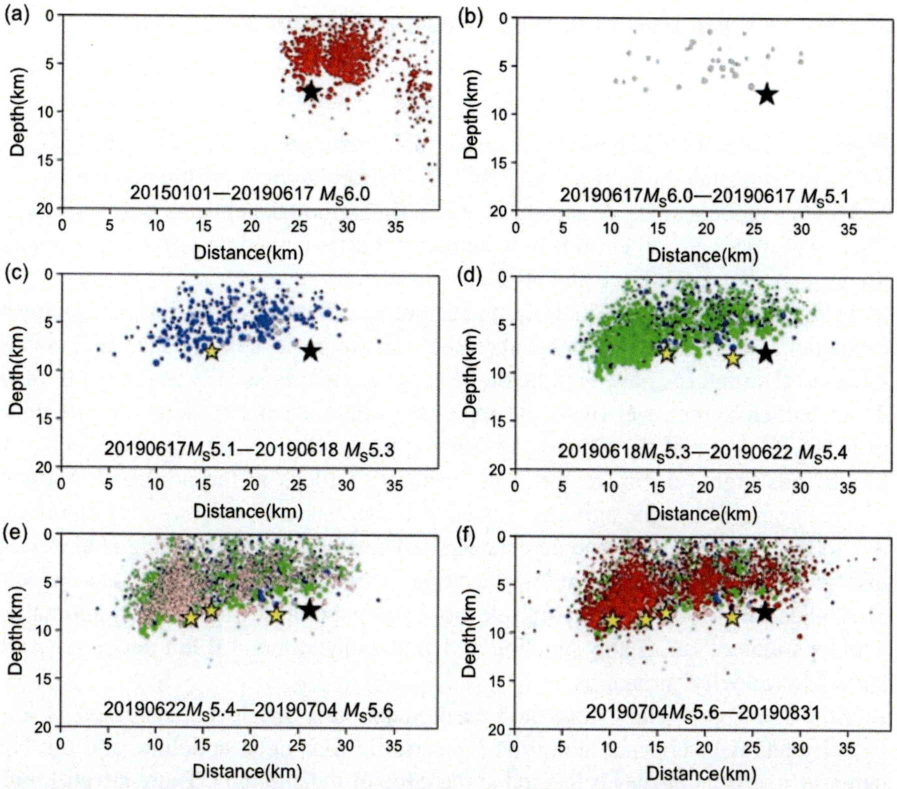

Fig. 8.10 Focal depth section along AA' profile in different time periods. **a** Red dots: January 1, 2015 to Changning M_S6.0 earthquake on June 17, 2019. Note: the western end of the profile contains too many blasting events before Changning M_S6.0, which are not shown in the figure. **b** Gray point: Changning M_S6.0 earthquake to M_S5.1 on June 17. **c** Blue dots: M_S5.1 on June 17 to M_S5.3 aftershock on June 18. **d** Green dots: M_S5.3 event on June 18 to M_S5.4 event on June 22. **e** Pink spots: M_S5.4 event on June 22 to M_S5.6 event on July 4. **f** Red dots: M_S5.6 event on July 4 to August 31

M_S6.0 on June 17 to the occurrence of M_S5.3 on June 18, and there is no significant difference in the number of earthquakes and focal depth between the western and eastern sides. During the period from the occurrence of M_S5.3 on June 18 to the M_S5.4 on June 22 (Fig. 8.10d), more events occurred in the western segment, with the maximum depth extending from 5 km on the eastern side to 10 km on the western side. After the M_S5.4 on June 22 and before the M_S5.6 on July 4 (Fig. 8.10e), lots of events occurred around the M_S5.4 hypocenter on the western segment. After July 4 (Fig. 8.10f), lots of events were still concentrated in the western segment, and since then the whole spatial distribution pattern has not changed significantly. In general, from east to west along the AA' profile, the focal depths of aftershocks gradually deepening with the maximum depth from 5 km to about 10 km, which is consistent with the relocated results of other studies (Yi et al., 2019; Lei et al., 2019a, Long et al. 2020).

8.4.2 Three-Dimensional Crustal V_P, V_S and V_P/V_S Structures

Figure 8.11 shows the horizontal tomographic images of V_P, V_S and V_P/V_S at different depth layers. Figures 8.12 and 8.13 show vertical profiles of the V_P, V_S and V_P/V_S models along or across several main earthquake clusters, respectively. In Figs. 8.11 and 8.12, all earthquakes before and after Changning M_S6.0 earthquake are projected on each layer and profile.

In order to discuss the distribution feature of aftershocks, in Fig. 8.13 only the earthquakes after Changning M_S6.0 earthquake are projected. The results show that the crustal structures have significant heterogeneities, especially at the depth of 0–7 km, which agrees well with the regional geological background. At the depth of 0 km, the core of the Changning-Shuanghe anticline system is associated with an obvious high-velocity region. The basins or valleys in the northeast and west of Changning-Shuanghe anticline system are characterized by low-velocity regions due to their younger sedimentary materials. The study results of Long et al. (2020) also show that the new strata at low elevation in this area corresponds to low velocity anomalies, and the old strata at high elevation corresponds to high velocity anomalies. The locations of Luochang syncline and Jianwu syncline at 0 km depth generally show low velocity anomalies.

At 4 km depth where relocated earthquakes are mainly concentrated, a low-velocity anomaly appears along the NE striking Shuanghe anticline, and the NE-extending seismic cluster is located at the edge of the anomaly. There are still high-velocity anomalies on the east and west sides of Shuanghe anticline, and low-velocity anomalies are mainly distributed in the northeast and west sides of Changning-Shuanghe anticline system, while the low-velocity anomalies around the Jianwu syncline are significantly weakened. At the depth of 7 km, the high-velocity anomaly on the east side of the Shuanghe anticline has weakened while the Luochang syncline area still maintains a strong low-speed anomaly. At the depth of 10 km and below, the lateral heterogeneity in the study area is obviously weakened.

For the V_P/V_S structure, at the depth of 0 km, to the east of the Shuanghe anticline there are low-V_P/V_S anomalies, and to the west there are high-V_P/V_S anomalies, consistent with the strata exposed on both sides. The Jianwu syncline area is also associated with the high-V_P/V_S anomaly. At the depth of 4 and 7 km, the west side of Shuanghe anticline shows low V_P/V_S anomaly, and its shape is consistent with Baixiangyan-Yutan anticline. The high-V_P/V_S anomaly in Jianwu syncline also disappears. At the depth of 10 km, a high-V_P/V_S anomaly appears in the Baixiangyan-Yutan anticline.

From the horizontal and vertical velocity and V_P/V_S images, we can see that the Changning M_S6.0 earthquake is located in the transition zone between high- and low-velocity regions of P-wave and S-wave, and is located in relatively low V_P/V_S region. The Gongxian M_S5.3 earthquake and Xingwen M_S5.7 earthquake are also located at the transition area of the high- and low-velocity anomaly. Wang and Kao (2019) found that strong earthquakes usually occur at the edge of velocity anomaly

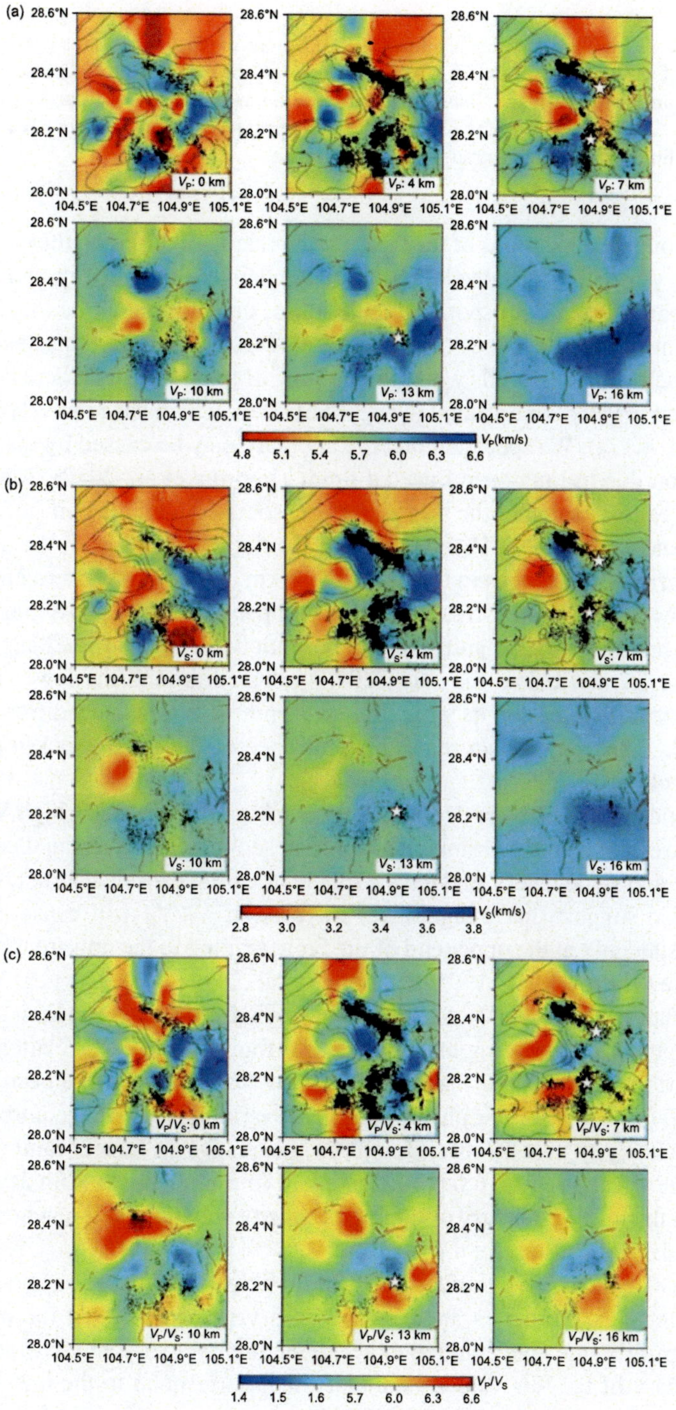

(a)
28.6°N
28.4°N
28.2°N
28.0°N
104.5°E 104.7°E 104.9°E 105.1°E
V_P: 0 km
V_P: 4 km
V_P: 7 km
V_P: 10 km
V_P: 13 km
V_P: 16 km
4.8 5.1 5.4 5.7 6.0 6.3 6.6
V_P(km/s)
(b)
V_S: 0 km
V_S: 4 km
V_S: 7 km
V_S: 10 km
V_S: 13 km
V_S: 16 km
2.8 3.0 3.2 3.4 3.6 3.8
V_S(km/s)
(c)
V_P/V_S: 0 km
V_P/V_S: 4 km
V_P/V_S: 7 km
V_P/V_S: 10 km
V_P/V_S: 13 km
V_P/V_S: 16 km
1.4 1.5 1.6 5.7 6.0 6.3 6.6
V_P/V_S

◂**Fig. 8.11** **a** V_P, **b** V_S and **c** V_P/V_S images at different depths. Black dots denote the vertical projection of earthquakes within 1.5 km on both sides of horizontal section. The white stars on the 7 km depth profile represent the Changning M_S6.0 earthquake and the Gongxian M_S5.3 earthquake, respectively. The white star on the 13 km depth profile represents the Xingwen M_S5.7 earthquake. Thin black lines indicate main stratigraphic boundaries

area. Although the locations of the three earthquakes are different, they all took place in the area where the crustal velocity structure changes dramatically, and the velocity gradient bands often correspond to fault zones, indicating that the crustal structure plays a controlling role in the occurrence of moderate-strong earthquakes in this area.

Right below the salt wells where Changning M_S6.0 is located, there is an obvious low-V_P and low-V_S zone around the depth of 4 km, which has relatively high V_P/V_S value (Fig. 8.12a). We speculate that the anomaly may be caused by the penetration or diffusion due to long-term injected fluids (Improta et al., 2017). Along the AA' profile, it can be seen that the seismicity is associated with a significantly high V_P, high V_S and very low V_P/V_S zone, which is relatively shallow at the southeastern end, and gradually deepens to the depth of 10 km at the northwestern end. The high V_S and low-V_P/V_S zone corresponds to the upper part of the pre-Sinian crystalline basement beneath the Changning-Shuanghe anticline system. The Changning M_S6.0 earthquake located below the salt wells is situated around the lower edge of the low-velocity anomaly, and its structural environment is obviously different from the aftershock. This may reflect that the occurrence of the main shock is induced by water injection.

We notice that at 0–2 km depth along the AA' profile there is a high V_P/V_S zone, which corresponds to the Ordovician–Silurian sedimentary layers that contain some mudstone and shale layers, and have a role in arresting earthquake rupture. Previous studies also suggest that a significant increase of V_P/V_S will cause shear forces decrease suddenly at the upper end of the fault, leading to the upward termination of the rupture (Thurber et al., 1995; Chiarabba et al., 2003). Doi et al. (2013) suggested that the relatively high V_P/V_S and low-velocity environment would limit the propagation of ruptures. Considering that no fault was found on the surface after Changning M_S6.0 earthquake, we infer that the Changning M_S6.0 earthquake been confined in the low-V_P/V_S area. Similar phenomenon that variation of V_P/V_S controls the depth of seismic activity has also been found on the south San Jacinto fault (Fang et al., 2019). The Changning M_S6.0 earthquake and four aftershocks with $M_S > 5$ are all located at the lower boundary of the high-velocity anomaly region, which is also the low boundary of earthquake distribution.

Figure 8.13 shows five profiles (BB'–FF') perpendicular to AA', and the distance between the five profiles is about 4 km. It is very obvious that the events along different profiles are concentrated either in the high-velocity anomaly area or along the boundary of the low-velocity anomaly, and correspond to the low-V_P/V_S area. Along the BB' profile through the M_S6.0 earthquake, the aftershocks are scattered. Along CC' and DD' profiles, the aftershocks are concentrated in a nearly vertical plane. On the EE' profile, the events are clearly distributed into two branches, with

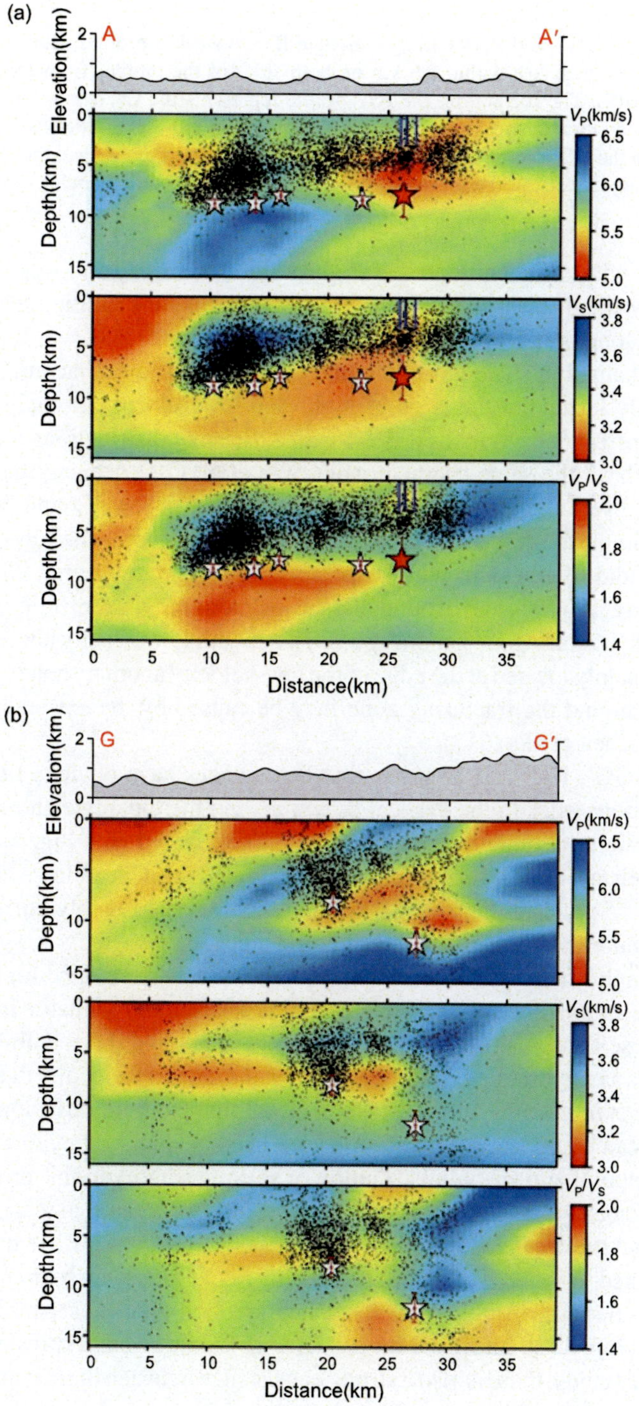
(a)
A
A′
Elevation(km)
Depth(km)
V_P(km/s)
6.5
6.0
5.5
5.0
V_S(km/s)
3.8
3.6
3.4
3.2
3.0
V_P/V_S
2.0
1.8
1.6
1.4
Distance(km)
(b)
G
G′
Elevation(km)
Depth(km)
V_P(km/s)
V_S(km/s)
V_P/V_S
Distance(km)

◂**Fig. 8.12 a–d** V_P, V_S and V_P/V_S images along different vertical profiles. Black dots are vertical projections of earthquakes within 2.5 km on both sides of the profile (from January 1, 2015 to August 31, 2019). The red star on the AA' profile represents the Changning $M_S6.0$ earthquake, white stars represent aftershocks with $M_S > 5$, and blue lines represent salt wells. The left and right white stars on the GG' profile represent the Xingwen $M_S5.7$ earthquake and the Gongxian $M_S5.3$ earthquake, respectively. Red error bars on $M_S > 5$ earthquakes mark the relocation errors

one branch with more earthquakes slightly tilted towards southwest and the other one slightly tilted towards northeast. The aftershocks distribution also dips to the southwest along the FF' profile.

In the Changning shale gas block, the GG' profile shows that small earthquakes are generally located at the edge of high- and low-velocity anomaly areas, which generally corresponds to the low-V_P/V_S area. Seismic distributions on the HH' and II' profiles have the same characteristics. Tan et al. (2020) found that earthquakes induced by hydraulic fracturing are usually located in low V_P and V_S zones, and the V_P/V_S in the earthquake source areas are relatively low. The low-V_P/V_S can be caused by fluid-filled cracks with large aspect ratio (Lin & Shearer, 2009; Tan et al., 2020). Chen et al. (2017) also found that the locations of microseismic events induced by hydraulic fracturing are generally associated with low velocity anomalies, and the events are mainly located at the edge of the low-velocity anomaly zones. Low velocity anomalies around the fracturing zone may be caused by the existence of fractures and fluids (Chen et al., 2017).

Profiles GG', HH' and II' show that earthquakes have obvious lineation distribution characteristics in the vertical direction, and the dip angle and strike of these seismic clusters are quite different (Fig. 8.12). Combined with Fig. 8.11, it is found that these seismic clusters are mainly located around the edge of low velocity anomalies and correspond to the low-V_P/V_S, and their strikes roughly correspond to the strikes of abnormal bodies. The relocated seismic clusters (Fig. 8.12) generally have large dip angles, which are consistent with the widely developed high angle faults or fractures in this area. We speculate that there are blind faults or fractures at the edge of these abnormal bodies. Tan et al. (2020) studied the relationship between earthquakes and hydraulic fracturing based on the detailed injection data of two shale gas well pads in the area, and the results reveal that the fluid diffusion by hydraulic fracturing can cause the reactivation of pre-existing faults. Previous studies show that the temporal and spatial distribution of seismic activity in this area is related to hydraulic fracturing (Lei et al., 2019b; Meng et al., 2019; Tan et al., 2020). Therefore, we infer that the frequent seismic activities in the Luochang-Jianwu syncline area may be related to the activation of pre-existing faults caused by hydraulic fracturing.

Previous research results show that water injection-induced seismicity is affected by local geological structure conditions (Bao & Eaton, 2016; Rubinstein & Mahani, 2015). In this study, there is also a strong consistency between the seismic distribution and the inverted crustal structure.

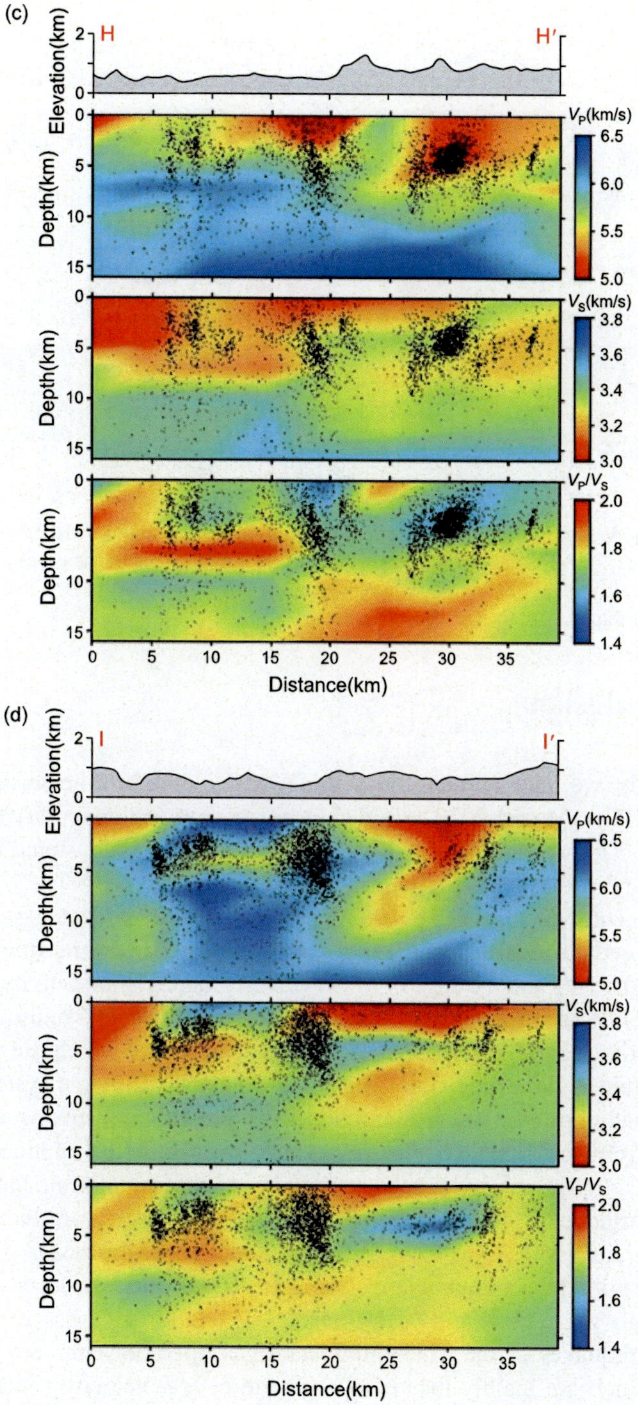

Fig. 8.12 (continued)

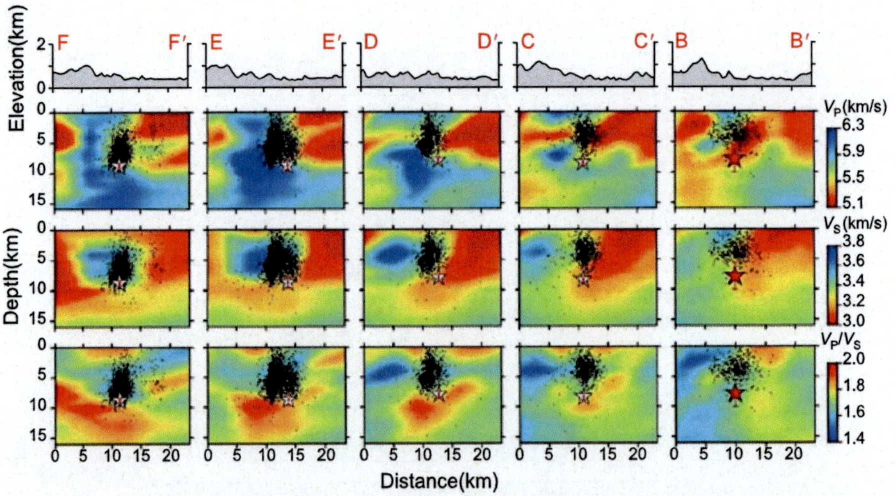

Fig. 8.13 V_P, V_S and V_P/V_S images along the profiles that perpendicular to the AA'. Black dots are the vertical projection of the earthquakes within 2 km on both sides of the profile after Changning M_S6.0 (denoted as red star). White stars represent the aftershocks with $M_S > 5$ and red error bars on $M_S > 5$ earthquakes mark the relocation errors

8.5 Conclusion

In this study, we have applied the TomoDDMC method to the seismic data from January 2015 to August 2019 recorded by permanent and temporary stations in the Changning-Xingwen area to obtain high-resolution three-dimensional V_P, V_S, V_P/V_S structures and earthquake locations in this area.

The epicenter of the Changning M_S6.0 earthquake is located at (104.894° E, 28.363° N) and the focal depth is 7.8 km, which is located in the transition zone of high and low V_P and V_S and the edge of low-V_P/V_S region. Seismic activity in Shuanghe-Yutan anticline area before the Changning M_S6.0 earthquake outlined several NE trending stripes, implying pre-existing small-scale faults which are perpendicular to major faults (NW-striking) in Changning-Shuanghe anticline system. We found that the Changning M_S6.0 broke through these pre-existing small-scale faults, and extended from the Shuanghe to Yutan anticline. Both the rupture process and aftershock activity were influenced by the pre-existing small-scale faults.

Most Earthquakes within the Changning salt mining area are located in a slant zone that gradually deepens from Shuanghe anticline on the east to Yutan anticline on the west with the maximum depth from 5 to 10 km. There are also associated with obvious high-V_S and low-V_P/V_S features.

The earthquakes in the Changning shale gas operation area are distributed in clusters, which are mainly located at the edge of low velocity anomalies and are associated with low-V_P/V_S areas. Combined with geological tectonic settings and previous research results, we infer that the frequent seismic activities in Luochang-Jianwu syncline area may be related to the activation of pre-existing faults or fractures caused by hydraulic fracturing process.

References

Bao, X., & Eaton, D. W. (2016). Fault activation by hydraulic fracturing in western Canada. *Science, 354*(6318), 1406–1409.

Chen, Y., Zhang, H., Miao, Y., Zhang, Y., & Liu, Q. (2017). Back azimuth constrained double-difference seismic location and tomography for downhole microseismic monitoring. *Physics of Earth Planetary Interiors, 264*, 35–46. https://doi.org/10.1016/j.pepi.2016.10.003

Chiarabba, C., & Amato, A. (2003). V_P and V_P/V_S images in the M_W 6.0 Colfiorito fault region (central Italy): A contribution to the understanding of seismotectonic and seismogenic processes. *Journal of Geophysics Research, 108*(B5). https://doi.org/10.1029/2001JB001665

Deng, Q., Zhang, P., Ran, Y., Yang, X., Min, W., & Chu, Q.. (2003). Basic characteristics of active tectonics of China. *Science of China, Series D: Earth Science, 46*(4), 356–372.

Doi, I., Noda, S., Iio, Y., Horiuchi, S., & Sekiguchi, S. (2013). Relationship between hypocentral distributions and V_P/V_S ratio structures inferred from dense seismic array data: A case study of the 1984 western Nagano Prefecture earthquake, central Japan. *Geophysical Journal International, 195*, 1323–1336.

Eberhart-Phillips, D. (1986). Three-dimensional velocity structure in northern California Coast Ranges from inversion of local earthquake arrival times. *Bulletin of the Seismological Society of America, 76*(4), 1025–1052.

Eberhart-Phillips, D. (1990). Three-dimensional *P* and *S* velocity structure in the Coalinga region, California. *Journal of Geophysical Research, 95*(B10), 15343–15363.

Efron, B., & Gong, G. (1983). A leisurely look at the bootstrap, the jackknife, and cross-validation. *American Statistician, 37*(1), 36–48.

Efron, B., & Tibshirani, R. (1991). Statistics data analysis in the computer age. *Science, 253*, 390–395.

Fang, H., Yao, H., Zhang, H., Thurber, C., Ben-Zion, Y., & van der Hilst, R. D. (2019). V_P/V_S tomography in the southern California plate boundary region using body and surface wave traveltime data. *Geophysical Journal International, 216*, 609–620.

Guo, H., & Zhang, H. (2017). Development of double-pair double difference earthquake location algorithm for improving earthquake locations. *Geophysical Journal International, 208*, 333–348.

Guo, H., Zhang, H., & Froment, B. (2018). Structural control on earthquake behaviors revealed by high-resolution V_P/V_S imaging along the Gofar transform fault, East Pacific Rise. *Earth Planet Science Letter, 499*, 243–255.

He, D., Lu, R., Huang, H., Wang, X., Jiang, H., & Zhang, W. (2019). Tectonic and geological setting of the earthquake hazards in the Changning shale gas development zone, Sichuan Basin, SW China. *Petroleum Exploration and Development, 46*(5), 1051–1064.

Huang, H. H., Wu, Y. M., Song, X., Chang, C. H., Lee, S. J., Chang, T. M., & Hsieh, H. H. (2014). Joint Vp and Vs tomography of Taiwan: Implications for subduction-collision orogeny. *Earth Planet Science Letter, 392*, 177–191.

Humphreys, E., & Clayton, R. W. (1988). Adaptation of back projection tomography to seismic travel time problems. *Journal of Geophysical Research, 93*(B2), 1073–1085.

Improta, L., Bagh, S., De Gori, P., Valoroso, L., Pastori, M., Piccinini, D., Chiarabba, C., Anselmi, M., & ButtinelliButtinelli, M. (2017). Reservoir structure and wastewater-induced seismicity at the Val d'Agri oilfield (Italy) shown by three-dimensional V_P and V_P/V_S local earthquake tomography. *Journal of Geophysical Research, 122*(11), 9050–9082. https://doi.org/10.1002/2017jb014725

Lei, X., Huang, D., Su, J., Jiang, G., Wang, X., Wang, H., Guo, X., & Fu, H. (2017). Fault reactivation and earthquakes with magnitudes of up to Mw4.7 induced by shale-gas hydraulic fracturing in Sichuan Basin, China. *Science of Reports, 7*, 7971. https://doi.org/10.1038/s41598-017-08557-y

Lei, X., Wang, Z., & Su, J. (2019a). Possible link between long-term and short-term water injections and earthquakes in salt mine and shale gas site in Changning, south Sichuan Basin, China. *Earth Planetary Physics, 3*(6), 510–525.

Lei, X., Wang, Z., & Su, J. (2019b). The December 2018 M_L 5.7 and January 2019 M_L 5.3 earthquakes in south Sichuan Basin induced by shale gas hydraulic fracturing. *Seismological Research Letters, 90*(3), 1099–1110.

Lin, G., & Shearer, P. M. (2009). Evidence for water-filled cracks in earthquake source regions. *Geophysical Research Letters, 36*, L17315. https://doi.org/10.1029/2009GL039098

Liu, S., Tang, C. C., Chen, C. H., & Xu, R. (2019). Spatiotemporal evolution of the 2018 M_W 6.4 Hualien earthquake sequence in Eastern Taiwan. *Seismological Research Letters, 90*(4), 1446–1456.

Long, F., Zhang, Z., Qi, Y., Liang, M., Ruan, X., Wu, W., Jiang, G., & Zhou, L. (2020). Three dimensional velocity structure and accurate earthquake location in Changning-Gongxian area of southeast Sichuan. *Earth Planetary Physics, 4*(2), 163–177.

Ma, X., Xie, J., & Yong, R. (2020). Geological characteristics and high production control factors of shale gas in Silurian Longmaxi Formation, southern Sichuan Basin, SW China. *Petroleum Exploration and Development, 47*(5). (in Chinese with English abstract). https://doi.org/10.11698/PED.2020.05.00

Meng, L., McGarr, A., Zhou, L., & Zang, Y. (2019). An investigation of seismicity induced by hydraulic fracturing in the Sichuan Basin of China based on data from a temporary seismic network. *Bulletin of the Seismological Society of America, 109*(1), 348–357.

Pang, J., Ding, X., Han, K., Zeng, Y., Chen, A., Zhang, Y., Zhang, Q., & Yao, D. (2017). The national 1:1000000 geological map spatial database. *Global Geology Data*. https://doi.org/10.23650/data.H.2017.NGA105570.T1.64.1

Ruan, X., Cheng, W., Zhang, Y., Li, J., & Chen, Y. (2008). Research of the earthquakes induced by water injections in salt mines in Changning, Sichuan. *Earthquake Research China, 24*(3), 226–234. (in Chinese with English abstract).

Rubinstein, J. L., & Mahani, A. B. (2015). Myths and facts on wastewater injection, hydraulic fracturing, enhanced oil recovery, and induced seismicity. *Seismological Research Letters, 86*(4), 1060–1067.

Shi, X., Tong, Y., Liu, W., Zhao, C., Liu, J., & Fang, J. (2019). Analysis of seismic-scale fracture system of shale reservoir and its petroleum significance: A case study of well Ning 201 area of Changning Block, Sichuan Basin. *Marine Origin Petrleum Geology, 24*(4), 87–96. (in Chinese with English abstract).

Sun, X., Yang, P., & Zhang, Z. (2017). A study of earthquakes induced by water injection in the changning salt mine area, SW China. *Journal of Asian Earth Sciences, 136*, 102–109.

Tan, Y., Hu, J., Zhang, H., Chen, Y., Qian, J., Wang, Q., Zha, H., Tang, P., & Nie, Z. (2020). Hydraulic fracturing induced seismicity in the southern Sichuan Basin due to fluid diffusion inferred from seismic and injection data analysis. *Geophysics Reseach Letters, 47*(4). https://doi.org/10.1029/2019GL084885

Tang, C. C., Lin, C. H., & Peng, Z. (2014). Spatial-temporal evolution of early aftershocks following the 2010 M_L 6.4 Jiashian earthquake in southern Taiwan. *Geophysical Journal International, 199*, 1772–1783.

Thurber, C. H., Atre, S. R., & Eberhart-Phillips, D. (1995). Three-dimensional Vp and Vp/Vs structure at Loma Prieta, California, from local earthquake tomography. *Geophysical Research Letters, 22*(22), 3079–3082.

Um, J., & Thurber, C. (1987). A fast algorithm for two-point ray tracing. *Bulletin of the Seismological Society of America, 77*(3), 972–986.

Wang, Y. M., Huang, J. L., Wang, S. F., Dong, D. Z., Zhang, C. C., & Guan, Q. Z. (2016). Dissection of two calibrated areas of the Silurian Longmaxi formation, changning and Jiaoshiba, Sichuan Basin. *Natural Gas Geoscience, 27*(3), 423–432. (in Chinese with English abstract).

Wang, Z., & Kao, H. (2019). The significance of tomographic edge zones for large earthquakes in Taiwan. *Journal of Geophysical Research, 124*(11), 11822–11839.

Wessel, P., & Smith, W. H. F. (1998). New, improved version of generic mapping tools released. *EOS, Transactions of the American Geophysical Union, 79*(47), 579.

Yi, G. X., Long, F., Zhang, Z. W., & Ruan, X. (2013). The relationship between earthquakes in Changning window and M≥5.5 earthquakes in Sichuang and its adjacent area. *Journal of Seismological Research, 36*(4), 427–432. (in Chinese with English abstract).

Yi, G. X., Long, F., Liang, M. J., Zhao, M., Wang, S. W., Gong, Y., Qiao, H. Z., & Su, J. R. (2019). Focal mechanism solutions and seismogenic structure of the 17 June 2019 M_S6.0 Sichuan Changning earthquake sequence. *Chinese Journal of Geophysics, 62*(9), 3432–3447. (in Chinese with English abstract).

Zhang, H., & Thurber, C. H. (2003). Double-difference tomography: The method and its application to the Hayward fault, California. *Bulletin of the Seismological Society of America, 93*(5), 1875–1889.

Zhang, H., & Thurber, C. H. (2006). Development and applications of double-difference seismic tomography. *Pure and Applied Geophysics, 163*, 373–403.

Zhang, H., & Thurber, C. H., & Bedrosian, P. (2009). Joint inversion for Vp, Vs, and Vp/Vs at SAFOD, Parkfield, California. *Geochemistry, Geophysics, Geosystem, 10*(11). https://doi.org/10.1029/2009GC002709

Zhang, Z. W., Long, F., Wang, S. Y., Gong, Y., Wu, P., Wang, H., & Wang, G. M. (2019). Earthquake location and velocity structure in Yibin Area, Sichuan. *Seismological Geology, 41*(4), 913–926. (in Chinese with English abstract).

Zhao, D., Hasegawa, A., & Horiuchi, S. (1992). Tomographic imaging of *P* and *S* wave velocity structure beneath northeastern Japan. *Journal of Geophysical Research, 97*(B13), 19909–19928.

Chapter 9
Estimated Casualty Risk for Disaster Preparation in Five Scenario Great Earthquakes, Sichuan-Yunnan Region, China

Jiawei Li, Max Wyss, Zhongliang Wu, and Shiyong Zhou

Abstract For earthquake disaster preparation and risk reduction, quantitative estimates of casualties caused by potential future ruptures is crucial. We estimate the numbers of casualties and the population strongly affected for five "worst-case scenarios" of devastating earthquakes in the Sichuan-Yunnan region, China. These earthquake scenarios are assumed to rupture the entire segments of the Xianshuihe, Anninghe, Zemuhe, Daliangshan and Xiaojiang faults. The source parameters of the scenario earthquakes are derived from fault models. Before calculating scenario losses however, we validate the algorithms and data sets of Quake Loss Assessment for Response and Mitigation (QLARM) for the 13 earthquakes of the mainland of China since 2003, which have caused fatalities, by matching observed intensities and casualty counts with calculated values. These near real-time fatality estimates are reviewed and found to have been correct and, for people injured, six are 100% correct and three have been within a factor of 2.5. We use an average attenuation and assume that earthquakes occur at 03:00 local time, which expects assumes maximum occupancy of buildings. The greatest disaster is derived from the scenarios of the Xiaojiang fault rupture, where we calculate the possible numbers of fatalities on average as 286,000 and injured as 751,000. The scenario of the maximum rupture along the Daliangshan fault results in approximately twice the casualties reported in

Supplementary Information The online version contains supplementary material available at (https://doi.org/10.1007/978-981-16-8607-8_9).

J. Li (✉)
Institute of Geophysics, China Earthquake Administration, Beijing, China
e-mail: lijw@cea-igp.ac.cn; lijw3@sustech.edu.cn

Institute of Risk Analysis, Prediction and Management, Academy for Advanced Interdisciplinary Studies, Southern, University of Science and Technology,, Shenzhen, China

M. Wyss
International Centre for Earth Simulation Foundation, Geneva, Switzerland

Z. Wu
Institute of Earthquake Forecasting, China Earthquake Administration, Beijing, China

J. Li · S. Zhou
Department of Geophysics, School of Earth and Space Sciences, Peking University, Beijing, China

© Higher Education Press 2022

Y.-G. Li et al. (eds.), *China Seismic Experimental Site*,
https://doi.org/10.1007/978-981-16-8607-8_9

the 2008 M_W 7.9 Wenchuan earthquake. The three scenarios along the Xianshuihe, the Anninghe and the Zemuhe faults result in estimates of casualties similar to those in the Wenchuan case. These estimates are significant not only for disaster preparation and relief by the government, but also for risk prevention and mitigation by the public.

Keywords Earthquake casualty risk · Human loss estimates · Sichuan-Yunnan region · China · Earthquake disaster preparation

9.1 Introduction

Quantitative estimates for casualty risk caused by potential future earthquakes are valuable for disaster preparation and risk reduction. The earliest tools for estimating losses due to earthquakes were EXTREMUM (Shakhramanjyan et al., 2001) and HAZUS (Federal Emergency Management Agency, 2003), the first of these responded to worldwide events, the second to USA earthquakes only. The tool we used, Quake Loss Assessment for Response and Mitigation (QLARM), is a second generation developed in part based on EXTREMUM, with elements of HAZUS incorporated. A summary of various approaches to earthquake risk estimates is presented by Erdik et al. (2014) and Wyss (2014). Many of these publications address a specific country, a region or a single city. Worldwide near real-time earthquake loss estimates for large events (e.g., M $\geq$ 5 for Europe and China) are provided by PAGER since 2006 (Jaiswal & Wald, 2010; Jaiswal et al., 2011) and QLARM since October 2002 (Wyss, 2014).

Near real-time loss estimates are based on earthquakes, that have happened and are deterministic. Estimates of monetary losses for insurance purposes are probabilistic as a concept (e.g. Salgado-Galvez et al., 2016a, 2016b, 2017). The quantitative scenario loss estimates we present here are deterministic, based on assumed ruptures of active faults with histories of large to great earthquakes in the Sichuan and Yunnan provinces of China.

A description of the engine of QLARM can be found in the *Appendix*. Based on QLARM, loss scenarios have been published for the Himalayan plate boundary (Wyss, 2005, 2006), Romania (Trendafiloski et al., 2009), Algeria (Rosset & Wyss, 2017; Rosset et al., 2009), Indonesia (Wyss, 2010), Chile (Wyss, 2010), Mexico (Wyss & Zuñiga, 2016), Nepal (Wyss, 2017; Wyss & Chamlagain, 2018, India (Wyss et al., 2018),Italy (Wyss and Rosset, 2020), Kyrgyzstan (Rosset et al., 2020), the Azores (Fontiela et al., 2020; Wyss et al., 2018), and Colombia (Wyss et al., 2021).

Here, we select five faults with high seismic risk in the Sichuan-Yunnan region, China, and construct "worst-case scenarios" of devastating earthquakes, although even greater earthquakes are possible along these faults. Before the calculation for scenario losses, we validate the algorithms and data sets of the QLARM software using the 13 earthquakes, which have recently caused fatalities on the Chinese mainland, by matching observed intensities and casualty counts with calculated ones.

All our calculations for scenarios are subject to assumptions and are affected by uncertainties, thus they should be considered order of magnitude estimates only. The assumptions we set forward are based on the validation work, and listed below:

(1) the fault zones are supposed to host the largest ruptures we can expect according to the length of the investigated fault segments;
(2) the energy radiated is assumed to come from lines with a depth of 15 km;
(3) we propose to use the attenuation of $C_3 = 3.9$ (refer to Eq. 1 in the *Appendix*);
(4) the occurrence time is assumed as 03:00 local time, which means maximum occupancy of buildings;
(5) the losses are calculated for 2020, and the regional population growth rate of the past few years is from the National Bureau of Statistics of China (see Data and Resources).

Approximately correct estimates of fatalities in near real-time (minutes after earthquakes) and good matches of earthquake casualties and observed intensities by theoretical values using line sources performed in this paper, give us confidence that our scenario results are realistic.

9.2 Validation of QLARM for the Mainland of China

We wish to validate the algorithms and data sets in QLARM for the mainland of China by using the final parameters for each earthquake for which intensity and fatality information exist since 2003. Theoretical intensities and numbers of casualties for the 13 earthquakes listed in Table 9.1 have been computed for comparison with observed values. Because QLARM is constructed to use the standard USGS M values, we therefore used the M_W from the U.S. Geological Survey (USGS; see *Data and Resources*) for our calculations.

We trust the epicenters provided by the China Earthquake Networks Center (CENC) rather than those by the USGS because they are derived from arrival times in regional networks. In addition, we consider macroseismic information for refining the epicenters. When the instrumental epicenters are offset from the center of the isoseismals, we assume that the difference is due to errors in the instrumental solution and use the latter for loss calculations because it represents the center of energy radiated.

Both of the USGS moment tensor depths and those based on the data from the regional networks are considered. However, depth estimates are uncertain, thus we use intensity observations to refine the depths, unless detailed source models with slip as a function of depth are available. Therefore, the depths shown in Table 9.1 are derived from the best fit of intensities calculated by QLARM matching the reported isoseismals.

Earthquake ruptures can be approximated as point sources when the magnitude is smaller than 6.5. For larger ruptures and cases with elongated isoseismal patterns, we assume that energy radiates from line sources. We estimate the lengths and widths of

Table 9.1 Earthquakes in the mainland of China with near real-time alerts from January 1, 2003 to April 1, 2017

Name	yyyy/mm/dd	hh:mm:ss	Lon	Lat	Depth	M_W	C_3
Dayao	2003/07/21	15:16:31	101.16	25.95	11	6.0	3.5
Zhaosu	2003/12/01	01:38:31	80.47	42.96	21	6.0	4.4
Wenchuan	2008/05/12	06:28:01	104.38	31.90	15	7.9	3.7
Haixi	2009/08/28	01:52:06	95.78	37.64	20	6.3	3.5
Yushu	2010/04/13	23:49:38	96.85	33.11	15	6.9	3.5
Xinyuan-Hejing	2012/06/29	21:07:33	84.69	43.32	25	6.3	4.5
Lushan	2013/04/20	00:02:47	102.97	30.25	26	6.6	4.0
Minxian-Zhangxian	2013/07/21	23:45:56	104.14	34.55	18	5.9	4.5
Ludian	2014/08/03	08:30:13	103.38	27.08	12	6.2	4.1
Jinggu	2014/10/07	13:49:39	100.50	23.37	18	6.1	4.1
Pishan	2015/07/03	01:07:47	78.15	37.56	20	6.4	4.0
Zaduo	2016/10/17	07:14:50	94.96	32.83	15	5.9	3.5
Arketao	2016/11/25	14:24:30	74.18	39.22	19	6.6	3.5

The epicenter coordinates and the depths are those used here for intensity calculations, M_w is the USGS magnitude, C_3 controls the attenuation according to Shebalin (1985). Times are UTC

the rupture from M, using established relations (Wells & Coppersmith, 1994; Wyss, 1979). In Table 9.2 we give the end-points (X_i, Y_i), dip, width (W), length (L) and depth (Z) of ruptures for $M \geq 6.5$ earthquakes in Table 9.1. In the case of line sources, the epicenter used for the calculation is artificially placed at the center of the rupture

Table 9.2 Fault model parameters of the $M \geq 6.5$ arthquakes given in Table 9.1

Name	Epi Lat	Epi Lon	Z	L	W	Dip	Y_1	X_1	Y_2	X_2	From
Wenchuan	31.90	104.38	15	305	25	50	31.10	103.45	32.70	105.31	Zhang et al. (2009)
Yushu	33.11	96.85	15	30	20	80	33.20	96.73	33.02	96.97	Zhang et al. (2010)
Lushan	30.24	102.97	26	25	30	30	30.14	102.93	30.35	103.01	Zhang et al. (2013)
Arketao	39.22	74.18	19	30	10	80	39.27	73.99	39.18	74.38	see *Data and Resources*

X_i and Y_i are the assumed end-points of the respective line ruptures. The epicenter is placed at the center of the line rupture to provide symmetrical calculations of intensities. The last column in Table 9.2 lists the detailed source parameters of the fault models.

We match the reported isoseismals by calculated intensities adjusting the depths (Z) and the attenuation (C_3 in the Eq. 1 of *Appendix*) when necessary. The best fitting depth is obtained once the intensity points calculated by QLARM for that depth meet the following criteria:

(1) I_{max}(calculated) equals I_{max}(reported);
(2) the distributions for calculated intensity ($\geq$ VI) points match the reported isoseismals well at high intensities, and within a deviations of 40–60 km in parts of the area shaken less strongly;
(3) the calculated casualties (fatalities and injured) approximately equal the reported numbers of fatalities and injured (Table 9.3).

We give max, middle and min weights to the criteria (1), (2) and (3), respectively. The number of reported casualties comes from the China Earthquake Administration (CEA) and news media.

The fractions of buildings and population in them used for the three size classes of settlements and the building classes are given in Table 9.4. We have modeled settlements in the mainland of China in three size classes, < 6000 inhabitants, between 6000 and 50,000, and > 50,000 in the data set of QLARM. These fractions of building classes that make up 100% of the buildings in one settlement vary in different countries. The European Macroseismic Scale (EMS-98) defines classes ranging from the weakest (A) to the most resistant building (F). This first order approximation will be refined in the future. At present, we accept this model for buildings and inhabitants, under the condition that the resulting validation results are correct. Although these data can and should be further refined, they are accepted in this study, because the overall validation results are correct. The degree of damage and the human losses are calculated for each building type in a settlement, based on the distributions of building stock and population of Table 9.4 and the intensity predicted using Eq. 1 of the Appendix.

The intensities reported in China are on a different scale from MMI (modified Mercalli intensity) used by QLARM. However, this difference is small (Hu, 2009; Wan, 2016; Yuan & Tian, 2012), thus we deem the Chinese intensity scale is approximately the same as MMI. Furthermore, intensities are usually reported as integers without fractions, so we define bands of intensities calculated by QLARM as rounded to the nearest integer.

In order to render this paper more readable, we explain in detail the validation of the 2003 Dayao earthquake as a point source, and the 2008 Wenchuan earthquake as well as the 2014 Jinggu earthquake as a line sources, but we refer the reader to the section on *Supplementary Material* for the validation of the remaining 9 earthquakes.

Table 9.3 The number of fatalities and injured reported compared to those calculated in near-real time using point sources and those in this paper using line sources for the earthquakes given in Table 9.1

Name	WAPMERR/ICES Foundation		This paper		Reported casualties	
	Fat^1(est)	Inj^1(est)	Fat^2(est)	Inj^2(est)	Fat^3	Inj^3
Dayao	95–260	–	30–370	150–1,000	16	584
Zhaosu	2–7	7–40	10–110	40–360	10	47
Wenchuan	40,000–100,000	95,730–414,259	55,000–140,000	146,000–378,000	87,149	374,643
Haixi	0–10	0–40	0–30	10–120	0	?
Yushu	200–4,000	500–10,000	450–3,000	2,000–10,000	2,968	10,701
Xinyuan-Hejing	0–50	10–200	30–350	130–1,000	0	52
Lushan	200–2,300	1,000–7,700	210–3,000	1,000–11,000	217	13,484
Minxian-Zhangxian	0–300	100–1,000	130–2,000	640–6,000	95	2,414
Ludian	80–635	316–2,233	190–1,000	650–4,000	617	3,143
Jinggu	0–400	100–1,400	70–1,000	390–4,000	1	331
Pishan	0–200	50–500	50–480	190–2,000	3	260
Zaduo	0–0	0–0	0–10	0–40	1	?
Arketao	0–50	20–170	10–170	60–600	1	?

Table 9.4 Distribution of building stock by EMS-98 classes and city sizes for the region of China studied

Settlement size	Urban (> 50,000 inh.)				Mid-urban (6000–50,000 inh.)				Rural (< 6000 inh.)		
EMS-98 classes	A	B	C	D	A	B	C	D	A	B	C
Building distribution	0.05	0.10	0.70	0.15	0.20	0.65	0.10	0.05	0.5	0.35	0.15
Population distribution	0.05	0.10	0.70	0.15	0.20	0.65	0.10	0.05	0.5	0.35	0.15

9.2.1 The Dayao Earthquake of 21 July 2003

For the 21 July 2003, Dayao earthquake located in the center of Yunnan Province, the magnitude was $M_S6.2/M_W6.0$, released by the CEA and USGS, respectively. We used a point source model for this earthquake. The isoseismal map for that earthquake is from Fei et al (2004), with I_{max}(reported) equal to VIII. The I_{max}, spatial distribution of intensities and the number of casualties match best when using a depth of 11 km (Fig. 9.1). The distance by which we adjusted the epicenter from CENC to the center of the isoseismals is about 7 km.

I_{max}(calculated) is VIII and the estimated ranges of fatalities and injured are 30–370 and 150–1000, respectively (Table 9.3). The ranges of the estimation of fatalities and injured are the formal limits for the result being within 95% probability. We take this wide range because there are additional uncertainties, such as precise location of the scenario earthquake, precise magnitude M, precise attenuation, and spectral distribution of radiated energy. The calculation for the mean degree of damage and the human losses depends on intensity I as elaborated by Trendafiloski et al. (2011).

The intensiy estimates, which depend on human perception, disturbance of objects, damage to buildings/structures and disruptions of the Earth's surface, are determined by macroseismic intensity surveys by the CEA. The site conditions and the type of buildings/structures have a significant influence on the isoseismals. Concerning these parameters, we have no relevant knowledge and cannot simulate the effects.

The model of QLARM that we preferred for the Dayao earthquake balances the difference between the observations and the calculations and makes them acceptable for each parameter. The parameters used and the constraints obeyed in the model we used for this earthquake and the subsequent ones (Table 9.1) are as follows.

(1) The depth we use is based on the results by CENC and the moment tensor solution of USGS, and is refined by intensity observations, unless seismic detailed source models are available.
(2) The occupancy rate is computed based on local time.
(3) The attenuation function used in our calculation is an average for the world, using the constant $C_3 = 3.5$ (Shebalin, 1985; *Appendix*) when we have no

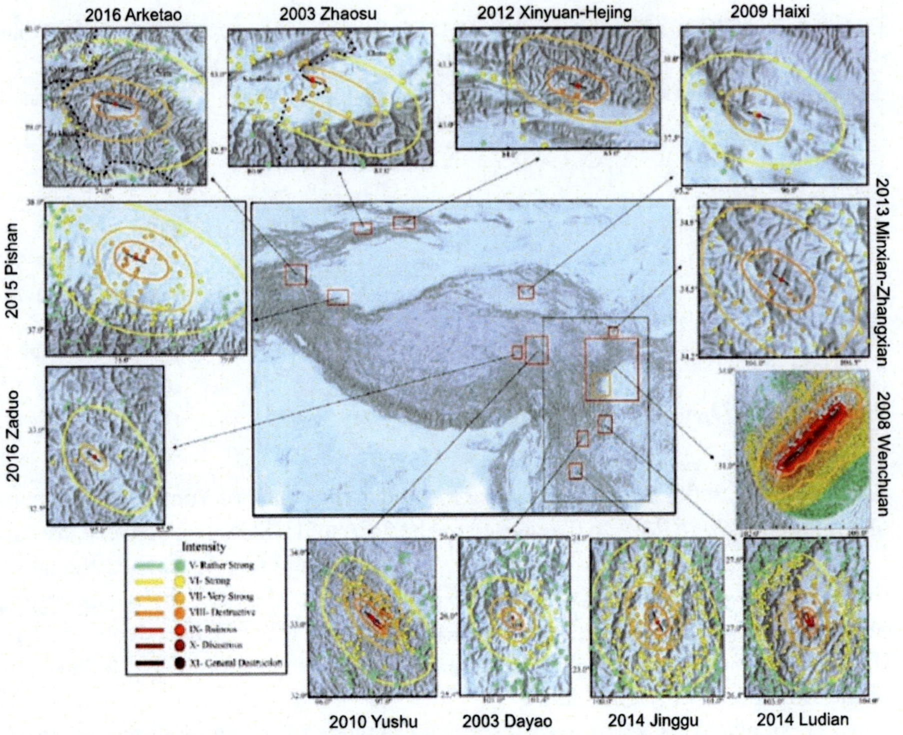

Fig. 9.1 Map of isoseismals and intensities calculated by QLARM (colored dots) for the events in Tables 9.1 and 9.2. The hexagrams indicate the macroscopic epicenters and the gray solid lines depict the line sources we used for calculations. Parameters used by QLARM for calculation are given in Tables 9.1 and 9.2. The yellow rectangle of the indexing figure shows the 2013 Lushan earthquake, in Sichuan province. The black rectangle shows the region of Fig. 9.3

other information. If the observed distribution of intensities demands it, we have used different factors for C_3 and discuss these in the cases concerned.

(4) The population data of settlements of QLARM in Sichuan-Yunnan region are from 2010. The number of computed fatalities given is corrected for the respective earthquake dates, using the annual population growth rate for China released by the National Bureau of Statistics of China (see *Data and Resources*).

(5) The farther a settlement is from the epicenter, the less reliable and significant is its comparison of reported with calculated intensities. So, for the matches we used mostly the settlements within the fields of high intensities.

9.2.2 The Wenchuan Earthquake of 12 May 2008

For the 12 May 2008, Wenchuan $M_S8.0/M_W7.9$ earthquake of Sichuan province, we used the isoseismals (Fig. 9.1) released by the CEA and collected by Chen and Booth

(2011). The isoseismal I_{max} equals XI and is elongated in the ENE direction. The highest intensities occurred in two separate areas, namely Wenchuan and Beichuan counties. On 25 August 2008, 87,149 people were reported killed (we consider 17,923 missing people to be dead) and at least 374,643 were injured (Table 9.3).

The spatio-temporal rupture process of the Wenchuan earthquake has been studied by several researchers e.g., (Ji, 2008; Wang et al., 2008; Du et al., 2009; Zhang et al., 2009; Huang et al., 2012; Source model of the Wenchuan rupture published by finite-source rupture model database of Source Inversion Validation project, see *Data and Resources*; Source model of this event released by USGS, see *Data and Resources*). However, in order to ensure consistency of the source models with other events used here, we only used results provided by Zhang Y. et al. (2009) for the Wenchuan earthquake .

We used a line source with a length of 305 km of which the end-points are given in Table 9.2. It is placed along the surface break and approximately within the center of the simplified source rectangle, which is based on asperities, and has an overall rupture width of 25 km and a dip of 50º. To evenly cover the area affected by strong shaking, we artificially placed the epicenter at the center of the line source. This artificial location is not the initial rupture point at the southern end.

The I_{max} and distribution of intensities of QLARM match the reported ones best if we assume a depth of 15 km (Fig. 9.1). The two sides of the rupture (mountains versus plains) have different attenuation, so we need to find a compromise. An attenuation with $C_3 = 3.7$ was used because the fields for IX and VIII are relatively matched within a deviation of 40 km and 60 km in this case. No model that is otherwise reasonable results in intensity XI, so we accept the calculated $I_{max} = X$, understanding that amplitudes of seismic waves are amplified by soil conditions locally. The ranges of the calculated number of fatalities and injured are 55,000–140,000 and 146,000–378,000, respectively (Table 9.3).

9.2.3 The Jinggu Earthquake of 7 October 2014

For the Jinggu $M_S6.9/M_W6.1$ earthquake of 7 October 2014 in Yunnan province, the isoseismal plot is shown in Fig. 9.1, and I_{max} equals VIII. The moment magnitude and focal mechanism of the Jinggu earthquake are similar to these of the 2014 Ludian earthquake. However, the damage caused by the Ludian and Jinggu earthquakes are different. Only 1 person in Jinggu, but 617 in Ludian were reported dead. The I_{max} values were not equal in the two cases and the number of people affected in the earthquake with more fatalities is higher.

The difference in damage may also have been due to a difference in the energy radiated (Choy & Boatwright, 2009; Choy et al., 2001; Song et al., 2016). Song et al (2016) reported different magnitudes for the Jinggu and the Ludian earthquakes. It can be seen from their results that although the M_W of the Jinggu earthquake approximately equals that of the Ludian earthquake, there are differences between them in the energy magnitude M_e and the broadband body-wave magnitude m_{Bc}, respectively. The differences in both M_e and m_{Bc} between the two earthquakes are about 0.3, meaning that the energy radiated at high frequencies is lower by a factor of 3 in the case of Jinggu earthquake compared to Ludian earthquake (Choy & Boatwright, 1995). This difference in the source spectra can explain partially the difference in the losses.

For the model calculation, we moved the epicenter from CENC to the macroepicenter. The adjusted distance is about 6 km. Owing to the isoseismals being elongated in the ESE direction, we assume that the end-point coordinates of this line source with length of 13 km are 23.44° N, 100.46° E and 23.30° N, 100.55° E. The match is good when we increase the depth of the hypocenter from those of CENC and USGS to 18 km along with $C_3 = 4.1$ (Fig. 9.1).

The calculated ranges of casualties, 70–1000 and 390–4000, respectively (Table 9.3), are overestimated. We propose that weak energy radiation in the Jinggu earthquake, as shown by the magnitudes sensitive to high frequencies, was the main cause of fewer casualties than expected.

9.2.4 *Loss Estimation for an Aftershock of the Second Kind: The Lushan Earthquake of 20 April 2013*

The epicenter of the Lushan earthquake, which happened about 5 years after the Wenchuan earthquake, was located 90 km southwest of the latter rupture (Wu et al., 2014). On 20 April 2013, the Lushan $M_S7.0/M_W6.6$ earthquake occurred in Sichuan province. The isoseismals (Fig. 9.2) and the losses (Table 9.3) have been published by the CEA (see *Data and Resources*). The isoseismals with I_{max} = IX are elongated in a NEN direction. The State Council of China released the following loss report on 24 April 2013: 217 deaths (196 deaths and 21 missing) and at least 13,484 injured.

According to the slip model of Zhang et al (2013), we used rupture end-points which are 30.14° N, 102.93° E and 30.35° N, 103.01° E to define a line source. The source properties are: length of 25 km, width of 30 km and dip of 30°. This calculation of casualties shows the smallest differences from the reported casualties when we used a depth of 26 km along with $C_3 = 4.0$. The distance by which we moved the epicenter from the CENC to the macroepicenter is about 6 km. The estimated number of fatalities and injured are 210–3000 and 1000–11,000, respectively (Table 9.3).

QLARM operators do not calculate damage and casualties for regular aftershocks of large earthquakes because the main shock has changed the properties of buildings in the area by eliminating weak ones (collapse) and by weakening strong ones (damage). In addition, people are likely staying in tents, which means the occupancy rate is different than normal. Therefore, calculations of consequences of aftershocks are meaningless. People who distribute such calculations based on automatic estimations are misleading the local population and rescuers.

However, the Lushan case is different from a regular aftershock. We introduce the term "aftershock of the second kind" for loss risk estimates of the Lushan earthquake

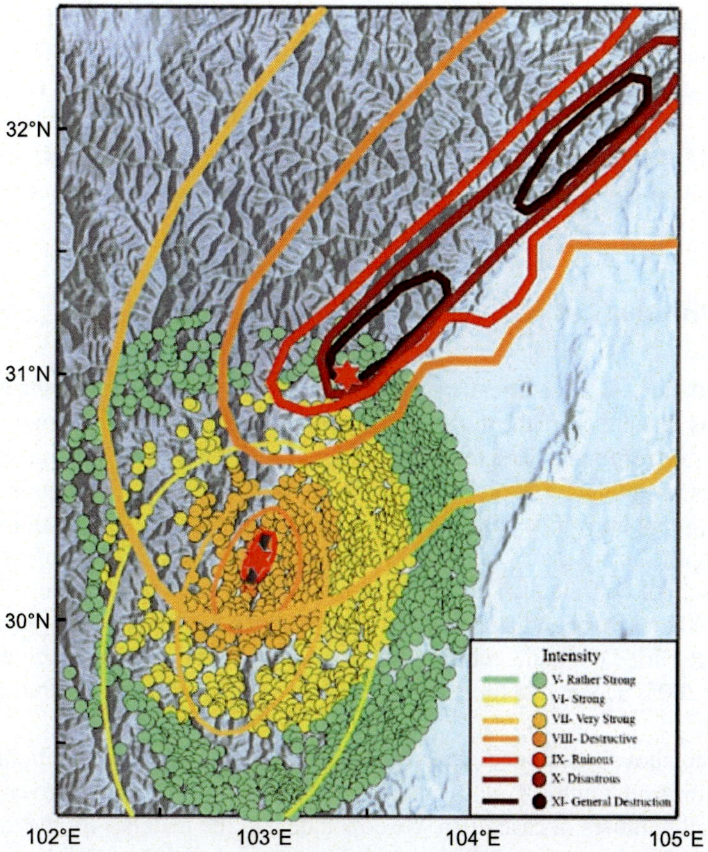

Fig. 9.2 Map of isoseismals of the Lushan earthquake (thin lines) and the Wenchuan earthquake (thick lines) compared with intensities calculated by QLARM (colored dots). The hexagrams indicate the macroscopic epicenter for the Lushan and the instrumental one for the Wenchuan earthquake. The gray solid line depicts the line source we used for calculating intensities for the Lushan earthquake. Parameters used by QLARM for the calculations are given in Tables 9.1 and 9.2.

because its area of strong shaking encroaches on that of the Wenchuan earthquake that happened five years earlier (Fig. 9.2). After the Wenchuan disaster, many new resistant buildings were constructed, and damaged buildings were retrofitted. This increase of building strength five years after the Wenchuan earthquake is opposite to the decrease of resistance in unmodified buildings in days and weeks immediately following a large earthquake. The losses caused by the Lushan earthquake in the area where the building stock has been strengthened cannot be calculated using the regular vulnerability model for the region. The losses in the Lushan earthquake depended on the changes to the building stock after the Wenchuan earthquake. This means that we should not use this earthquake as a validation case.

We observe that fewer fatalities, but more injuries than those expected for the Lushan earthquake were reported. In the near real-time sense, however, the calculation of QLARM was still approximately correct. During the Lushan earthquake, the southern part where buildings had remained unchanged, were affected severely. This part produced the usual number of casualties and injured. In contrast, in the northern part where buildings are now stronger, thus a portion of the expected fatalities were shifted into the category of "injured". Therefore, the observed number of injured was larger than expected in the near real-time calculation, and the observed fatalities were less than the average expected (Table 9.3).

9.3 Summary of Validation

For validation of theoretical casualties, the large magnitude earthquakes with numerous casualties are the most important ones, as they are also the most important ones for first responders and for mitigation plans. Matches for earthquakes with few casualties are subject to large variations because the collapse of a single apartment building, school or office building can cause in the order of 200 casualties. In 2008, for example, during the Wenchuan earthquake, main teaching building of Beichuan middle school of Beichuan county collapsed during the school hours. There were about 1,000 teachers and students in 21 classrooms, except for a few who escaped, most were buried under the rubble. Therefore, an over-estimation of theoretical casualties by 200 is acceptable for small events, for which near zero casualties have been reported.

For the above mentioned reasons, we give a high weight for calibration events with significant numbers of casualties, whereas a low weight is given for cases with small numbers of casualties. We conclude that the matches for the three events with large numbers of casualties are within a factor of 1.1 of the reported values (Fig. 9.1, Table 9.3). These excellent matches include the Wenchuan, the Yushu, and the Ludian earthquakes (Table 9.3). The details of 10 validation matches of calculated with observed casualties and isoseismals are discussed in detail in the section of *Supplemental Material.*

The overall conclusion mainly based on the large catastrophes for which QLARM works correctly÷ is that the QLARM program for estimating casualties in future scenario earthquakes is valid for the Chinese mainland.

Five cases among the validations are not far from the target region, in which we estimate the consequences of scenario earthquakes. They are the Dayao ($C_3 = 3.5$), Wenchuan ($C_3 = 3.7$), Lushan ($C_3 = 4.0$), Ludian ($C_3 = 4.1$) and Jinggu ($C_3 = 4.1$) earthquakes. So, we propose to use the average attenuation that fits best the validation cases, namely $C_3 = 3.9$, in the scenario estimations conducted subsequently.

9.3.1 Near Real-Time Earthquake Loss Alerts for the Mainland of China

Some near real-time loss estimations for the earthquakes of China's mainland have been released by the World Agency for Planetary Monitoring and Earthquake Risk Reduction (WAPMERR, Geneva, Switzerland) and the International Centre for Earth Simulation Foundation (ICES Foundation, Geneva, Switzerland; see *Data and Resources*) since QLARM was used to calculate casualties and distribute loss alerts starting in October 2002. The average delay time between the origin time of the earthquake and the loss alerts is 29 min, which depends in part on the time that USGS, GFZ (the Geoforschungszentrum) and EMSC (European Mediterranean Seismological Center) release the earthquake source parameters because all alerts based on QLARM are calculated and released manually (Wyss, 2014).

Precise parameters of the fault which caused the earthquake are not available in the first minutes after the event. For instance, the azimuth of the fault rupture is not known at that time, and therefore point source approximations are used for near real-time loss calculations. Also, epicenters available in near real-time are usually incorrect with errors of at least 10 km, and the depths is not well known. The macroseismic distribution is especially informative, but it is not known within weeks and sometimes months after earthquakes. All of this renders the near real-time loss estimations of QLARM strongly dependent on uncertainties. Therefore differences between rapidly calculated and final reported casualties must be expected.

However, we show that all the near real-time fatality estimates for mainland of China are correct. For estimates of the numbers of injured, six cases are hundred percent correct and three cases are within a factor of 2.5. Table 9.3 shows the loss alerts calculated by QLARM for the mainland of China and distributed by WAPMERR and ICES Foundation within the period from 1 January 2003 to 1 April 2017, along with the calculated results of this paper and the reported casualties. For small numbers, a difference less than 200 is not wrong if we take into account the collapse of one single building that can kill several hundred, or more people, as seen in Beichuan in the 2008 Wenchuan earthquake.

9.4 Earthquake Scenario Casualty Calculations

Predicting earthquakes, is extremely difficult because many factors, may influence the occurrence time and magnitude. For a valid earthquake prediction, origin time, magnitude and location should be specified with uncertainties, and in addition, the probability of the event to occur should be given.

Here, we do not predict future earthquakes. Instead, we identify active faults that have generated large earthquakes in the past, and which will rupture again in the future. We do not know when they will rupture, but suggest strongly that residents and authorities should prepare for these earthquakes, which could happen in their respective lifetimes or that of the next generations.

We model only the largest ruptures one can expect, according to the length of the investigated fault segments.. Although small and even medium earthquakes occurring frequently along some of them may generate losses that are tragic, here we focus only on large earthquakes, which would be of national important disaster.

We assume 15 km to be the depth of energy release, because the final results from validations tell us that 15 km is an appropriate choice for modelling the main release of energy, on average.

9.4.1 Reasons for Selecting Scenario Earthquake Source Parameters

We selected the Xianshuihe, Anninghe, Zemuhe, Daliangshan and Xiaojiang faults for earthquake scenario calculations (Fig. 9.3). This choice is mainly based on the following three considerations:

(1) These faults are included in the National Earthquake Monitoring and Prediction Experiment Site of Sichuan-Yunnan (23°–32° N, 98.5°–104.5°E; see *Data and Resources*) and China Seismic Experimental Site (21°-32°N, 97.5°-105.5°E; see Data and Resources);
(2) The scenarios are located in the southern part of the central China north–south seismic belt, which has been selected to be one of the testing regions of the Collaboratory for the Study of Earthquake Predictability (CSEP) project (Mignan et al., 2013; Zhang et al., 2016a, 2016b);
(3) This choice is based on outcomes of the annual consultation meetings of the CEA, which has focused on this area in recent years.

We also consulted the fifth-generation seismic hazard zoning map of China, see *Data and Resources.* The peak ground acceleration with 10% probability of exceedance in 50 years is 0.2–0.4 g along the scenario faults and in surrounding regions. The displacement accumulation rates are 10–20, 6–7, 6–7, 3–4 and 10 mm/yr for the Xianshuihe, Anninghe, Zemuhe, Daliangshan and Xiaojiang faults, respectively (Xu et al., 2003, 2005). The assumed source parameters of these faults to rupture in the future possible earthquakes are shown in Table 9.5.

Frequently earthquakes rupture the Xianshuihe fault. At least eight $M > 7.0$ earthquakes have occurred there since the year 1725 (Fig. 9.3). We measure the total length of the Xianshuihe fault as being about 320 km (Xu et al., 2003, 2005), with an assumed width of 20 km (Fig. 9.3, Table 9.5).

The Anninghe fault starts at Shimian in the north and ends at Xichang in the south (Fig. 9.3, Table 9.5). Ran et al. (2008) investigated the recurrence interval of historical major earthquakes and suggested that the Anninghe fault alternates between strong and weak activity. We measure the length of this fault as 140 km (Xu et al., 2003, 2005), the width of the fault plane is assumed to be 20 km (Fig. 9.3, Table 9.5).

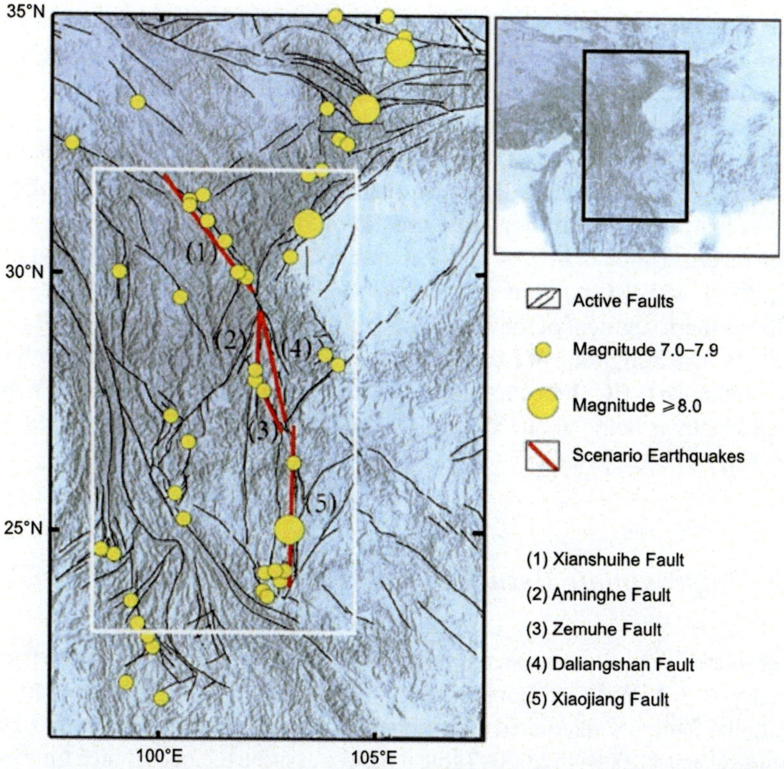

Fig. 9.3 Map of the scenario ruptures (heavy red lines), historical earthquakes (yellow dots) (Department of Earthquake Disaster Prevention of the China Earthquake Administration, 1999; Department of Earthquake Disaster Prevention of the State Seismological Bureau, 1995) on record from 0 BC to AD 2017 with $M \geq 7.0$ and active faults (Deng et al., 2003) in the target region (21.0°–35.0° N, 97.5°–107.5° E). The white rectangle shows the region of the National Earthquake Monitoring and Prediction Experiment Site of Sichuan-Yunnan (23°–32° N, 98.5°–104.5°E). The indexing figure at the top right shows the region under discussion

Table 9.5 Source parameters of the hypothetical earthquakes

Fault	Epi Lat.	Epi Lon.	Z	L	W	Dip.	M(W&C)	Y_1	X_1	Y_2	X_2
Xianshuihe	30.75	101.13	15	320	20	70	7.9	31.89	100.10	29.61	102.16
Anninghe	28.60	102.27	15	140	20	70	7.5	29.25	102.29	27.95	102.25
Zemuhe	27.43	102.53	15	130	20	60	7.5	27.95	102.23	26.90	102.82
Daliangshan	28.18	102.62	15	250	20	60	7.8	29.25	102.33	27.10	102.90
Xiaojiang	25.45	103.05	15	350	20	70	7.9	27.03	103.09	23.87	103.01

X_i and Y_i are the assumed end points of the respective line ruptures. M is calculated from the rupture area. The epicenter is placed at the center of the line rupture

The Zemuhe fault links with the Anninghe fault in the Xichang region and extends to Qiaojia (Fig. 9.3, Table 9.5). We measure its length as 130 km (Xu et al., 2003, 2005) and assume that the width is 20 km (Fig. 9.3, Table 9.5).

The Daliangshan fault starts at Shimian, and extends to Ningnan (Fig. 9.3, Table 9.5). According to the Chinese historical literature since 0 BC, no large earthquake has occurred along this fault. However, paleoseismological excavations of old slips suggests that there occurred at least nine $M > 7$ earthquakes before the advent of written records (Song et al., 2002). We measure its length as being 250 km (Xu et al., 2003, 2005), while its width is assumed to be 20 km (Fig. 9.3, Table 9.5).

The northern segment of the Xiaojiang fault links with the Zemuhe fault (Fig. 9.3, Table 9.5). The Songming $M8.0$ earthquake is the largest earthquake along the Xiaojiang fault since 0 BC. It occurred in 1833 and caused severe damage. We measure its total length as being about 350 km (Xu et al., 2003, 2005); the assumed width is 20 km (Fig. 9.3, Table 9.5).

9.4.2 Approximate Results of Casualty Scenarios

The estimated maximum expected magnitude for the Xianshuihe rupture from fault geometry is 7.9 (Wells & Coppersmith, 1994, Table 9.5). For the scenario of the Xianshuihe fault, we calculated the numbers of possible fatalities as 68,000-167,000, and injured as 181,000-452,000 (Table 9.6). We assume the occurrence time as 19:00 UTC, which equals 03:00 local time, and which therefore means that the occupancy of buildings is at a maximum. The approximate numbers of people in the area of minor damage (VI + VII) and in the devastated area (VIII+) are 19,860,000 and 5,955,000, respectively (Table 9.7). We defined the sum of these numbers as the strongly affected population.

The estimated maximum expected magnitude for the Anninghe rupture is 7.5, based on the assumed rupture area (Wells & Coppersmith, 1994). For the scenario of a rupture with $M = 7.5$ of the Anninghe fault along with a maximum occupancy of houses, our calculations show that the possible fatalities and injured are 49,000-134,000 and 131,000-364,000, respectively (Table 9.6). The approximate numbers of people in the intensity VI + VII and VIII+ areas are 16,253,000 and 5,348,000 respectively (Table 9.7).

The estimated maximum expected magnitude for the Zemuhe rupture is 7.5, based on the assumed rupture area (Wells & Coppersmith, 1994). With this earthquake occurring at 03:00 local time, the range of numbers of fatalities and injured estimated by QLARM are 38,000-111,000 and 104,000-302,000, respectively (Table 9.6). The approximate numbers of the population in the area of minor damage and in the devastated area are 14,350,000 and 4,376,000, respectively (Table 9.7).

The estimated maximum expected magnitude for the Daliangshan rupture is 7.8, according to the rupture area we specified (Wells & Coppersmith, 1994). For this earthquake occurring at 19:00 UTC the results calculated by QLARM show that the possible numbers of fatalities and injured are 111,000-259,000 and 292,000-693,000,

Table 9.6 The range of numbers of casualty estimated by QLARM for each earthquake scenario

Fault	Xianshuihe	Anninghe	Zemuhe	Daliangshan	Xiaojiang
Fatality	68,000–167,000	49,000–134,000	38,000–111,000	111,000-259,000	176,000-400,000
Injured	181,000-452,000	131,000-364,000	104,000-304,000	292,000-693,000	459,000-1,053,000

Table 9.7 The approximate numbers of people in the areas of intensitis (VI + VII) and (VIII+) level in each earthquake scenario, the sum of these numbers are the strongly affected people

	Xianshuihe	Anninghe	Zemuhe	Daliangshan	Xiaojiang
People in the minor damage area (VI + VII)	19,860,000	16,253,000	14,530,000	27,047,000	24,433,000
People in the devastated area (VIII+)	5,995,000	5,348,000	4,376,000	9,258,000	16,361,000

respectively (Table 9.6). The approximate numbers of people in the area of minor damage and in the devastated area are 27,047,000 and 9,258,000, respectively (Table 9.7).

The estimated maximum expected magnitude for the Xiaojiang rupture from fault geometry is 7.9 (Wells & Coppersmith, 1994). For this earthquake, we calculated the possible numbers of fatalities as 176,000-400,000 and injured as 459,000-1,053,000 (Table 9.6) at 19:00 UTC, when there would be a maximum occupancy of dwellings. The approximate numbers of people in the area of minor damage and in the devastated area are 24,433,000 and 16,361,000, respectively (Table 9.7).

9.5 Discussion

Scenario analyses could be done in a more useful way if we had more detailed data on buildings and their uses. For example, we should have a separate data set of locations and sizes of schools and another data set of health facilities, to estimate their probable damage in future earthquakes. This would help with both decision-making for reinforcement and for first responders (e.g. the rescue team). We have proposed to accumulate these data by an Open Building Map (OBM) project (Danijel Schorlemmer, personal communication; the OBM project see Data and Resources). In addition, it would be helpful to construct more detailed and more useful city models.

There are several source properties that influence damage and casualty numbers, which are currently not modeled. The impact of source complexity on estimation of earthquake casualties is not possible to be evaluated at this stage of development of alert tools. Using line sources for validations and scenarios is the best that can be done at this time, and it yields more accurate results than point sources (Li et al., 2020).

The spectrum of energy radiated by a rupture is unknown and not modeled, although it can influence the response. Earthquakes that cause more casualties radiate more than average energy at high frequencies, as suggested in the Ludian and Jinggu cases described above. The moment magnitude Mw is a low frequency concept, that does not measure the radiation of energy (Song et al., 2016), it would be desirable to estimate the impact of source spectra.

Earthquakes in high stress environments such, as thrusts, may cause more casualties than low stress events along normal faults, but this has not been investigated. In addition, the flip effect on the hanging-wall of thrust faults may augment accelerations (Brune, 1996, 2001a, 2001b).

Earthquakes with super-shear ruptures can lead to more serious disasters than normal earthquakes (Das, 2007). Analyses of past disasters indicate that super-shear ruptures can occur on strike slip faults (Xu et al., 2015; Zhang et al., 2017) and the damage caused by earthquakes depends in part on how fast the fault ruptures. The rupture speed is usually not known and it is not modeled in existing earthquake loss estimating tools.

A ubiquitous source of uncertainty is the inaccuracy of the location of epicenters and that of asperities along the fault plane. Epicenter errors in near real-time can map into large errors in casualty estimates (Wyss & Kossobokov, 2012; Wyss et al., 2011). Even final epicenter estimates by regional seismograph networks can deviate from those derived from macroseismic observations by up to 10 km, as shown in this paper. In the case of relatively small earthquakes in areas with uneven population distribution, these uncertainties can lead to significant errors in casualty estimates.

The extent and direction of ruptures could be defined accurately by analyses of satellite images. The use of this technique might dramatically increase the accuracy of near real-time loss estimates, as shown by Wyss et al. (2006). The frequency of satellite passes and the automatization of analysis of satellite data may have moved this approach into the realm of reality.

Local intensities of shaking are not known in near real-time. They are estimated months after an earthquake and are poor approximations to actual ground accelerations. In the future we expect that in a region with a high density MEMS (micro-electro-mechanical systems) for rapid intensity reports, casualties may be estimated more accurately. For example, in the mainland of China after the completion of the National System for Fast Report of Intensities and Earthquake Early Warning project, we will be able to use the near real-time reported intensity distribution to calculate casualties.

The Longmenshan fault zone consists of four parallel faults, including the front-range Longmenshan fault, the central Longmenshan fault, the back-range Long menshan fault and the mountain front blind fault. The Wenchuan earthquake ruptured on the Beichuan-Yingxiu fault in the middle section of the central Longmenshan fault and the Guanxian-Jiangyou fault in the middle part of the front-range Longmenshan fault. We should not ignore the seismic hazard of the Longmenshan fault zone although the 2008 Wenchuan earthquake (Chen & Booth, 2011) and the 2013 Lushan earthquake (Wu et al., 2014) have recently occurred nearby on it. Assessing the seismic hazard along the Longmenshan fault zone, Wang et al. (2010) considered seismic moment deficits (unreleased moment) or surplus (over-released moment) and found that the Wenchuan earthquake did not release all seismic moment on the fault. However, the buildings that would be affected in a Longmenshan scenario are probably mostly those that have been rebuilt after the 2008 Wenchuan earthquake and are stronger than they were before. Therefore, we do not have the correct building stock for this scenario. In the future, the losses should be further studied for this case, with a new model used for the building stock.

The Yutian M_S7.3/M_W6.8 earthquake of 12 February 2014 in Xinjiang Uygur Autonomous Region does not qualify for a calibration case. The fact that nearly all calculations result in 0 or very few fatalities is due to the absence of people in the affected region. There is a settlement in the field of I = IX and another settlement at the edge of the fields VIII and VII. Our calculations depend on many settlements contributing to the loss because the soil conditions and other variations are approximately averaging out. We cannot rely on the Yutian event with near zero input as a pillar for documenting the performance of QLARM, thus we did not use it.

9.6 Engineering Measures to Reduce Casualties

Although we are not specialists in engineering, a few remarks about how to reduce casualties are offered in this article. There exists a great opportunity to save lives by strengthening buildings. Local standards for seismic precautionary intensity should be reassessed and raised to a new reasonable level appropriate for buildings to withstand the shaking of intensities VII–IX.

Some potentially dangerous houses, i.e. those constructed of brick and adobe, should be reinforced with lateral resisting systems (e.g. shear walls, frames with in-fill walls and constructional columns), or reconstructed with reinforced concrete materials. The reinforced concrete structures should obey the principle of "strong column and weak beam" and, for those buildings which already exist, but do not meet the earthquake resistance level, carbon fiber sheet and steel-bonded reinforcement technique should be introduced (Hu, 2009).

If we assume that 70–90% of fatalities and injured could be saved by strengthening buildings, this would mean that the number of fatalities and injured could be reduced by 30,000–350,000 and 70,000–940,000, respectively, in each earthquake, depending on the particular case. This assumption of a reduced casualty rate is realistic. Xinjiang Uygur Autonomous Region of China is an earthquake-prone region. The earthquake of M6.8 occurred in the Bachu-Jiashi region on February 24, 2003, and killed 268 and injured 4,853. Since 2004, the local government has launched a rural earthquake-resistant project. More than 50 earthquakes larger than M5 occurred in Xinjiang during the 10 years from 2004 to 2014, but no one died and only 73 people were injured.

In addition, for people who cannot afford retrofitting the whole house, an earthquake closet that could improve the resident's chance of survival and not being injured by factors of 1000–30,000 (Wyss, 2012) is recommended. This factor depends on the ratio of the resistances to shaking by the building to that of the earthquake protection unit (EPU). For a house built poorly of heavy material that will collapse, the improvement of surviving in an EPU that is constructed as strong as a bunker, is almost infinity. In case of strong shaking, a family has about 5–10 s time to get into the EPU because this time span is equal to the difference between the first arriving P-wave with small amplitudes, and the later-arriving, destructive S-wave.

9.7 Conclusions

We validate the algorithms and data sets of QLARM for the 13 earthquakes of the mainland of China since 2003 for which intensity and fatality information exist (Fig. 9.1). It appears isoseismals are more accurate than instrumental locations for pinpointing the centriod. The distance on average from the instrumental epicenter to the macroepicenter is about 9 km for these validations.

Considering an earthquake with small casualty numbers, a loss difference less than 200 is acceptable because the collapse of one single building can kill many people, or more than the expected death toll. Therefore, we concluded that 13 near real-time fatality estimates are correct. For the results of injured, six cases are 100% correct and three cases are within a factor of 2.5.

For the three events with large numbers of casualties (the Wenchuan, the Yushu, the Ludian earthquakes) to which we give the most weight, the matches of theoretical intensities and numbers of the casualties are within a factor of 1.1 of the reported values (Fig. 9.1, Table 9.3).

Although several improvements in loss estimates may become available in the future, the near real-time verification calculations reported herein show that earthquake losses can be estimated within factors of 2–3 in the mainland of China using the currently available methods and data sets. It gives us confidence that the QLARM program for estimating casualties in future scenario earthquakes is applicable for the mainland of China.

We select the Xianshuihe, Anninghe, Zemuhe, Daliangshan and Xiaojiang faults (Fig. 9.3) for possible ruptures in their entire length in the Sichuan-Yunnan region, China, and construct "worst-case scenarios" of devastating earthquakes. We assume the occurrence time at 03:00 local time, which implies maximum occupancy of buildings. For the five scenarios, the numbers of possible casualties are calculated and listed in Table 9.6.

The greatest disaster is expected for the scenario of the Xiaojiang fault for which we calculated the possible numbers of fatalities on average as being 286,000 and injured as being 751,000. These expected numbers of casualties are about a factor of 3 larger than those reported in the Wenchuan earthquake of 2008. Part of the reason for this is that we assume better transmission of seismic waves in the Xiaojiang scenario cases than in Wenchuan.

The scenario of the maximum rupture along the Daliangshan fault results in approximately twice the casualties as reported in the Wenchuan earthquake. The three scenarios along the Xianshuihe, Anninghe and Zemuhe faults result in estimates of casualties similar to those in the Wenchuan case. Although our calculations are approximate, they are realistic. Earthquakes similar to the ones modeled here are inevitable, only their occurrence time is unknown. We hope that the numbers of casualties we forecast here motivate the government and the local inhabitants to take preventive measures.

9.8 Data and Resources

The earthquake catalog of USGS was searched via https://earthquake.usgs.gov/earthquakes/search/ (last accessed on 20 September 2017). The annual population growth rate for China is taken from the National Bureau of Statistics of China, available at http://data.stats.gov.cn/search.htm (last accessed on 21 September 2017). The different rupture models of the 2008 Wenchuan earthquake of China can be also found via the finite-source rupture model database (http://equakerc.info/SRCMOD/searchmodels/allevents/ , last accessed 21 September 2017) and the U.S. Geological Survey (https://earthquake.usgs.gov/earthquakes/eventpage/usp000g650#finite-fault, last accessed 21 September 2017). The rupture model of the 2016 Arketao earthquake are taken from the Institute of Geophysics, China Earthquake Administration, available at http://www.cea-igp.ac.cn/tpxw/275080.html (last accessed on 20 September 2017). The reported numbers and isoseismals for Chinese earthquakes can be found at the disaster assessment notice of the website of CEA (http://www.cea.gov.cn/publish/dizhenj/468/549/index.html, last accessed on 21 September 2017). The alerts released by the International Centre for Earth Simulation Foundation can be accessed at http://www.icesfoundation.org/Pages/CustomPage.aspx?ID=122 (last accessed 21 September 2017). The information about the National Earthquake Monitoring and Prediction Experiment Site of Sichuan-Yunnan can be accessed at http://www.ncepe.cn/syccn/index.php (last accessed 24 September 2017). The new generation (fifth generation) seismic hazard zoning map of China can be found at http://www.gb18306.cn/ (last accessed 22 September 2017). The information of Open Building Map (OBM) project are accessed at http://www.openbuilingmap.org (last accessed 12 January 2018). The news of the rural earthquake-resistant project of Xinjiang can be found at http://www.chinanews.com/sh/2014/02–14/5838784.shtml (last accessed 2 April 2018).

Acknowledgements We thank Robert Bishop, Philippe Rosset, and Yonggang Li for helpful remarks on the text. We also thank Changsheng Jiang for supporting us with the program to map the terrain and Zengping Wen, Sanhong Chen, Zhengya Si for providing suggestions about earthquake engineering.This project is supported by the Guangdong Basic and Applied Basic Research Foundation under Grant No. 2020A1515110844.

References

Brune, J. N. (1996). Particle motions in a physical model of shallow angle thrust faulting. *Proceedings of the Indian Academy of Sciences - Earth and Planetary Sciences, 105*(2), L197–L206.

Brune, J. N. (2001a). Erratum: Shattered rock and precarious rock evidence for strong asymmetry in ground motions during thrust faulting. *Bulletin of the Seismological Society of America, 91*(6), 1937–1940. https://doi.org/10.1785/0120010283

Brune, J. N. (2001b). Shattered rock and precarious rock evidence for strong asymmetry in ground motions during thrust faulting. *Bulletin of the Seismological Society of America, 91*(3), 441–447. https://doi.org/10.1785/0120000118

Chen, Y., & Booth, D. C. (2011). *The Wenchuan earthquake of 2008: Anatomy of a disaster*. Science Press Cooperating with Springer.

Choy, G. L., & Boatwright, J. (1995). Global patterns of radiated seismic energy and apparent stress. *Journal of Geophysical Research, 100*(B9), 18205–18228.

Choy, G. L., Boatwright, J., & Kirby, S. (2001). The radiated seismic energy and apparent stress of interpolate and intraplate earthquakes at subduction zone environments: Implications for seismic hazard estimation. *USGS Open-File Report, 01–0005*, 1–17.

Choy, G. L., & Boatwright, J. (2009). Differential energy radiation from two earthquakes in Japan with identical M_W: The Kyushu 1996 and Tottori 2000 earthquakes. *Bulletin of the Seismological Society of America, 99*(3), 1815–1826.

Das, S. (2007). The need to study speed. *Science, 317*, 905–906.

Deng, Q. D., Zhang, P. Z., Ran, Y. K., Yang, X. P., Min, W., & Chen, L. C. (2003). Active tectonics and earthquake activities in China. *Earth Science Frontiers, 10*(Supplement), 66–73. (in Chinese with English abstract).

Department of Earthquake Disaster Prevention of the State Seismological Bureau. (1995). *The catalogue of Chinese historical strong earthquakes*. Seismological Press. (in Chinese with English abstract).

Department of Earthquake Disaster Prevention of the China Earthquake Administration. (1999). *The catalogue of Chinese modern earthquakes*. China Science and Technology Press. (in Chinese with English abstract).

Du, H. L., Xu, L. S., & Chen, Y. T. (2009). Rupture process of the 2008 great Wenchuan earthquake from the analysis of the Alaska-array data. *Chinese Journal of Geophysics, 52*(2), 372–378. (in Chinese with English abstract).

Erdik, M., Şeşetyan, K., Demircioğlu, M. B., Zülfikar, C., Hancılar, U., Tüzün, C., Harmandar, E. (2014). Rapid earthquake loss assessment after damaging earthquakes. In A. Ansal (Ed.), *Perspectives on european earthquake engineering and seismology, geotechnical, geological and earthquake engineering* (pp. 53–95). https://doi.org/10.1007/978-3-319-07118-3_2

Fei, M. L., Zhou, G. Q., Shi, H. W., Lu, Y. K., & Chen, K. H. (2004). Seismic intensity and disaster analysis on the Dayao *M* 6.2 earthquake. *Journal of Seismological Research, 27*(Supplement), 70–74. (Chinese with English abstract).

Federal Emergency Management Agency (2003). Multi-hazard loss estimation methodology-earthquake model: HAZUS-MH MR4 technical manual. Washington D C. www.fema.gov/plan/prevent/hazus

Hu, Y. X. (2009). *Earthquake engineering: Second edition*. Seismological Press. (in Chinese). ISBN 978-7-5028-2852-3/P·1264(3271).

Fontiela, J., Rosset, P., Wyss, M., Bezzeghoud, M., Borges, J., Rodrigues, F. C., 2020. Human losses and damage expected in future earthquakes on Faial Island-Azores. *Pure and Applied Geophysics, 177*, 1831-1844.

Huang, B. S., Chen, J. H., Liu, Q. Y., Chen, Y. G., Xu, X. W., Wang, C. Y., Lee, S. J., & Yao, Z. X. (2012). Estimation of rupture processes of the 2008 Wenchuan earthquake from joint analyses of two regional seismic arrays. *Tectonophysics, 578*(2012), 87–97.

Jaiswal, K. S., & Wald, D. J. (2010). An empirical model for global earthquake fatality estimation. *Earthquake Spectra, 26*(4), 1017–1037.

Jaiswal, K. S., Wald, D. J., Earle, P. S., Porter, K. A., & Hearne, M. (2011). Earthquake casualty models within the USGS prompt assessment of global earthquakes for response (PAGER) system. In R. Spence, E. So, & C. Scawthorn (Eds.), *Human casualties in earthquakes: Progress in modeling and mitigation* (pp. 83–94). Springer.

Ji, C. (2008) Preliminary result of the May 12, 2008 M_W 7.9 Shichuan earthquake. http://www.geol.ucsb.edu/faculty/ji/big_earthquakes/2008/05/12/ShiChuan.html. Last access: 21 September 2017.

Li, J. W., Böse, M., Wyss, M., Wald, D. J., Hutchison, A., Clinton, J. F., Wu, Z. L., Jiang, C. S., Zhou, S. Y., 2020. Estimating rupture dimensions of three major earthquakes in Sichuan, China, for early warning and rapid loss estimates. *Bulletin of the Seismological Society of America, 110*(2): 920-936

Mignan, A., Jiang, C. S., Zechar, D. J., Wiemer, S., Wu, Z. L., & Huang, Z. B. (2013). Completeness of the China earthquake catalog and implications for the setup of the mainland of China earthquake forecast testing center. *Bulletin of the Seismological Society of America, 103*(2A), 845–859. https://doi.org/10.1785/0120120052

Ran, Y. K., Chen, J. W., Gong, H. L., & Chen, L. C. (2008). Late quaternary geomorphic deformation and displacement rates of the Anninghe fault around Zimakua. *Seismology & Geology, 30*(1), 86–98. (Chinese with English abstract).

Rosset, P., Trendafiloski, G., Yelles, K., Semmane, F., & Wyss, M. (2009). *Application of the loss estimation tool QLARM in Algeria*. European Geosciences Union General Assembly.

Rosset, P., & Wyss, M. (2017). Seismic loss assessment in Algeria using the tool QLARM. *Civil Engineering Research Journal,* 2, 555583. https://doi.org/10.19080/CERJ.2017.02.555583

Rosset, P., Tolis, S., Wyss, M., 2020. The use of QLARM to estimate seismic risk in Kirghizstan at the regional and city scale. *Acta Geophysica, 68*, 979-991

Salgado-Gálvez, M. A., Romero, D. Z., Velásquez, C. A., Carreño, M. L., Cardona, O.-D., & Barbat, A. H. (2016a). Urban seismic risk index for Medellín, Colombia, based on probabilistic loss and casualties estimations. *Natural Hazards, 80*(3), 1995–2021.

Salgado-Gálvez, M. A., Bernal, G. A., Barbat, A. H., Carreño, M. L., & Cardona, O.-D. (2016b). Probabilistic estimation of annual lost economic production due to premature deaths because of earthquakes. *Human and Ecological Risk Assessment: An International Journal, 22*(2), 543–557. https://doi.org/10.1080/10807039.2015.1095072

Salgado-Gálvez, M. A., Bernal, G. A., Zuloaga, D., Marulanda, M. C., Cardona, O.-D., & Henao, S. (2017). Probabilistic seismic risk assessment in Manizales, Colombia: Quantifying losses for insurance purposes. *International Journal of Disaster Risk Science, 8*(3), 296–307.

Shakhramanjyan, M. A., Nigmetov, G. M., Larionov, V. I., Nikolaev, A. V., Frolova, N. I., Sushchev, S. P., & Ugarov, A. N. (2001). Advanced procedures for risk assessment and management in Russia. *International Journal of Risk Assessment and Management, 2*(3/4), 303–318.

Shebalin, N. V. (1985). Regularities of the natural disasters (Russian). *Nauki o Zemle, Znanie, 11*, 48.

Song, F. M., Li, R. C., & Xu, X. W. (2002). Preliminary results of the investigation of paleo-earthquakes along the Daliangshan fault zone, Sichuan Province, China. *Seismology & Geology, 24*(1), 27–34. (Chinese with English abstract).

Song, X. X., Wu, Z. L., Jiang, C. S., & Li, J. W. (2016). Difference in energy radiation from earthquakes with similar moment magnitude and focal mechanism—A comparison of the broadband body-wave magnitudes of the 2014 Ludian and Jinggu, Yunnan Province, China, earthquake. *Earthquake Research in China, 32*(4), 653–662. (Chinese with English abstract).

Trendafiloski, G., Wyss, M., Rosset, P., & Marmureanu, G. (2009). Constructing city models to estimate losses due to earthquakes worldwide: Application to Bucharest Romania. *Earthquake Spectra, 25*, 665–685.

Trendafiloski, G., Wyss, M., & Rosset, P. (2011). Loss estimation module in the second generation software QLARM. In R. Spence, E. So, & C. Scawthorn (Eds.), *Human casualties in earthquakes: Progress in modeling and mitigation* (pp. 95–106). Springer.

Wan, Y. G. (2016). Introduction to seismology. Science Press. (in Chinese). ISBN 978-7-03-047926-6.

Wang, H., Liu, M., Shen, X. H., & Liu, J. (2010). Balance of seismic moment in the Songpan-Ganze region, eastern Tibet: Implication for the 2008 great Wenchuan earthquake. *Tectonophysics, 491*(1–4), 154–164. https://doi.org/10.1016/j.tecto.2009.09.022

Wang, W. M., Zhao, L. F., Li, J., & Yao, Z. X. (2008). Rupture process of the M_S 8.0 Wenchuan earthquake of Sichuan, China. *Chinese Journal of Geophysics, 51*(5), 1403–1410. (Chinese with English abstract).

Wells, D. L., & Coppersmith, K. J. (1994). New empirical relationships among magnitude, rupture length, rupture width, rupture area, and surface displacement. *Bulletin of the Seismological Society of America, 84*(4), 974–1002.

Wu, Z. L., Jiang, C. S., Li, X. J., Li, G. J., & Ding, Z. F. (2014). *Earthquake phenomenology from the field: The April 20, 2013, Lushan earthquake*. Springer. https://doi.org/10.1007/978-981-4585-15-6

Wyss, M. (1979). Estimating maximum expectable magnitude of earthquakes from fault dimensions. *Geology, 7*(7), 336–340. https://doi.org/10.1130/0091-7613(1979)7

Wyss, M. (2005). Human losses expected in Himalayan earthquakes. *Natural Hazards, 34*(3), 305–314.

Wyss, M. (2006). The Kashmir *M*7.6 shock of 8 October 2005 calibrates estimates of losses in future Himalayan earthquakes. In B. van de Walle, & M. Turoff (Eds.), *Proceedings of the 3rd International ISCRAM Conference*. Newark.

Wyss, M. (2010). Predicting the human losses implied by predictions of earthquakes: Southern Sumatra and central Chile. *Pure and Applied Geophysics, 167*(8–9), 959–965.

Wyss, M. (2012). The earthquake closet: Rendering early-warning useful. *Natural Hazards, 63*(2), 761–768.

Wyss, M. (2014). Ten years of real-time earthquake loss alerts. In M. Wyss (Ed.), *Earthquake hazard, risk, and disasters* (pp. 143–165). Elsevier.

Wyss, M. (2017). Four loss estimates for the Gorkha *M*7.8 earthquake, April 25, 2015, before and after it occurred. *Natural Hazards, 86*(1 Supplement), 141–150.

Wyss, M., Wang, R., Zschau, J., & Xia, Y. (2006). Earthquake loss estimates in near real-time. *EOS Transactions American Geophysical Union, 87*(44), 477–492.

Wyss, M., Elashvili, M., Jorjiashvili, N., & Javakhishvili, Z. (2011). Uncertainties in teleseismic earthquake locations: Implications for real-time loss estimates. *Bulletin of the Seismological Society of America, 101*(3), 1152–1161.

Wyss, M., & Kossobokov, V. (2012). Errors in expected human losses due to incorrect seismic hazard estimates. *Natural Hazards, 62*(3), 927–935.

Wyss, M., & Zuñiga, F. R. (2016). Estimated casualties in a possible great earthquake along the Pacific coast of Mexico. *Bulletin of the Seismological Society of America, 106*(4), 1867–1874.

Wyss, M., Gupta, S., & Rosset, P. (2018). Casualty estimates in repeat Himalayan earthquakes in 725 India. Bulletin of the Seismological Society of America (submitted).

Wyss, M., & Chamlagain, D. (2018). Estimated casualties in possible future earthquakes south and 723 west of the M7.8 Gorkha earthquake of 2015. Natural Hazards (submitted).

Wyss, M., Rosset, P., 2020. Near-real-time loss estimates for future Italian earthquakes based on the M 6.9 Irpinia example. In The November 23rd, 1980 Irpinia-Lucania, Southern Italy Earthquake: Insights and Reviews 40 Years Later, Porfido, S., Alessio, G., Gaudiosi, G., Nappi, R., Michetti, A. M., (eds.), no. 165

Wyss, M., Rosset, P., Triveno, L., 2021. The ratio of rural/urban people killed in earthquakes needs to be assessed for countries separately, the example of Colombia. *Seismological Res. Lett., 92*, 1036-1051

Xu, X. W., Wen, X. Z., Zheng, R. Z., Ma, W. T., Song, F. M., & Yu, G. H. (2003). Pattern of latest tectonic motion and its dynamics for active blocks in Sichuan-Yunnan region, China. *Science China Earth Science, 33*(Supplement), 151–162. (Chinese with English abstract).

Xu, X. W., Zhang, P. Z., Wen, X. Z., Qin, Z. L., Chen, G. H., & Zhu, A. L. (2005). Features of active tectonics and recurrence behaviors of strong earthquakes in the western Sichuan province and its adjacent regions. *Seismology & Geology, 27*(3), 446–461. (Chinese with English abstract).

Xu, J. K., Zhang, H. M., & Chen, X. F. (2015). Rupture phase diagrams for a planar fault in 3-D full-space and half-space. *Geophysical Journal International, 202*(3), 2194–2206.

Yuan, Y. F., & Tian, Q. W. (2012). Engineering seismology. Seismological Press. (in Chinese). ISBN 978-7-5028-4130-4/P·(4781).

Zhang, Y., Feng, W. P., Xu, L. S., Zhou, C. H., & Chen, Y. T. (2009). Spatio-temporal rupture process of the 2008 great Wenchuan earthquake. *Science China Earth Science, 52*(2), 145–154.

Zhang, Y., Xu, L. S., & Chen, Y. T. (2010). Fast inversion of rupture process of the 14 April 2010 Yushu, Qinghai, earthquake. *Acta Seismologica Sinica, 32*(3), 362–365. (Chinese with English abstract).

Zhang, Y., Xu, L. S., & Chen, Y. T. (2013). Rupture process of the Lushan 4.20 earthquake and preliminary analysis on the disaster-causing mechanism. *Chinese Journal of Geophysics, 56*(4), 1408–1411. (Chinese with English abstract).

Zhang, S. F., Wu, Z. L., & Jiang, C. S. (2016a). The central China north–south seismic belt: Seismicity, ergodicity, and five-year PI forecast in testing. *Pure & Applied Geophysics, 173*(1), 245–254. https://doi.org/10.1007/s00024-015-1123-9

Zhang, S. F., Wu, Z. L., & Jiang, C. S. (2016b). Reducing false alarms of annual forecast in the central China north-south seismic belt by reverse tracing of precursors (RTP) using the pattern informatics (PI) 'hotspots.' *Pure & Applied Geophysics, 174*(6), 2401–2410.

Zhang, Z. G., Xu, J. K., Huang, H. Q., & Chen, X. F. (2017). Seismic characteristics of supershear and sub-rayleigh earthquakes: Implication from simple cases. *Geophysical Research Letters, 44*(13), 6712–6717. https://doi.org/10.1002/2017GL074158

Chapter 10
China Seismic Experimental Site: Seismicity, Ergodicity and Nowcasting Earthquakes

Shengfeng Zhang and Yongxian Zhang

Abstract China Seismic Experimental Site (CSES) has been launched on May 12th, 2018 at the International Conference for the Decades Memory of the Wenchuan Earthquake and served as the second step of testing center for the Collaborative for Study of the Earthquake Predictability (CSEP2.0). To comprehensively understanding the earthquake preparation and occurrence process, the statistical methods provides several new prospects to describe the features of seismotectonics, earthquake cycle and rupture dynamics. In this work, we outline the background seismicity of CSES region using the catalog provided by the China Earthquake Networks Center. Analysis on ergodicity under the assumptions of meta-stable equilibrium shows that this behavior exists for most of the time period from 1980 to the present, and shows a difference in two sub-regions, Sichuan and Yunnan region. At the same time, the ergodicity was apparently disrupted by the 2008 Wenchuan earthquake sequence. Finally, Nowcasting method based on the 'natural time' sequence was used to describe the earthquake potential above $M6.0$ in CSES and its two sub-regions. The Earthquake Potential Score (EPS) of CSES region above 80% for target earthquake within $M6.0 \sim 7.5$ reveals a long time absence of target events from the last large event within the magnitude range in this area. The Sichuan region has a relatively high EPS value than the Yunnan region. Finally, after the forward forecasting given by Nowcasting method, a retrospective experiment for the evaluation shows that the Nowcasting method outperforms random guesses. We also obtain the effective threshold of EPS value for whole CSES and its two sub-regions, Sichuan and Yunnan region, is 30%, 25% and 60%, respectively. The combination of the Nowcasting method and the Annual Consultation Conference in China may provide a measure to determine the probability of the next strong earthquake.

Keywords China seismic experimental site (CSES) · Ergdicity · Nowcasting method · Annual consultation conference

S. Zhang (✉) · Y. Zhang
Institute of Earthquake Forecasting, China Earthquake Administration, Beijing 100036, China
e-mail: yxzhseis@sina.com

S. Zhang
School of Earth Science and Engineering, Nanjing University, Nanjing 210023, Jiangsu, China

© Higher Education Press 2022
Y.-G. Li et al. (eds.), *China Seismic Experimental Site*,
https://doi.org/10.1007/978-981-16-8607-8_10

10.1 Introduction

In the reduction of earthquake disaster risk in China, the North–South Seismic Region (21.0°–41.5° N, 97.5°–107.5° E) in central China is one of the focuses of special attention. Several disastrous earthquakes occur in this belt, including the 1920 Haiyuan $M8.5$ great earthquake (Department of Earthquake Disaster Prevention of the State Seismological Bureau, 1995, 1999), 1970 Tonghai $M_S7.8$ earthquake, 2008 Wenchuan $M_S8.0$ earthquake (Chen & Booth, 2011), and 2013 Lushan $M_S7.0$ earthquake (Wu et al., 2014). Tectonically, this seismic region lies among the Tibetan Plateau, the Ordos block, the Sichuan Basin, and the South China block (Zhang et al., 2003). The region is characterized by the variation of crust thickness from 30 km to 46 km in the east to 46–74 km in the west, reflecting the rising and spreading of the Tibetan Plateau with the collision between the India plate and the Eurasia plate (Li et al., 2006). Due to the reality of this region, many field experiment sites in history have been located there, such as the West-Yunnan Earthquake Prediction Experiment Site since 1980 and the National Experiment Site for Earthquake Monitoring and Forecast since 2014 (Sun & Wu, 2007). Directly related to statistical seismology, this region is featured by complicated fault systems and heterogeneous seismicity (Xu & Deng, 1996; Xu et al., 2005; Yi et al., 2002), so that it is hard to simply apply the existing recurrence models such as the time-predictable and/or magnitude-predictable models. It is somehow challenging, therefore, this region is suggested (Mignan et al., 2013) as being one of the first steps of testing regions of the Collaboratory for the Study of Earthquake Predictability (CSEP1.0), which is suggested by SCEC after the Reginal Earthquake Likelihood Models (RELM) workshop and proposed that different countries and regions should join the test process for different forecast models using religious evaluation methods (Jordan, 2006). As a 'baseline' analysis of the predictability of seismicity, a series of retrospective forecasts were made (Mignan et al., 2013) using the 'simple smoothed seismicity model', or the 'TripleS model' of Zechar and Jordan (2010), smoothening the locations of past earthquakes to construct a predictive density of future seismicity. The spatial distribution of earthquakes from 1970/01/01 to 2020/10/10 in the North–South Seismic Region is shown in Fig. 10.1.

The field experiments of China Seismic Experimental Site (CSES) (21°–32° N, 97.5°–107.5° E) has been established from 12 May 2018, at the opening ceremony of the International Conference for the Decade Memory of the Wenchuan $M_S8.0$ earthquake with the 4th International Conference on Continental Earthquakes (4th ICCE) (Wu et al., 2019b; Wu, 2020). The spatial region of CSES only contains the Sichuan and Yunnan provinces in China and seems smaller than the previous field site. After the detailed top-level design for the CSES research plan, it learns from the concept of ecology and environmental science and proposed a new cooperated research consideration called Coordinated Distributed Experiments (CDEs) (Wu et al., 2019a). To significantly understanding the test blocks of scientific hypotheses for the field experiment, this concept and cooperated framework will renew the links between the field observations and the indoor experiments and data processing. The

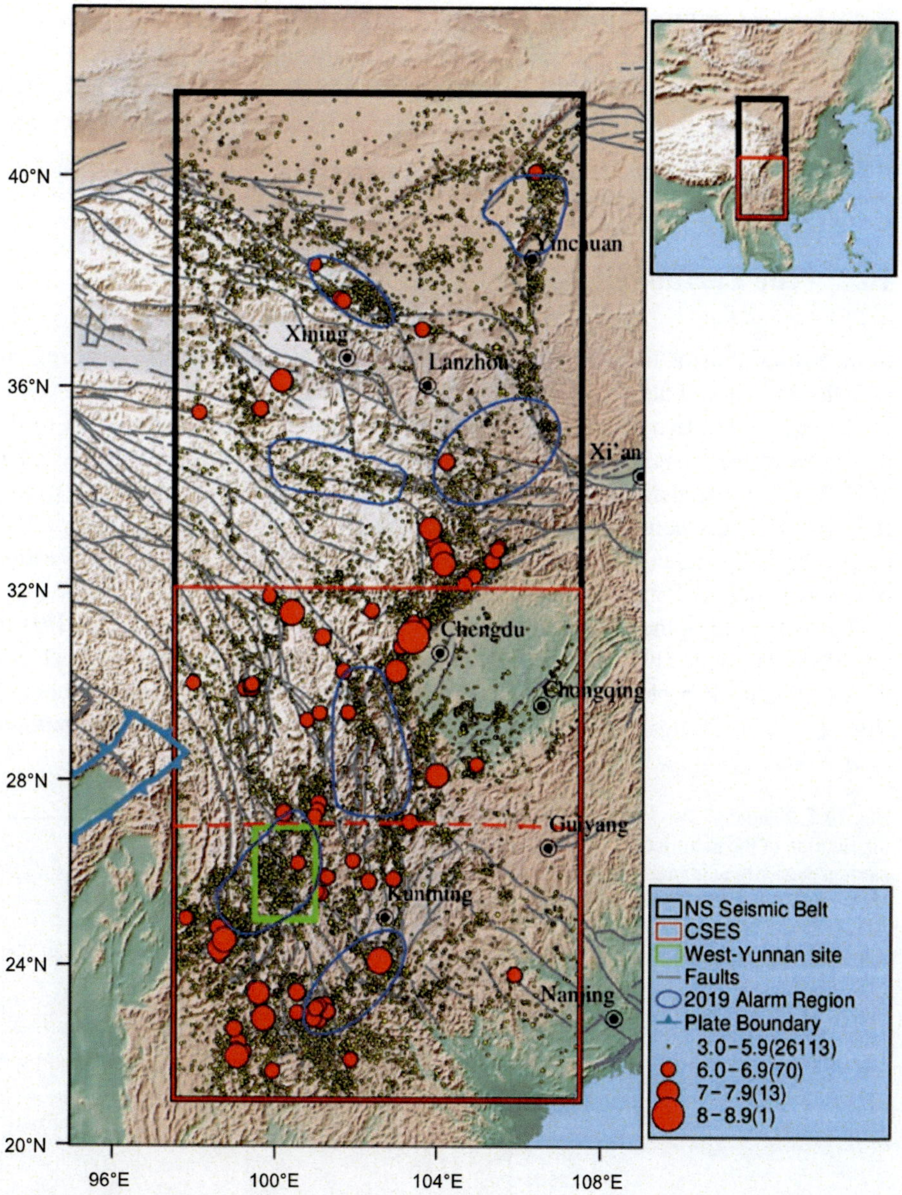

Fig. 10.1 Spatial distribution of earthquakes for the period from 1970/01/01 to 2020/10/10 in the central China North–South Seismic Region. The black, red overlaid on black and green rectangular frames denote the North–South Seismic Region, China Seismic Earthquake Site (CSES) which is the study region in this work, west Yunnan site. The CSES region is divided into two sub-regions, namely Sichuan region (top) and Yunnan region (bottom), separated by a dashed red line with latitude 27° N. The active faults are indicated by the lines with deep color. The blue irregular circles are the alarm regions that were outputted on the 2019 Annual Consultation Conference in China. The magnitude size and events number within the different ranges are also shown in the legends. The indexing figure to the top right is the location of this boundary in the world map

second step of CSEP (CSEP2.0) has been announced in 2019 (Michael & Werner, 2018; Savran et al., 2018) and will build the testing center in CSES region (Zhang et al., 2019a, b). This work will give an outline of seismicity features, ergodicity behavior and the Nowcasting (Rundle et al., 2016, 2017, 2019) experiment for CSES and its sub-regions.

10.2 The Earthquake Catalog

Somehow arbitrarily, the study region is delimitated by the ranges of CSES region, with the reference of the CSEP2.0 testing region (see Fig. 10.1). We divided the whole CSES region into two sub-regions, Sichuan region and Yunnan region, due to the difference in the seismicity between the south and north parts. While, Figs. 10.2 and 10.3 show the spatial distribution and magnitude-time plot of historical strong events in CSES region, where ancient Chinese literature recorded several magnitude 8 + earthquakes (Earthquake data from: Department of Earthquake Disaster Prevention of the State Seismological Bureau, 1995, 1999).

The data used in the present study is the earthquake catalog from 1970/01/01 to 2020/10/10 provided by the China Earthquake Networks Center (CENC). As shown in Fig. 10.4, the completeness magnitude given by different methods (Woessner and Wiemer, 2005) changed with time, depending on the development of observational

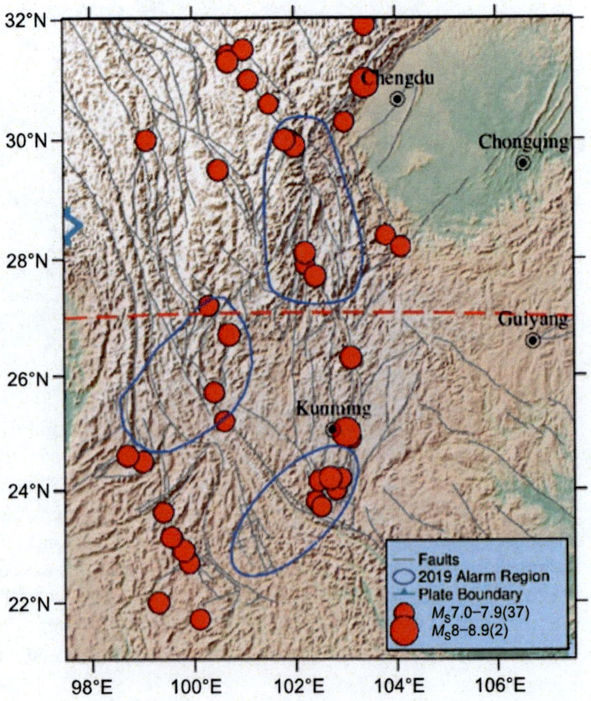

Fig. 10.2 Spatial distribution of the historical earthquakes with magnitude above M_S7.0 from 0 BC to AD 2020 in CSES region. The marks are the same with Fig. 10.1

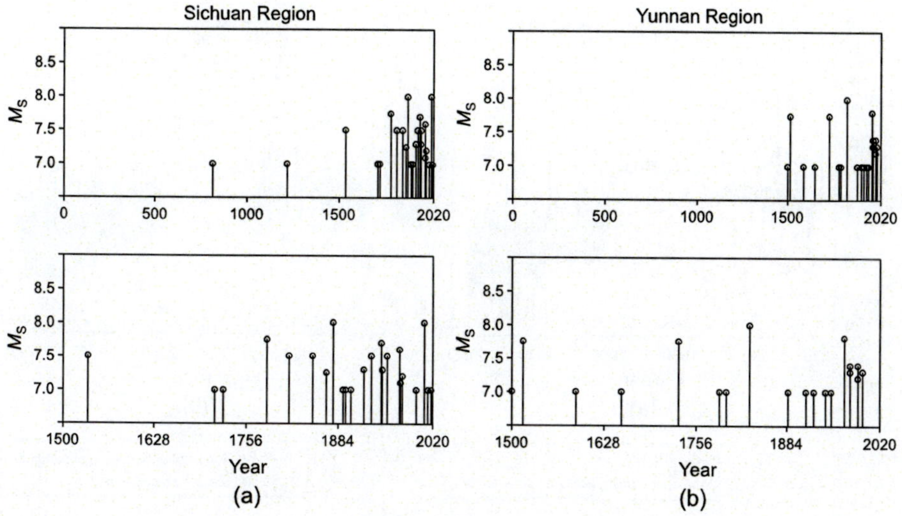

Fig. 10.3 Magnitude-Time diagram of two sub-regions of CSES within different periods. **a** Sichuan region; **b** Yunnan region

facilities. The 2008 Wenchuan earthquake sequence and 2013 Lushan earthquake sequence exhibit a significant influence on the overall completeness of the catalog, due to the limitation of routine seismological observation and interpretation, which is to be dealt with in the future. As an overall estimate, a homogeneous monitoring capability of completeness magnitude 3.0 (Mignan et al., 2013; Su et al., 2003) can be estimated since 1970. Figure 10.5 gives the magnitude-time plot of the small and strong earthquake in CSES region from 1970 to 2020, which shows the complex feature from the small events and in average ten-years recurrence period from the strong events above M_S7.0.

The catalog uses a 'mixed' magnitude system. For earthquakes below 4.0 the Chinese local magnitude M_L is used. For earthquakes above 5.5 the Chinese surface wave magnitude M_S is used. For the events in between, the transition from the Chinese M_L to the Chinese M_S is not clearly defined. Figure 10.6 shows the magnitude in the present catalog and that in the NEIC catalog. In the comparison, the events in both catalogs, with the differences of latitude and longitude less than 0.1° and origin time less than 5 s, are regarded as the identical events with different parameters determined by different networks. Similar to Bormann et al. (2007), the color bar is used to indicate the number of events, and the diagonal line indicates the equality of the two magnitudes.

Considering the ambiguity in the magnitude, in this study, we use only 'magnitude' for the size of the events in the catalog. For the analysis based on number-counting, such as the calculation of the Earthquake Potential Score (EPS) of Nowcasting method (Rundle et al., 2016, 2017, 2019), this ambiguity does not affect the result. This is the case especially when the cutoff magnitude is taken as below 4.0 (that is,

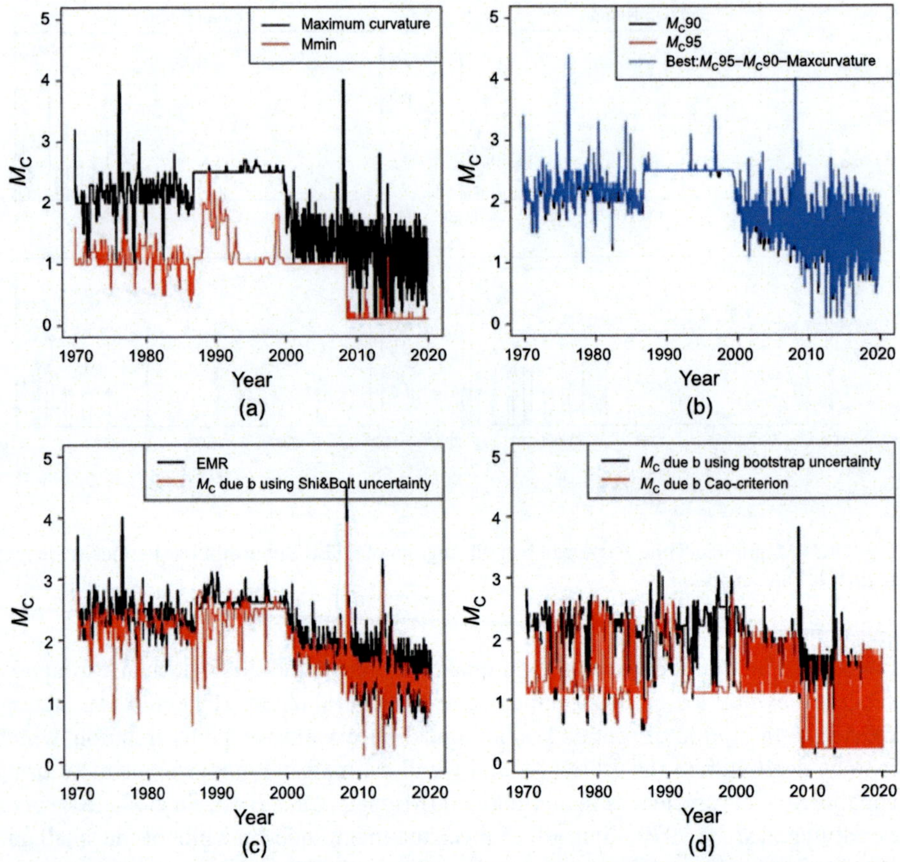

Fig. 10.4 Completeness magnitude calculated using different statistical seismology methods (Woessner & Wiemer, 2005) for earthquake catalog of CSES. **a** Maximum curvature method and Mmin method; **b** M_C90 and M_C95 methods; **c** EMR method and Shi&Bolt uncertainty method; **d** bootstrap uncertainty method and Cao-criterion method

the magnitude is simply the local M_L). According to a convention, the cutoff magnitude of the earthquake catalog under study should be no less than the completeness magnitude and is generally taken as about 2 magnitude units less than that of the 'target' earthquakes (Jiang & Wu, 2010; Rundle et al., 2016). In our analysis, we take the cutoff magnitude as 3.0 and the 'target' magnitude as above 6.0, the same with the work in Rundle et al. (2016). However, for other algorithms which consider not only the number-counting but also the size of earthquakes, the issues of magnitude transition, as well as of magnitude uncertainty, have to be taken into account.

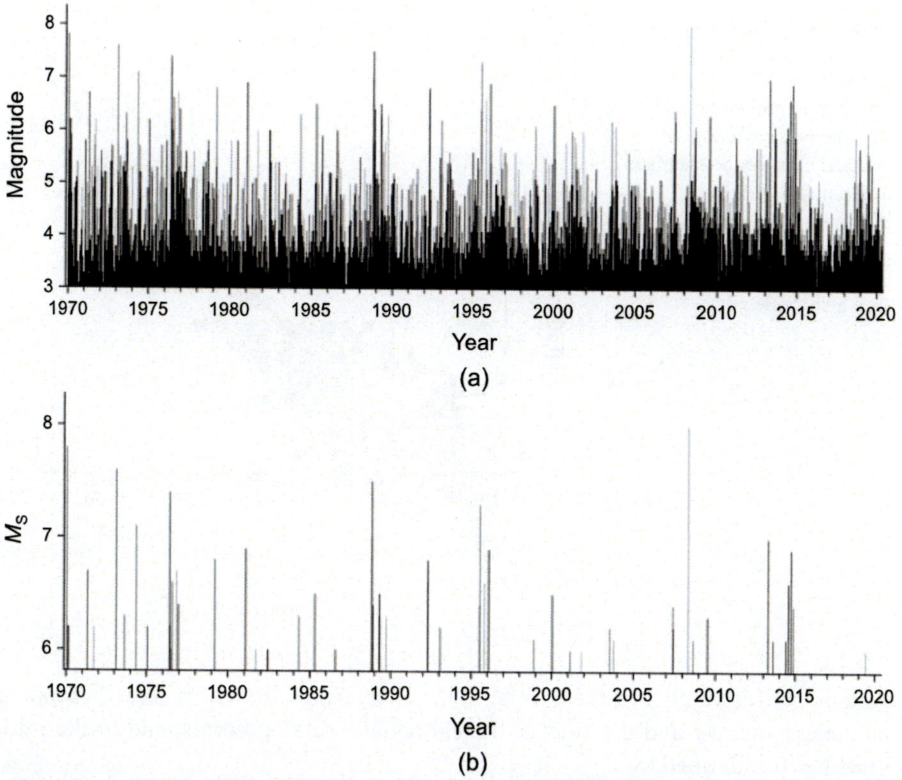

Fig. 10.5 Magnitude-Time plot of the earthquakes in CSES region during 1970 to 2020. **a** Small earthquakes in CSES region with magnitude above $M_L3.0$; **b** Strong earthquakes in CSES region with magnitude above $M_S6.0$

10.3 Ergodicity of CSES

Many studies have shown that the driven dissipative mean-field systems appear to show effective ergodic dynamics and the models of statistically stationary tend to reside in a sequence of physical states that are similar to equilibrium. Large events have the ability to drive the system out of equilibrium for a period of time, after which this state is reestablished (Egolf, 2000; Tiampo et al., 2003, 2007). Due to the temporal complexity of the small events in this region, which is denoted in Fig. 10.5a, we choose the approach of Tiampo et al., (2004, 2007) to describe the degree of its heterogeneity and in turn the statistical predictability of strong events. In general, the ergodicity behavior can be measured using the Thirumalai-Mountain diagram (the TM diagram) (Thirumalai & Mountain, 1993; Thirumalai et al., 1989). Initially, the TM metrics can measure effective ergodicity or the difference between the time average of a quantity, generally related to the energy at each sub-region and the ensemble average level over the entire system. For a simple statistical model, like slide block model (Tiampo et al., 2002), the TM metrics can be calculated not for the

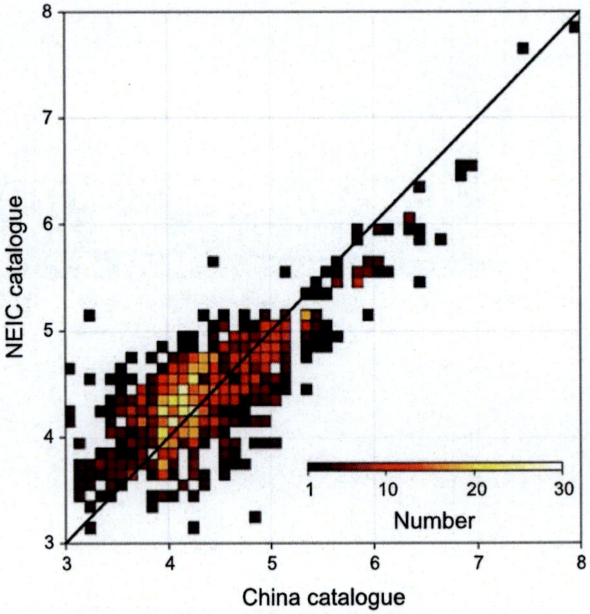

Fig. 10.6 Comparison of the magnitude in the CENC catalog (noted as China catalog) in horizontal axis, and the NEIC catalog in vertical axis for the events in the central China North–South Seismic Belt. The time span is from 1980/01/01 to 2021/01/01. See text for details

energy but for number of the events. In the analysis, the function $TM_n(t)$ computed at creasing time t and the start of the earthquake catalog correspond to the initial time $t = 0$ is defined by

$$TM_n(t) = \frac{1}{L}\sum_{i=1}^{L}[n_i(t) - \overline{n}(t)]^2 \tag{10.1}$$

in which L is the number of the grids in the spatial range under consideration, $n_i(t)$ the number of events in the ith grid within time 0–t, and $\overline{n}(t)$ the mean number of events located in all of the grids within time 0–t,

$$n_i(t) = \frac{1}{t}\int_0^t n_i(\acute{t})\mathrm{d}\acute{t} \tag{10.2}$$

$$\overline{n}(t) = \frac{1}{L}\sum_{i}^{L} n_i(t) \tag{10.3}$$

According to Tiampo et al., (2004, 2007), the deviation of time-average quantity from its ensemble average value in Eq. 5.1 is decreasing as the time and all grids in the system are statistically similar in physical. It means if the system exhibits the 'effective ergodic behavior' in a long period, then the function $1/TM$ will be proportional to time t, or, $TM_n(t) \propto 1/t$.

Figure 10.7 shows the TM diagrams of the seismicity in CSES region and its two sub-regions (defined in Figs. 10.1and 10.2), with grid size 0.2° × 0.2° and cutoff magnitude 3.0. The results of the top and below indicate the case of containing and not containing the 2008 Wenchuan earthquake sequence. We can see that its two sub-regions show different ergodic behavior and both of them have not good ergodicity before 1980. Previous results (e.g., Tiampo et al., 2004, 2007) concluded that, once variations in the catalog data resulting from technical and network issues are considered, natural seismicity displays ergodicity. In this regard, and considering the history of seismic networks in China, the ergodicity before 1980 in CSES region and its two sub-regions may be attributed to the regionalized characteristics of seismological observation and interpretation.

In general, since about 1980, the seismicity in the whole CSES region, the Sichuan region and the Yunnan region show good ergodicity. The 2008 Wenchuan earthquake seems to have a significant influence on the ergodicity, as observing the case of the Sichuan region, in which the disruption of the TM diagram around 2008 is predominately contributed by the Sichuan region (with Wenchuan events), and disappears when the Wenchuan sequence is removed (with no Wenchuan events). This is consistent with the previous results (e.g., Tiampo et al., 2007) that a large event disrupts the ergodicity. But here it may be argued that such an influence may simply be the contribution of the aftershock sequence which obeys a different statistic (that is, the Omori's law).

In the statistical physics of earthquakes, the importance of ergodicity lies in that effective ergodicity generally implies stationary periods of meta-stable equilibrium, which guaranteed the applicability of the analysis models developed based on the theoretical assumptions such as mean-field and spatio-temporal correlations, being applied to (applicable for) a large class of physical systems, driven threshold systems

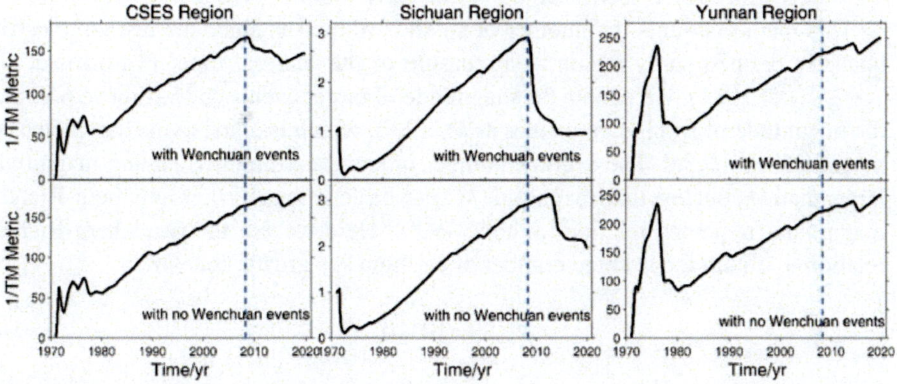

Fig. 10.7 The TM diagram to show the ergodicity of the seismicity in CSES region and two sub-regions. **Left:** the whole CSES region; **Middle:** Sichuan region; **Right:** Yunnan region. **Top:** the result computed using the earthquake catalog with Wenchuan earthquake sequence; **Bottom:** the result without Wenchuan earthquake sequence. The vertical dashed lines indicate the time of 2008 Wenchuan earthquake

in general and earthquake fault systems in particular (Tiampo et al., 2004, 2007). Our result shows that for the North–South Seismic Region in central China, the availability of seismic data to test such models (for example, the Pattern Informatics (PI) algorithm) (Zhang et al., 2015) is from 1980.

10.4 Nowcasting Method

The item "Nowcasting" has been used for a long time in meteorology or economic discipline and extended into many fields, such as used as standard measures to assess the state of the economy in the tendency prediction of the stock market,[1] reveal the real-time scene of the cloud movement used in the fact-based weather data and was finally applied into the hazard evacuation of the earthquake by seismologists, called "Nowcasting Earthquakes" (Luginbuhl et al., 2017, 2018; Rundle et al., 2016, 2017, 2019). It usually refers to the determination of the current state of a system via some indirect means, which is different from the item "forecasting" that is usually used to output a probability result of future activity in the hazard (earthquake) cycle. In seismology study field, this method used to be applied to the area in which there is a relatively strong seismic activity and has a long history earthquake record, for example, South California in America. "Nowcasting" earthquakes used the number of small earthquakes in the time intervals of large earthquakes that occurred in the same or nearly same study area having the same dynamic evolution history, to calculate the Earthquake Potential Score (EPS). The selection of the magnitude threshold of small earthquakes should ensure that all small earthquakes in the catalog were a completeness sequence. The main advantage of this method is that it can be applied more easily than the direct way in the hazard assessment in earthquake prediction and forecasting which seems difficult to be carried out in practice.

This method counts the number of small earthquakes since the last large earthquake in defined study region as a measure of the 'natural time' (Varotsos et al., 2002, 2005, 2011). We denote the magnitude of target events as M_λ (above 6.0) and the magnitude of small earthquakes as M_σ (3.0), which is equal to the completeness magnitude as M_L3.0. The average number of small earthquakes having magnitude larger than M_σ but less than magnitude M_λ can be described by the Gutenberg-Richter magnitude-frequency relation (Scholz, 1990). Here we use the Gutenberg-Richter relation to obtain the average number of earthquakes greater than M:

$$N_{avg} = 10^a 10^{-bM} \tag{10.4}$$

where the b value is typically near (around) 1 and a value can give the seismicity level of the region.

[1] https://en.wikipedia.org/wiki/Nowcasting_(economics).

If we denote N_σ the average number of the small earthquakes greater than M_σ and N_λ the average number of the large earthquakes greater than M_λ,

$$N_\sigma = 10^a 10^{-bM_\sigma} \tag{10.5}$$

$$N_\lambda = 10^a 10^{-bM_\lambda} \tag{10.6}$$

then we can obtain the average number of the small earthquakes N between-in large earthquakes:

$$N = \frac{N_\sigma - N_\lambda}{N_\lambda} = 10^{b(M_\lambda - M_\sigma)} - 1 \tag{10.7}$$

here N does not depend on a value.

Mathematically, the process of Nowcasting calculation can be expressed through the potential for large earthquakes (target events) by computing the cumulative distribution function (CDF) using the small events with magnitude large than M_σ but less than M_λ. We use the standard method (Bevington & Robinson, 2003) to calculate the probability density function (PDF) and CDF of small earthquakes of magnitude larger than M_σ but less than M_λ in each larger earthquake cycle. After we get the CDF of the small earthquakes, we can compute the current value of CDF, $p[n \le n(t)]$, using the current count of small earthquakes $n(t)$ at time t, here t is the calendar time since the last large earthquake. This value can be defined as the earthquake potential score (EPS) at time t:

$$\text{EPS} = p[n \le n(t)] \tag{10.8}$$

We then use the EPS as the potential for the occurrence of the next large earthquake greater than M_λ in our study region. From Eqs. 10.7 and 10.8, we can see that the EPS value will increase with time since the last large earthquake and will quickly reset to EPS = 0 once the next large earthquake occurs. After that, it will again begin to increase until the next large event occurs. In the whole process, there is no fitting calculation for the model parameters and just a direct way to interpret the earthquake data. Because the definition of the EPS has no relation with the seismicity rate in study region, so the "Nowcasting" method can be applied to not only a large region but also a small region somewhere like a regional (district) area of a city.

10.5 Nowcasting for Earthquake Potential

Figure 10.8 shows the EPS result calculated through Nowcasting method for the larger events with $M6.0$ and $M6.5$ in CSES and its two sub-regions using the small

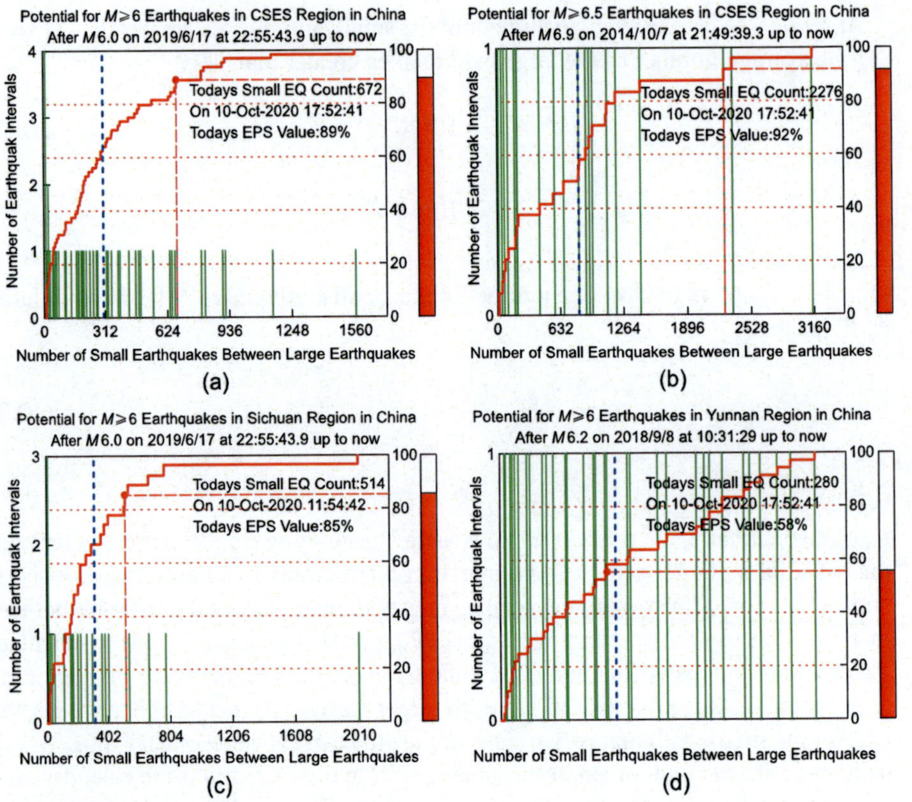

Fig. 10.8 Earthquake Potential Score (EPS) calculated by Nowcasting method up to the end time of the earthquake catalog (here is Oct. 10, 2020) for CSES and its sub-regions. **a** Potential for $M \geq 6.0$ in CSES region; **b** Potential for $M \geq 6.5$ in CSES region; **c** Potential for $M \geq 6.0$ in Sichuan region; **d** Potential for $M \geq 6.0$ in Yunnan region. The green bars are the histograms of the number of small events ($M_\sigma \geq 3.0$) between the large events. The red curve is the CDF $p[n \leq n(t)]$ of the histogram values. The red points correspond to the number of small events since the last large events and the relative EPS is shown on right. The mean number of small events ($M_\sigma \geq 3.0$) between large events is indicated by the vertical dashed line in blue color. The information each case gives is also texted in every plot

events between the large events as the earthquake sample. For example, Fig. 10.8a gives the information that it has occurred 672 small events in CSES region since the last large event with magnitude above 6.0 which occurred in 2019/6/17 and the EPS value reaches 89%. Similarly, Fig. 10.8b–d suggest that the EPS value is 92% in CSES region for the large events with magnitude above 6.5, and 85, 56% in Sichuan and Yunnan region for the large events with magnitude above 6.0. To describe the earthquake potential for different sizes of the target earthquake, Fig. 10.9 gives the computation results of the larger earthquake within magnitude range M6.0–7.5. Due to the number of small events or samples separated by the larger events, the smoothness of the EPS curve in each diagram will be influenced by the number of existed larger events. Despite this, the result of CSES region can still reveal a

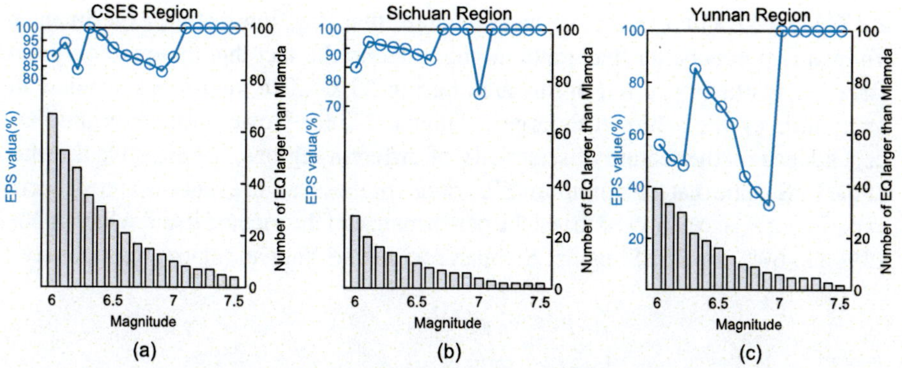

Fig. 10.9 EPS value calculated by Nowcasting method up to Oct. 10, 2020 for the target earthquake in CSES region, Sichuan region and Yunnan region, within magnitude range *M*6.0– 7.5. The left axis uses the blue circles and line to indicate the EPS value, the right axis uses the grey bars to note the number of large events used in each computation

relatively high potential of larger earthquakes. The Sichuan region has a higher EPS value for the target earthquake within *M*6.0–7.5, comparing with the Yunnan region.

Figure 10.10 shows the retrospective experiment we carried out to evaluate the performance of Nowcasting method for the target earthquakes above 6.0 during 2010–2020 in CSES region and its sub-regions, and the calculation for the ratio of the EPS cases above 0.5, which can be seemed as the performance of random guess. The ratio of EPS value above 0.5 in CSES region, Sichuan region and Yunnan region

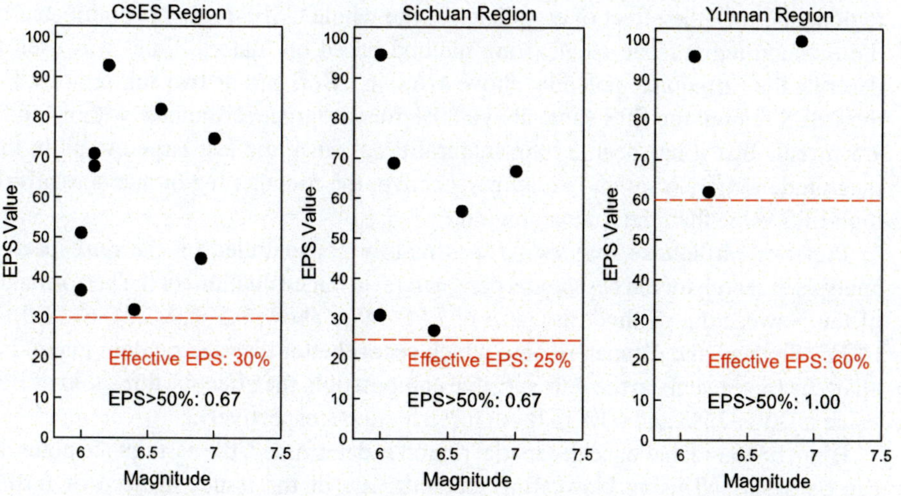

Fig. 10.10 Evaluation for Nowcasting method used to forecast the target events above 6.0 during 2010–2020 in CSES and its sub-regions. The ratio of cases that the EPS value larger than 50% is shown, comparing this method with the random guess. The effective threshold of the EPS value relative to the large events that occurred in practice is also noted by the horizontal dashed line and text with red color

are 0.67, 0.67 and 1. Here we can conclude that the Nowcasting experiment for Yunnan region is better than random guess due to the fact that there are only three target events that occurred in this area during 2010–2020, and it has accumulated many little events before each large earthquake. Since Nowcasting algorithm has a dependence on the feature of seismicity of different regions, the effective threshold of the EPS value that the minimum EPS value for these target events in a retrospective experiment is also analyzed to test the performance of the method itself, which is 30%, 25% and 60% for CSES region, Sichuan region and Yunnan region, respectively.

10.6 Conclusion and Discussion

The experiment of earthquake forecast is a fundamental and essential work in CSES region. After some regional forecast with time-independent and/or time–dependent models in the framework of CSEP1.0 applied in the central North–South Seismic Region (Jiang & Wu, 2008; Jiang & Zhuang, 2010; Zhang et al., 2015, 2016), discussion using different statistical methods of forecast and algorithms to analysis earthquake catalog is also needed in CSES region.

In this work, we analyzed the earthquake catalog in CSES region and its sub-regions to describe the basic characteristics of seismicity. As an indicator for the assumptions of meta-stable equilibrium, ergodicity was shown to exist for most of the time period from 1980 to the present. Such ergodicity is apparently disrupted by the 2008 Wenchuan earthquake sequence. Due to the clustering of aftershocks of the Wenchuan earthquake, the ergodicity of the Yunnan region is better than the Sichuan region, and thus the effect of ergodicity of the whole CSES region is dominated by the Sichuan region. The Nowcasting method based on 'natural time' was used to describe the earthquake potential above $M6.0$ in CSES and its two sub-regions. For the CSES region, the EPS value above 80% for the target earthquake within $M6.0$–7.5 reveals that it has been a long 'natural time' since the last large events in this magnitude range. From an overall perspective, the Sichuan region has a relatively high EPS value than the Yunnan region.

Previous earthquake forecasting was mainly concentrated on the retrospective analysis to search for the best model parameters. In our evaluation for the performance of the Nowcasting method, the ratio of EPS value above 0.5 is 0.67, 0.67, 1.0 for CSES, Sichuan and Yunnan region, which seems better than the random guess. For all of the large events in the retrospective computation, the effective threshold of EPS value is 30%, 25% and 60% in these three regions, respectively.

It can be found that once the spatial region is determined, the earthquake potential can be described using Nowcating method. One of the issues in need of further discussion is, how long the forecast window from now this method can give in practice. For this question which might determine the application to the detailed earthquake work in different regions, the combination of the Nowcasting method with the Annual Consultation Conference in China might be a good chance. The Annual Consultation Conference has been a persistent forward forecast test organized by

China Earthquake Administration (CEA) since the 1970s (Zhao et al., 2010; Wu et al., 2013; Zhang et al., 2019a, 2019b). Previous studies showed that the performance of such annual forecast outperforms random guess (Shi et al., 2001), such as the alarm regions in Figs. 10.1 and 10.2, while to much extent contributed by the statistics of seismicity (Zhuang & Jiang, 2012). But for the Nowcasting method, the detailed study on the data analysis using clustered or non-cluster catalog is still in need. On the other hand, a possible query is that the Nowcasting method used in the area where there is weak seismicity might not give a convincing result, because there are not enough earthquake samples to take part in the calculation. But from another view, the actual seismic recording is the basic requirement for the Nowcasting method, especially for the region having weak seismicity.

Acknowledgements The earthquake catalog is provided by the China Earthquake Networks Center (CENC). The authors thank Prof. Wu Zhongliang in IEFCEA for his support and guide for this work. Prof. John B. Rundle in UC Davis and Jiang Changsheng in IGPCEA also gave guidance to Nowcasting method and the ergodicity analysis, respectively. Prof. Li Yonggang of Department of Earth Sciences, University of Southern California, help to make corrections for the text. This work has been supported by the National Key Research and Development Program of China (Grant No. 2018YF E0109700), the National Science Foundation of China (Grant No. 42004038) and the Special Fund of the Institute of Earthquake Forecasting, China Earthquake Administration (Grant No. 2020IEF0501).

References

Bevington, P. R., & Robinson, D. K. (2003). Data reduction and error analysis for the physical sciences, 3rd ed. McGraw-Hill.

Bormann, P., Liu, R., Ren, X., Gutdeutsch, R., Kaiser, D., & Castellaro, S. (2007). Chinese national network magnitudes, their relation to NEIC magnitudes, and recommendations for new IASPEI magnitude standards. *Bulletin of the Seismological Society of America, 97*, 114–127.

Chen, Y., & Booth, D. C. (2011). *The Wenchuan earthquake of 2008: Anatomy of a disaster*. Springer Science & Business Media.

Department of Earthquake Disaster Prevention of the State Seismological Bureau. (1995). *The catalogue of Chinese historical strong earthquakes*. Seismological Press.

Department of Earthquake Disaster Prevention of the China Earthquake Administration. (1999). *The catalogue of Chinese modern earthquakes*. China Science and Technology Press.

Egolf, D. A. (2000). Equilibrium regained: From nonequilibrium chaos to statistical mechanics. *Science, 287*(5450), 101–104.

Jiang, C. S., & Wu, Z. L. (2008). Retrospective forecasting test of a statistical physics model for earthquakes in Sichuan-Yunnan region. *Science in China Series d: Earth Sciences, 51*, 1401–1410.

Jiang, C. S., & Wu, Z. L. (2010). PI Forecast for the Sichuan-Yunnan Region: Retrospective test after the may 12, 2008, Wenchuan Earthquake. *PAGEOPH, 167*, 751–761.

Jiang, C. S., & Zhuang, J. C. (2010). Evaluation of background seismicity and potential source zones of strong earthquakes in the Sichuan-Yunan region base on the space-time ETAS model. *Chinese Journal of Geophysics-Chinese, 53*, 305–317.

Jordan, T. H. (2006). Earthquake predictability, brick by brick. *Seismological Research Letters, 77*, 3–6.

Li, S. L., Mooney, W., & Fan, J. C. (2006). Crustal structure of the mainland of China from deep seismic sounding data. *Tectonophysics, 420*, 239–252.

Luginbuhl, M., Rundle, J. B., & Turcotte, D. L. (2018). Natural time and nowcasting earthquakes: Are large global earthquakes temporally clustered? *PAGEOPH, 175*, 661–670.

Luginbuhl, M., Rundle, J. B., Hawkins, A., & Turcotte, D. L. (2017). Nowcasting earthquakes: A comparison of induced earthquakes in Oklahoma and at the Geysers, California. *PAGEOPH, 175*, 49–65.

Michael, A. J., & Werner, M. J. (2018). Preface to the focus section on the collaboratory for the study of earthquake predictability (CSEP): New results and future directions. *Seismological Research Letters, 89*, 1226–1228.

Mignan, A., Jiang, C. S., Zechar, J. D., Wiemer, S., Wu, Z. L., & Huang, Z. (2013). Completeness of the mainland of China earthquake catalog and implications for the setup of the China earthquake forecast testing center. *Bulletin of the Seismological Society of America, 103*, 845–859.

Rundle, J. B., Luginbuhl, M., Giguere, A., & Turcotte, D. L. (2017). Natural time, nowcasting and the physics of earthquakes: Estimation of seismic risk to global megacities. *PAGEOPH, 175*, 647–660.

Rundle, J. B., Luginbuhl, M., Giguere, A., & Turcotte, D. L. (2019). *Natural time, nowcasting and the physics of earthquakes: Estimation of seismic risk to global megacities* (pp. 123–136).

Rundle, J. B., Turcotte, D. L., Donnellan, A., Grant Ludwig, L., Luginbuhl, M., & Gong, G. (2016). Nowcasting earthquakes. *Earth and Space Science, 3*, 480–486.

Savran, W. H., Maechling, P. J., Werner, M. J., Jordan, T. H., Schorlemmer, D., Rhoades, D. A., Marzocchi, W., Yu, J., & Vidale, J. E. (2018). The collaboratory for the study of earthquake predictability version 2.0 (CSEP2.0): New capabilities in earthquake forecasting and testing. Poster Presentation at 2018 SCEC Annual Meeting.

Scholz, C. H. (1990). *The mechanics of earthquakes faulting*. Cambridge University Press.

Shi, Y. L., Liu, J., & Zhang, G. M. (2001). An evaluation of Chinese annual earthquake predictions, 1990–1998. *Journal of Applied Probability, 38A*, 222–231.

Su, Y. J., Li, Y. L., Li, Z. H., Yi, G. X., & Liu, L. (2003). Analysis of minimum complete magnitude of earthquake catalog in Sichuan-Yunnan Region. *Journal of Seismological Research, 26*, 10–16. (in Chinese with English abstract).

Sun, Q. Z., & Wu, S. G. (2007). *Development of the earthquake monitoring and prediction in China during 1966–2006*. Seismological Press. (in Chinese with English abstract).

Thirumalai, D., & Mountain, R. D. (1993). Activated dynamics, loss of ergodicity, and transport in supercooled liquids. *Physical Review E, 47*, 479–489.

Thirumalai, D., Mountain, R. D., & Kirkpatrick, T. R. (1989). Ergodic behavior in supercooled liquids and in glasses. *Physical Review A, 39*, 3563–3574.

Tiampo, K. F., Rundle, J. B., McGinnis, S., Gross, S. J., & Klein, W. (2002). Mean-field threshold systems and phase dynamics: An application to earthquake fault systems. *Europhysics Letters, 60*(3), 481–487.

Tiampo, K. F., Rundle, J. B., Klein, W., Martins, J. S. S., & Ferguson, C. D. (2003). Ergodic dynamics in a natural threshold system. *Physics Review Letters, 91*(23), 238501.

Tiampo, K. F., Rundle, J. B., Klein, W., & Martins, J. S. S. (2004). Ergodicity in natural fault systems. *PAGEOPH, 161*, 1957–1968.

Tiampo, K. F., Rundle, J. B., Klein, W., Holliday, J., Sá Martins, J. S., & Ferguson, C. D. (2007). Ergodicity in natural earthquake fault networks. *Physical Review E, 75*, 066107.

Varotsos, P. A., Sarlis, N. V., & Skordas, E. S. (2002). Long-range correlations in the electric signals that precede rupture. *Physical Review E, 66*, 011902.

Varotsos, P. A., Sarlis, N. V., & Skordas, E. S. (2011). *Natural time analysis: The new view of time*. Springer.

Varotsos, P. A., Sarlis, N. V., Tanaka, H. K., & Skordas, E. S. (2005). Some properties of the entropy in the natural time. *Physical Review E, 71*, 032102.

Woessner, J., & Wiemer, S. (2005). Assessing the quality of earthquake catalogues: Estimating the magnitude of completeness and its uncertainty. *Bulletin of the Seismological Society of America, 95*, 684–698.

Wu, Z. L. (2020). Seismic experimental sites: Challenges and opportunities. *Journal of the Geological Society of India, 95*, 113–116.

Wu, Z. L., Zhang, Y., & Li, J. W. (2019a). Coordinated distributed experiments (CDEs) applied to earthquake forecast test sites. In L. Yong-Gang (Ed.), *Earthquake and disaster risk: Decade retrospective of the Wenchuan earthquake* (pp. 107–115). Higher Education Press and Springer Nature Singapore Pte Ltd.

Wu, Z. L., Zhang, X. D., & Sun, K. (2019b). China seismic experiment site: Scientific challenges. *Acta Geologica Sinica—English Edition, 93*, 273–273.

Wu, Z. L., Ma, T. F., Jiang, H., & Jiang, C. S. (2013). Multi-scale seismic hazard and risk in the China's mainland with implication for the preparedness, mitigation, and management of earthquake disasters: An overview. *International Journal of Disaster Risk Reduction, 4*, 21–33.

Wu, Z. L., Li, G. J., Li, M., Li, X. J., Jiang, H., & Zhang, S. F. (2014). *The field investigation of the earthquake, earthquake phenomenology from the field: The April 20, 2013, Lushan earthquake* (pp. 13–22). *Springer*

Xu, X. W., & Deng, Q. D. (1996). Nonlinear characteristics of paleoseismicity in China. *Journal of Geophysical Research: Solid Earth, 101*, 6209–6231.

Xu, X. W., Zhang, P. Z., Wen, X. Z., Qin, Z. L., Chen, G. H., & Zhu, A. L. (2005). Features of active tectonics and recurrence behaviors of strong earthquake in the western Sichuan province and its adjacent regions. *Seismology and Geology, 27*, 446–461. (in Chinese with English abstract).

Yi, G. X., Wen, X. Z., & Xu, X. W. (2002). Study on recurrence behaviors of strong earthquakes for several entireties of active fault zones in Sichuan-Yunnan Region. *Earthquake Research in China, 18*, 267–276. (in Chinese with English abstract).

Zechar, J. D., & Jordan, T. H. (2010). Simple smoothed seismicity earthquake forecasts for Italy. *Annals of Geophysics, 53*, 99–105.

Zhang, Y. X., Wu, Z. L., Zhang, X. D., Yu, H. Z., Zhang, S., Zhang, X.T., Yu, C., & Hu, C. Z. (2019a). CSEP2 from China Perspective. Poster Presentation at 2019 SCEC Annual Meeting.

Zhang, Y. X., Wu, Z. L. Zhang, X. T, Li, G. (2019b). Annual earthquake potential consultation: A real forward prediction test in China. In L. Yong-Gang (Ed.), *Earthquake and disaster risk: Decade retrospective of the wenchuan earthquake* (pp. 117–134). Higher Education Press and Springer Nature Singapore Pte Ltd. https://doi.org/10.1007/978-981-13-1085-0_5

Zhang, P. Z., Deng, Q. D., Zhang, G. M., Ma, J., Gan, W. J., Min, W., Mao, F. Y., & Wang, Q. (2003). Active tectonic blocks and strong earthquakes in the continent of China. *Science in China Series d: Earth Sciences, 46*, 13–24.

Zhang, S. F., Wu, Z. L., & Jiang, C. S. (2015). The Central China North-South seismic belt: Seismicity, ergodicity, and five-year PI forecast in testing. *PAGEOPH, 173*, 245–254.

Zhang, S. F., Wu, Z. L., & Jiang, C. S. (2016). Reducing false alarms of annual forecast in the Central China North-South seismic belt by reverse tracing of precursors (RTP) Using the pattern informatics (PI) 'hotspots.' *PAGEOPH, 174*, 2401–2410.

Zhao, Y. Z., Wu, Z. L., Jiang, C. S., & Zhu, C. Z. (2010). Reverse tracing of precursors applied to the annual earthquake forecast: Retrospective test of the annual consultation in the Sichuan-Yunnan Region of Southwest China. *PAGEOPH, 167*, 783–800.

Zhuang, J. C., & Jiang, C. S. (2012). Scoring annual earthquake predictions in China. *Tectonophysics, 524–525*, 155–164.

Chapter 11
Characterization of the San Andreas Fault by Fault-Zone Trapped Waves at Seismic Experiment Site, Parkfield, California: A Review

Yong-Gang Li

Abstract The San Andreas Fault (SAF) at Parkfield, California has been taken as a seismic experimental site since the 1970s. The San Andreas Fault Observatory at Depth (SAFOD) measured a ~200-m-wide damage zone with a seismic velocity reduction of ~25–30% within the mature SAF at ~3 km depth. Observations and 3-D finite-difference simulations of fault-zone trapped waves (FZTWs) recorded in a sequence of seismic experiments conducted at the Parkfield SAF characterized a low-velocity waveguide in the SAF, likely extending from the surface to ~7-km depth with a width taper from ~200 m to ~100 m, within which seismic wave velocities are reduced by 20–40% and Q is 15–50. The FZTW data recorded before and after the 2004 M6 Parkfield earthquake show that the SAF co-seismically weakened with seismic wave velocities reduced by ~2.5% and consequently healed with seismic velocity recovery by ~1.25% within approximate three months after the mainshock. This chapter is a retrospective review of the results from our previous experiments at the Parkfield SAF, California, we expect that it will be valuable for researchers who are carrying out seismic experiments at the active faults to develop the community models of seismic wave velocity, fault structure and earthquake forecasting in the China Seismic Experimental Site (CSES) and global earthquake regions.

Keywords San Andreas fault · Fault-zone trapped (guided) wave · Waveguide effect · Dispersion · Strong motion · Earthquake hazard and mitigation · Propagator matrix · Finite-difference simulation · Rupture zone width and depth · Co-seismic damage · Post-seismic healing · Seismic experimental site · SAFOD · CSES

Y.-G. Li (✉)
Department of Earth Sciences, University of Southern California, Los Angeles, CA 90089, USA
e-mail: ygli@usc.edu

© Higher Education Press 2022

Y.-G. Li et al. (eds.), *China Seismic Experimental Site*,
https://doi.org/10.1007/978-981-16-8607-8_11

11.1 Finding and Modeling of FZTWs Generated by Microearthquakes at the SAF, Parkfield California

Field and laboratory studies indicate that fault zones appear to undergo high, fluctuating, stresses and pervasive cracking during earthquakes (Aki, 1984; Mooney and Ginzburg, 1986; Chester et al. 1993; Andrews, 2005; Cochran et al., 2009). Numerous seismological studies in the past decades have revealed a low velocity zone surrounding the surface trace of the San Andreas fault (SAF) in central California (e.g., Lees and Malin, 1990; Eberhart-Phillips and Michael, 1993; Catchings et al., 2002; Thurber et al., 2004, 2006). Body-wave arrival-time inversions on the SAF at Parkfield show this zone to be 1–2 km wide and with S-wave velocities of 2.0–3.0 km/s and Vp/Vs ratios of 2.0–2.3 (Michelini and McEvilly, 1991). A possible cause for the low S-wave velocities and corresponding high Vp/Vs ratios within the fault zone is dilatant fracturing due to high pore-fluid pressures (Nur, 1972). Byerlee (1990) and Rice (1992) note that the high pore pressures within a fault zone at seismogenic depths may be due in part to greater permeability than in the adjacent blocks. Rice (1992) further presumes a supply of fluids at the ductile roots of the crustal fault zone. The high pore pressure within the fault zone may be caused by porosity reduction due to progressive volumetric compacting (Blanpied et al., 1998), thermal expansion of fluids due to frictional heating (Lachenbruch, 1980), or the dehydration of clay minerals (Raleigh, 1977).

However, due in part to their depth of burial, the magnitude, depth and lateral extent of the resulting rock damages are not well constrained (Hickman and Evans, 1992, 1995; Sleep et al., 2000). The thickness and depth of such zones are critical ingredients in the understanding of fault mechanics (Sibson, 1977; Dieterich, 1978; Scholz, 1990; Rice, 1992; Das and Aki, 1997) and evaluate strong ground motion near the faults (e.g. McBride and Brown, 2001; Catchings et al., 2016; Li et al., 2016) for mitigation of earthquakes. Ground-motion variability typically controls hazard at levels of engineering interest (Anderson and Brune, 1999). Shaw (2006) found that rupture nucleation, termination and propagation in multicycle earthquake simulations are highly sensitive to fault geometries (including endpoints, connecting structures and geometric complexities) which are poorly constrained at seismogenic depths (Papageorgiou and Aki, 1983). In addition, as with prior UCERF2-based CyberShake studies, characterization of ground-motion path effects due to low-velocity fault zones, is highly dependent on their subsurface velocity structure. Recently, observations and computer modeling suggest that mature faults, long-lived faults along major plate boundaries like the San Andreas Fault (SAF) in California, host large earthquakes at much lower levels of stress than the expected static strength. In order to better understand the physics that drive the major earthquakes occurring on it, (Lambert et al., 2021) have examined the relationships between the size of an earthquake, the energy it radiates out, and the heat generated by movement along the SAF, using computer modeling. Their potential explanations are that the faults are quasi-statically strong but experience considerable weakening during earthquake propagation as self-healing pulses (e.g. Heaton, 1990), or that the faults are persistently weak, because of fluid overpressure (e.g. Noda et al., 2009; Perry et al., 2020). Both types of fault rupture models yield large earthquakes, but the forces creating those earthquakes are different. Understanding the differences will be key to developing more accurate earthquake forecast maps and to expect shaking from large earthquakes. A better knowledge of detailed damage and rheology structure of mature faults at seismogenic depth would help with distinguishing two physics-based models that estimate shaking and seismic hazard from future earthquakes.

As a result of the seismic velocity reduction due to intense fracturing, brecciation, liquid-saturation, and possibly high pore-fluid pressure, the fault zone forms a natural low-velocity waveguide. When an earthquake occurs within the fault zone, some seismic energy is trapped in the waveguide and propagates as normal modes that are formed by the constructive interference of multiple reflections at boundaries between the low velocity fault zone and the high-velocity surrounding rock (Li, 1988; Li et al., 1990; Li and Leary, 72 1990). Because of the dispersion and concentration of trapped (or call guided) waves within and near the low-velocity fault zone, these waves are able to depict the fault zone width of tens to several hundreds of meters. These waves have been used as a probe of the state of the SAF in California, (e.g. Li et al., 1994a, 1994b; Li and Vidale, 1996; Ben-Zion, 1998; Korneev et al., 2003; Malin et al., 2006; Ellsworth and Malin, 2011). In this Chapter, we review our previous researches for spatial and temporal characterizations of the SAF at Parkfield, using the fault-zone trapped wave (FZTW) since middle 1980s (Li, 1988; Li and Leary, 1990; Li et al., 1990; Li et al., 1997, 2004, 2007a, 2008). Parkfield, California has been taken as a scientific seismic experimental site in USA since 1970s, and the SAF is the target fault to investigate earthquake physics and forecasting. Since then, more than ten types of field experiments (including seismic, geophysical, geochemical, geodetic ones) have been carried out at this experimental site. In the fall of 2003, a pair of scientific wells was drilled at the San Andreas Fault Observatory at Depth (SAFOD) site; the main-hole (MH) passed a ~200-m-wide low-velocity zone (LVZ) and fault core with highly fractured rocks of the SAF at the depth of ~ 3.2 km below the drilling wellhead on the ground (Hickman et al., 2005; Zoback, 2007, 2008; Lockner et al., 2011). Borehole seismographs were installed in the SAFOD MH in 2004, which were located within the LVZ of the fault at ~ 3-km depth to recorded microearthquakes occurring on the SAF. On September 28 2004, a M6 earthquake occurred ~15 km southeast of the town of Parkfield. Combination of the data recorded in our field experiments conducted before and after the 2004 M6 earthquake provides a unique opportunity to insight the co-seismic damage and post-mainshock heal of the SAF associated with this strong earthquake (Li et al., 2006, 2007b). We expect that the retrospective of the results from our previous experiments at the Parkfield SAF, California, will be valuable for on-going experiments conducted at active faults by the Southern California Earthquake Center (SCEC) and the China Seismic Experimental Site (CSES) to develop the community velocity and fault models as well as earthquake hazard and forecasting models for earthquake mitigation.

Seismic waves trapped in fault zone were first discovered in a three-dimensional vertical seismic profiling experiment in 1986 at the area surrounding a borehole drilled into fault zone of the Oroville, California, earthquake of 1975, and consequently similar waveforms were found at the Middle Mountain station of Parkfield borehole seismic network located within the San Andreas Fault (SAF) for local earthquakes (Leary et al., 1987; Li et al., 1987, 1990; Li, 1988; Li & Leary, 1990). Records of borehole seismographs placed near the fault zone showed an unusually long-period wavetrain when the active seismic source or micro-earthquakes were located within the fault zone. At the early stage, we attributed these characteristic waveforms to a Love wave type mode trapped in a low-velocity, low-Q zone. The Parkfield ten-station borehole seismic network of three-component seismographs was installed at depths of 200–300 m and operated by the U.S. Geological Survey and University of California at Santa Barbara, which recorded numerous instances of pronounced dispersive wavetrains following the S-wave arrivals. These wavetrains are first time called fault-zone trapped waves (Li et al., 1990, Science). Station MM is located ~ 8 km northwest of the tiny town of Parkfield.

For example, Fig. 11.1a shows seismograms recorded at two borehole network stations MM and JN. We recognized prominent FZTWs at station MM, but not at station JN located ~ 3 km away from the SAF for a M0.6 earthquake occurring

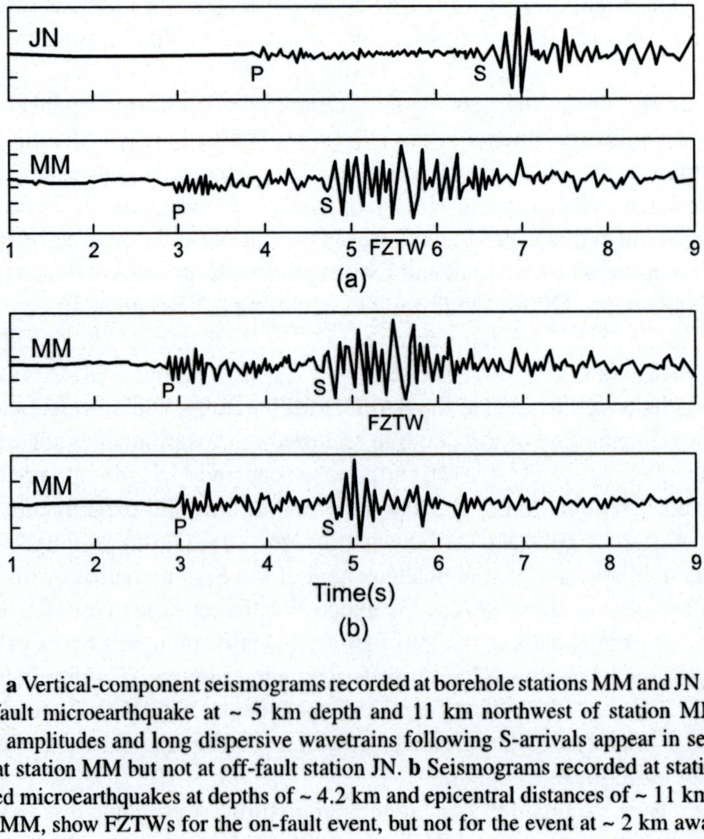

Fig. 11.1 **a** Vertical-component seismograms recorded at borehole stations MM and JN at Parkfield for a on-fault microearthquake at ~ 5 km depth and 11 km northwest of station MM. FZTWs with large amplitudes and long dispersive wavetrains following S-arrivals appear in seismograms recorded at station MM but not at off-fault station JN. **b** Seismograms recorded at station MM for two repeted microearthquakes at depths of ~ 4.2 km and epicentral distances of ~ 11 km northwest of station MM, show FZTWs for the on-fault event, but not for the event at ~ 2 km away from the SAF

within the fault zone at depth of 5 and 11 km northwest of station MM. Figure 11.1b shows seismograms recorded at station MM for two microearthquakes occurring within and away from the fault zone, respectively. Prominent FZTWs appear in seismograms for the on-fault event. In contrast, body is dominant in seismograms for the off-fault event. These observations show existence of a low-velocity waveguide on the SAF between event epicenters and station MM, and the sensitivity of FZTW's generating and recording to positions of the source and receiver relative to the fault-zone waveguide. Figures in this section updated from Li (1988), Li et al. (1990) and Li and Leary (1990).

The FZTWs recorded at station MM were first modeled with the simplest possible physical theory (*SH-Love* trapped waves), the simplest component of motion (vertical), and the simplest waveguide model (infinite, uniform, parallel planar layers, non-attenuation media) to obtain an initial estimate of the fault-zone waveguide properties (Li & Leary, 1990; Li et al., 1990). At Parkfield, the SAF separates two crustal blocks. The Franciscan block lies to the northeast, which contains fluid-rich metamorphic oceanic sedimentary rocks. The southwest crustal block is primarily harder granite, in which the seismic velocities are faster. Seismic tomography shows the asymmetry in seismic velocity variations across the SAF (Thurber et al., 2004). Therefore, we assumed a four-part vertical velocity structural model. The model contains a fault zone with lowest seismic velocity and transition zones with mild low velocity, which are sandwiched by the high-velocity half-space on the southwest and northeast sides with the lower velocity in the NE than in the SW half-space (Fig. 11.2). Model parameters include velocities of multi-layers, widths of the fault zone and transition zone, and locations of source and receivers.

We employed a modified propagator matrix method to compute the vertical wavenumbers that lead to propagating SH-Love energy at specified frequencies for a planar multilayer velocity model to simply synthesize FZTWs (Li, 1988; Li & Leary, 1990). The number of layers can be expanded arbitrarily as required for the purpose. In the isotropic medium, the equation of Love-type motion in the jth layer may be written by

$$\mu_j \frac{\partial^2 U_j}{\partial x^2} + \mu_j \frac{\partial^2 U_j}{\partial z^2} = \rho_j \frac{\partial^2 U_j}{\partial t^2} \tag{11.1}$$

where U_j is the y-component displacement in the jth layer. μ is the rigidity of isotropic rock with material density ρ_j. The displacement $U_j(x, z, t)$ as a function of time t and position in the x–z plane at $y = 0$ is expressed by

$$U_j(x, z, t) = A_j(x) \exp[ik(z - ct)] \tag{11.2}$$

where $A_j(x)$ is the x-dependent amplitude term, k is the wavenumber and c is the phase velocity. Putting (11.2) into the equation of motion (11.1), we obtain

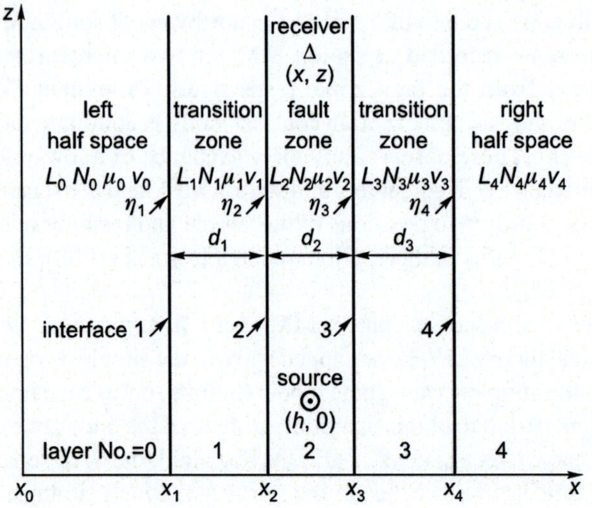

Fig. 11.2 The simple fault zone model with multiple vertical planar layers in between two half spaces. The fault core zone is sandwiched by two transition zones. The layers are numbered $j = 0$, 4 for the two half spaces of high speed rocks, $j = 1$, 3 for transition zones of moderate velocities, $j = 2$ for a fault core zone of the lowest velocity. The interface x coordinates are x_1, x_2, x_3, and x_4. Thicknesses of transition zones and fault core zone are d_1, d_2 and d_3. The shear rigidity μ_j for isotropic medium, L_j and N_j for transversely isotropic medium, and vertical wavenumber ν_j are assigned to each layer. The phase shift η_j is assigned to each interface. A *SH* source with motion along *y*-axis is located at coordinates (*h*, 0) within the fault zone. A receiver is located at position (*x*, *z*) in the model

$$\frac{\partial^2 A_j(x)}{\partial x^2} = -k_j^2\left(\frac{c^2\rho_j}{\mu_j} - 1\right)A_j(x) = -\nu_j^2 A_j(x) \tag{11.3}$$

where ν_j is the vertical wavenumber (normal to the fault zone) for the jth layer.

$$\nu_j = k\sqrt{\left(\frac{c^2}{V_{sh}^2} - 1\right)_j} = k\nu_j^\circ \tag{11.4}$$

We define ν_j as a pseudo-vertical wavenumber for the *j*th layer. Introducing the right-going and left-going plane waves with wavenumber ν_j and amplitudes $\grave{S}_j$ and $\acute{S}_j$, respectively, into Eq. (11.3), the total displacement amplitude $A_j(x)$ of the *SH-Love* wavefield in the *j*th layer is defined by

$$A_j(x) = \grave{S}_j\exp[-i\nu_j(x - x_j)] + \acute{S}_j\exp[+i\nu_j(x - x_j)] \tag{11.5}$$

The right-ward and left-ward plane waves may yield a constructive interference in the *j*th layer if the difference between their phases are $2n\pi$; n corresponds the order of trapped modes. Following Rader et al. (1985), we introduce the complex phase shift η_j at the *j*th interface between the *j*th layer and the $j + 1$th layer, relating amplitude $\grave{S}_j$ of the right-going wave to amplitude $\acute{S}_j$ of the left-going wave by

$$\acute{S}_j = \grave{S}_j \exp[2i\eta j] \tag{11.6}$$

Then (11.5) may be written by

$$\begin{aligned} A_j(x) &= \grave{S}_j \exp[i\eta_j]\{\exp[-i(\nu_j(x - x_j) + \eta_j)] \\ &\quad + \exp[i(\nu_j(x - x_j) + \eta_j)]\} \\ &= B_j \cos[\nu_j(x - x_j) + \eta_j] \end{aligned} \tag{11.7}$$

and

$$B_j = 2\grave{S}_j \exp[i\eta_j] \tag{11.8}$$

The *SH-Love* type trapped waves radiate within the two half spaces with the amplitudes exponentially decaying

$$A_o(x) = B_o \exp[i\nu_o(x - x_1)] \tag{11.9}$$

and

$$A_4(x) = B_3 \exp[i\nu_3(x - x_4)] \tag{11.10}$$

Because the displacement and stress must be continuous at the interface, the continuity of displacement $A_j(x)$ and stress across the interface between layer j and $j + 1$ can be expressed by the condition

$$\begin{pmatrix} \cos(\nu_j d_j + \eta_j) & \cos(\eta_{j+1}) \\ \mu_j \nu_j \sin(\nu_{j+1} d_j + \eta_j) & -\mu_{j+1} \nu_j \sin(\eta_{j+1}) \end{pmatrix} \begin{pmatrix} B_j \\ B_j \end{pmatrix} = 0 \tag{11.11}$$

where d_j is the thickness of the jth layer. The first matrix on the left side of this homogeneous equation is called the layer matrix. It forms a standard eigenvalue problem. For a non-trivial solution, the displacement of the layer matrix must be equal to zero, it yields

$$\mu_{j+1}\nu_{j+1} \tan[\eta_{j+1}] = \mu_j \nu_j \tan[\eta_j + \nu_j(d_j)] \tag{11.12}$$

Instead of the standard Thomson-Haskell propagator matrix method, we used a rapid phase-shift recursion algorithm to compute *SH-Love* trapped wave dispersion and amplitude response in the model. From the recursion Eq. (11.10), a set of unknown phase shifts η_j at successive interfaces ($j = 1, 2, 3, 4$) can be recursively computed with the starting phase shift η_1 at the interface between the half space and transition zone left of the fault zone (refer to Fig. 12.1),

$$\eta_1 = \tan^{-1}\left(-i\frac{v_0}{\mu_1 v_1}\right) \tag{11.13}$$

The phase shift at the interface between the transition zone and half space right of the fault zone is computed by

$$\eta_4 = \tan^{-1}\left(-i\frac{\mu_4 v_4}{\mu_3 v_3}\right) = \eta_3 + v_3 d_3 \tag{11.14}$$

For a given frequency ω, the phase velocities for different modes may be computed by recursively solving (11.12), (11.13) and (11.14) as a root-finding condition. After knowing a set of phase shift values at interfaces and phase velocities in multiple layers and the two half spaces for the specified mode n, the amplitudes of *SH-Love* trapped waves can be computed as a function of the location in the model using Eqs. (11.7)–(11.10). With phase velocities c at each frequency for each layer being known via the above procedure, group velocities of *SH-Love* type trapped waves can be stably computed from integral relations (Aki & Richards, 1980),

$$V = \frac{1}{c}\frac{I_2}{I_1} \quad I_1 = \frac{1}{2}\int_{-\infty}^{\infty} \rho(x)A^2(x)\mathrm{dx}$$

$$I_2 = \frac{1}{2}\int_{-\infty}^{\infty} \mu(x)A^2(x)\mathrm{dx} \tag{11.15}$$

The simulated seismograms for particle motion at receiver position (*x, z*) due to a source at position (*h,* 0) may be computed using the *Love*-wave Green's function (Aki & Richards, 1980)

$$Gyy(x, z, h, 0; \omega) = \sum_N \frac{A(\omega, \nu_n, h)A(\omega, \nu_n, x)}{4\nu_n I_2}\exp[i(\nu_n z + \pi/2)] \tag{11.16}$$

where ω and ν_n, are respectively the frequency and the vertical wavenumber, and n is the order of the mode of trapped waves, $n = 0$ for fundamental mode, $n = 1$ for the first-high mode and so on. z is the distance between the source and receiver along the vertical fault zone. $A(\omega, \nu_n, h)$ and $A(\omega, \nu_n, x)$ are displacement contribution terms dependent on the source location (*h,* 0) and the receiver location (*x, z*) in the model for a given frequency ω. They can be computed by Eqs. (11.7)–(11.10). The energy integral I_2 is obtained from Eq. (11.15).

A wave packet of the *SH-Love* trapped waves for a single mode is then computed by the integration of the spectral density $|F(\omega)|$ given by (11.16) in the interesting frequency range,

$$f(z,t) = \frac{1}{2\pi}\int_{-}^{+}|F(\omega)|\exp[-i\omega t + i\nu_n z + i\omega(\omega)]\mathrm{d}\omega. \tag{11.17}$$

where $\phi(\omega)$ is the initial phase at the source position.

Figure 11.3 illustrates computed dispersion curves of the fundamental and first-higher modes and wave component amplitudes as a function of distance normal to the fault zone for a series of frequencies, using formulas given above and specific model parameters. The dispersion curves show slower phase and group velocities

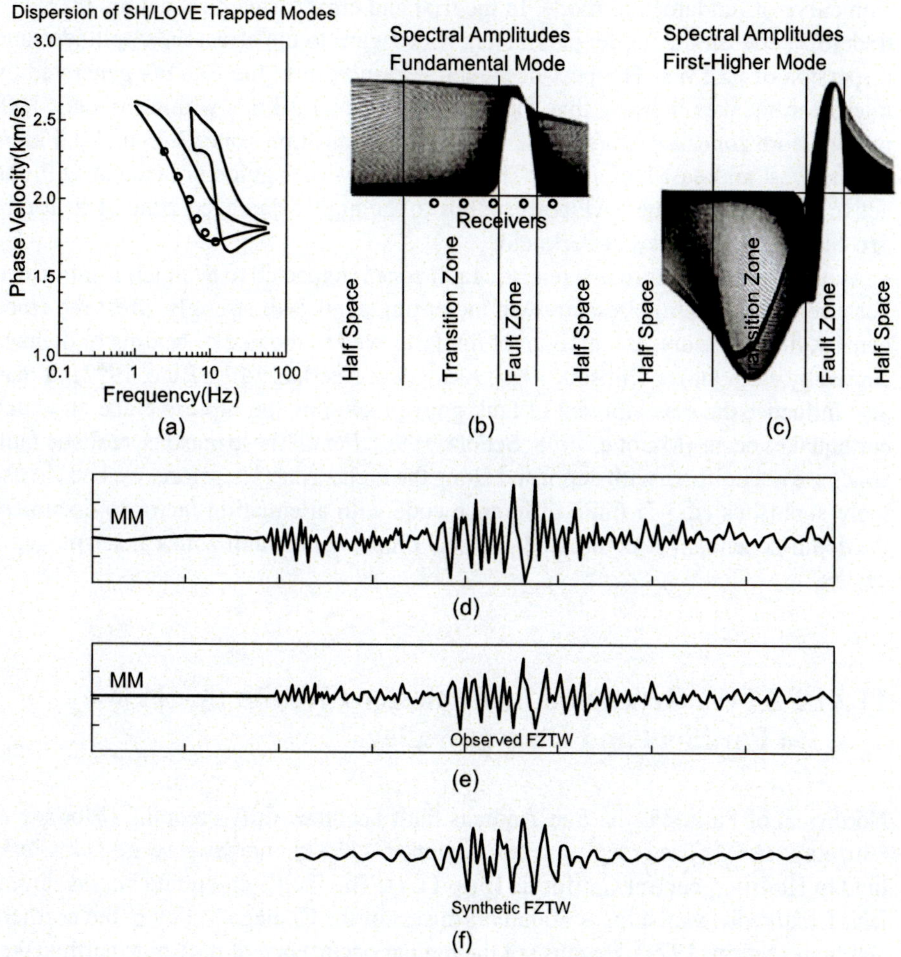

Fig. 11.3 **a** Computed dispersion curves for the fundamental and first-higher modes of SH-Love type FZTWs in the fault-zone model in (**b**). Circles denote group velocities measured from recorded FZTWs at six specific frequencies. **b** and **c** Spectral amplitudes for the fundamental and first-higher modes at 3.5–30 Hz. **d** and **e** FZTWs in vertical-component seismograms recorded at MM station for two microearthquakes on the SAF at 5 km depth and 11 km northwest of MM (Fig. 11.4). **f** Synthetic SH-Love type trapped waves within fault zone using the model in (**b**), matching recorded FZTWs

at higher frequencies, and the Airy-phase with the minimum group velocity. The maximum spectral amplitude for the fundamental mode appears at the center of the low-velocity fault-zone waveguide while the first-higher mode has highly diagnostic amplitude extinction in the center of the fault zone due to phase reversal. The simulated *SH/Love* trapped waveforms at station within the fault core zone for a source ~ 11 km from receivers are comparable with the FZTWs with large amplitudes and dispersive wavetrains following S-arrivals recorded at station MM located within the SAF near Parkfield (Fig. 11.1). Group velocities measured from recorded FZTWs using a narrow-band filter with 1 Hz band, are comparable for the synthetic dispersion curve of fundamental mode. In the trial and error forward modeling, there is a trade-off between the model parameters with regard to the effect on amplitudes and dispersion of FZTWs. The best-fit modeling results from the FZTWs generated by micro-earthquakes indicate that shear velocities are 1.8 km/s within the fault 160-m-wide core zone, 2.4 km/s in the 400-m-wide transition zone, 2.75 and 2.6 km/s for the west and east half space. The low-velocity waveguide coincident with the active fault plane of the SAF corresponds to the highly damaged zone of recurrent $M6$ earthquake episodes at Parkfield.

However, the structure of a realistic fault zone is expected to be much complicated because the increasing pressure with increasing depth will strongly affect the crack density, fluid pressure, and amount of fluids, as well as the rate of healing of damage caused by earthquakes (Sibson, 1977; Aki, 1984; Byerlee, 1990; Rice, 1992). It may also influence the development of fault gouge including the rupture plane on which earthquakes occur (Marone, 1998; Scholz, 1990). For all these reasons, realistic fault zones are not uniform with depth and along the fault strike. Therefore, we need to use more sophisticated 3-D finite difference code with attenuation factor to document the depth-dependent structure and physical properties of fault zones at seismogenic depths.

11.2 FZTWs Generated by Explosions Within the SAF at Parkfield and Cienega Valley

Northwest of Parkfield the San Andreas fault becomes fully creeping (Mooney & Ginzburg, 1986). The creeping section extends ~ 150 km northwestward from Parkfield to Hollister, central California (Fig. 11.4). The fault's creep rate begins diminishing markedly, vanishing ~ 30 km northwest of the Cienega Valley in the northern Gabilans (Nason, 1971). Because of the regular occurrence of moderate earthquakes on the SAF at Parkfield and in the northern Gabilan range, both of them have been identified as potential sites for the SAFOD (Hickman et al, 1994; Zobark et al., 2007, 2008). In order to determine the upper limit of the Parkfield seismogenic zone for this project, the U.S. Geological Survey conducted an active seismic reflection and refraction experiment near Parkfield. Similarly, a larger active experiment was carried out at Cienega Valley by the University of Wisconsin-Madison, and the USGS

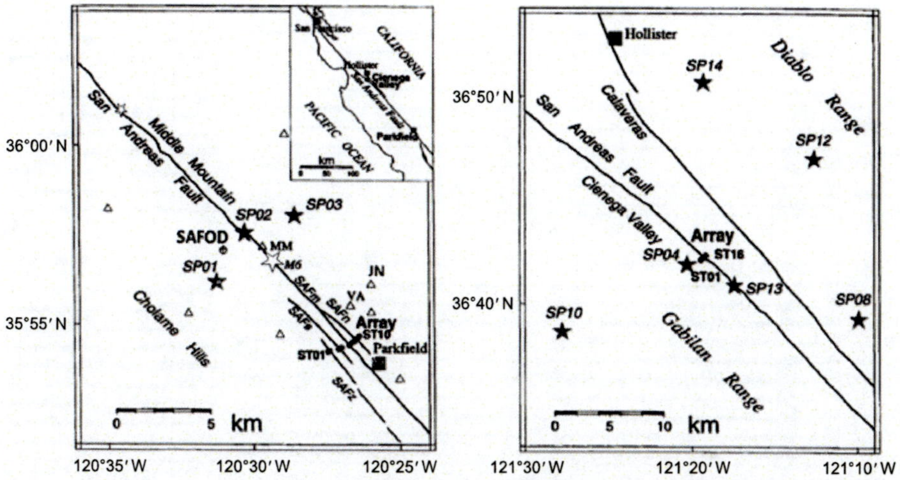

Fig. 11.4 Maps showing locations of explosions SP01, SP02 and SP03 (stars) and 10-station (ST01 to ST10) array at Parkfield (left) and 16-station (ST01 to ST16) array at Cienega Valley (right). The inset shows locations of working areas in California. Large open stars denote the *M*6 earthquake in 1966 and a microearthquake which generated the FZTWs (see Fig. 11.3d). Open triangles denote Parkfield Borehole Seismic Network stations. MM—Middle Mountain; JN—Joaquin North. Lines indicate surface traces of the SAF, including the main fault strand SAFm, northern strand SAFn, southern strand SAFs, and southwestern fracture zone SAFz. The circle with cross denotes the San Andreas fault drilling site (SAFOD). Figures from Li et al. (1997)

(Thurber et al., 1996). A group from University of Southern California participated in the experiments and deployed linear seismic arrays across the SAF to record FZTWs excited by the explosions detonated within the fault zone (Li et al., 1997).

At Parkfield, the USGS detonated three shots SP01, SP02, and SP03 (~ 200 kg of explosive gel set at a depth of about 25 m) at Parkfield. Shot SP02 was detonated within the San Andreas fault zone about 0.75 km southeast of station MM. It was at MM that the FZTWs generated by earthquakes were first observed on the SAF (Li et al., 1990). Shots SP01 and SP03 were about 2.5 km southwest and northeast of station MM, respectively. We deployed a 10-station array of portable, 1-Hz three-component seismic recorders (Fig. 11.4) along a 2-km-long line crossing multiple fault strands of the SAF. The array was ~ 7 km southeast of MM and ~ 2 km north-west town Parkfield. Stations ST02, ST03, ST05, and ST08 were deployed at the surface breaks of SAFz, SAFs, SAFm, and SAFn, respectively. Six other stations were deployed between these fault traces. Since fault-zone guided waves propagate smoothly only along connected fault segments, this configuration of stations was designed to see if any of the fault strands were coupled to the SAFm at shot point SP02. Figure 11.5a shows FZTWs generated by explosion SP02, which are similar to those generated by earthquakes (Fig. 11.1) but have lower frequencies and travel more slowly, suggesting that FZTWs travel at lower velocities within the shallower portion of the SAF. The coda-normalized ratios show maximum 2-Hz peaks centered at stations ST05 and ST08 located on SAFm and SAFn due to guided waves on theses

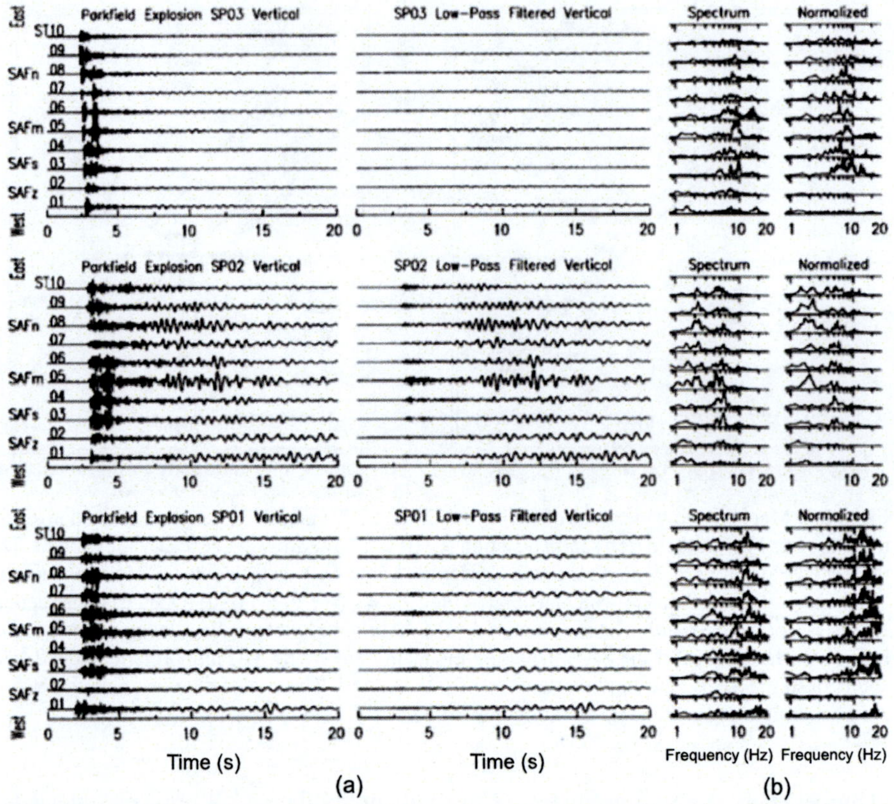

Fig. 11.5 **a** Vertical components of raw and low-pass (< 4 Hz) filtered seismograms recorded at the 10-station array (ST01 through ST10) for three explosions SP01, SP02, and SP03 at Parkfield, show FZTWs arriving at stations ST05 to ST08 at ~ 7 s after the shot time for SP02 detonated within fault zone. SAFm, SAFn, SAFs, and SAFz indicate the corresponding locations of the SAF strands. **b** The amplitude spectra of raw seismograms computed in a 10-s window starting from the P arrival (left), and the ratio of amplitude spectra in a 5-s window starting from the S arrival to the spectra in a window with the same length starting at 30 s after the origin time (right). Normalized spectra are plotted using a fixed scale

strands. (Fig. 11.5b). In contrast, FZTWs were not recorded for SP01 and SP03 away from the fault zone. Because SP02 was detonated within the main fault northwest of the array (Fig. 11.4), this implies that the northern fault strand should connect to the main fault at shallow depth. To the southwest of SAFm, the coda-normalized spectral amplitude ratio at 2-Hz decreases more sharply as the distance from the fault trace increases than that to the northeast of SAFn, consistent with a northeastern transition zone of intermediate velocity detected in Fig. 11.2.

At Cienega Valley south of Hollister, thirteen shots were detonated, including one explosion SP13 within the SAF adjacent to the northern Gabilan range (Fig. 11.4). We deployed a 16-station (ST01 to ST16) array along a 625-m-long line across the fault trace near SP13. The station spacing was 25 m in the central part of the array and 100 m at the ends of the array. The surface trace of the SAF lies between

ST07 and ST08. Figure 11.6a shows three-component seismograms at the cross-fault array for explosion SP13 at Cienega Valley. We observed distinct FZTWs with large amplitudes and long post-S durations beginning ~ 5 s after the shot origin time for stations between ST05 and ST11 located close to the SAF fault trace. The amplitude spectra show the peak at 3 Hz, which decays rapidly as the distance from the fault trace increases (Fig. 11.6b). These distinct wavetrains are similar to those at Parkfield but with slightly higher frequencies than those at Parkfield. We computed amplitude spectra in a 5-s window length starting from the S-arrivals. The spectral peak at 3 Hz decreases rapidly as the distance from the fault trace increases.

We computed synthetic *Love-type* guided modes in a simple one-dimensional model composed of a uniform low-velocity planer layer sandwiched between two

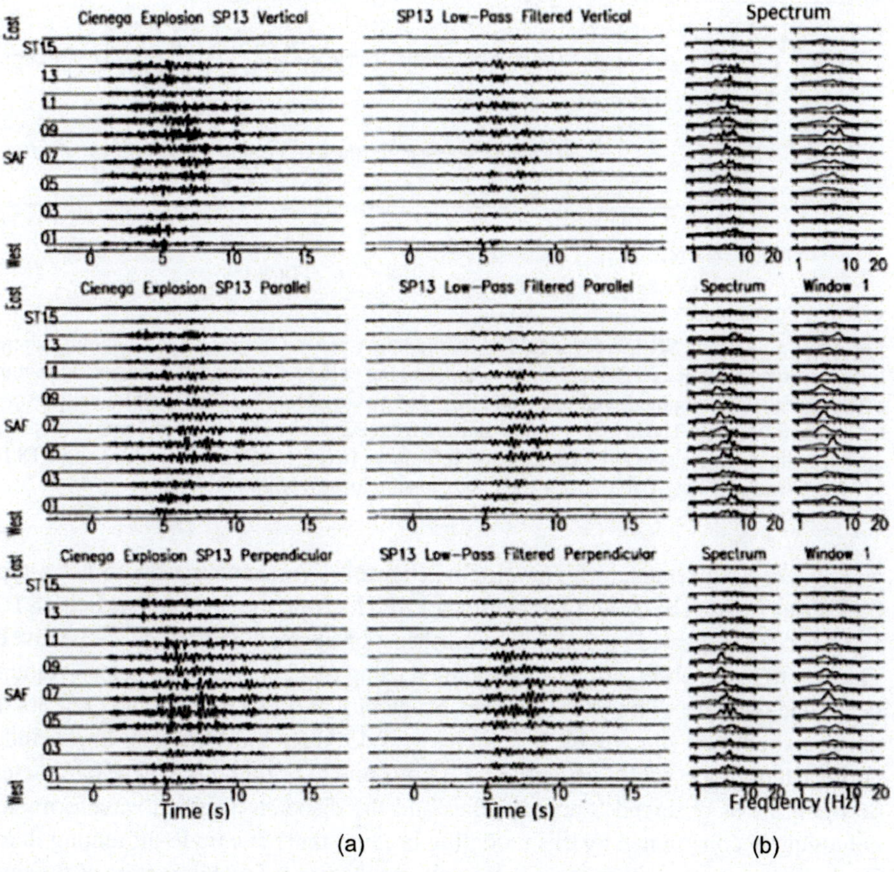

Fig. 11.6 **a** Vertical-components raw and < 4 Hz filtered seismograms recorded at the 16-station array for explosion SP13 at Cienega Valley. Station ST07 was located on the SAF surface trace. FZTWs appear between 5 s and 10 s at stations ST05 to ST11 within the distances span of 150 m. **b** The amplitude spectra of seismograms computed in a 10-s window starting from the P arrival (left), and amplitude spectra computed in a 5-time windows starting from the S arrival are plotted using a common scale

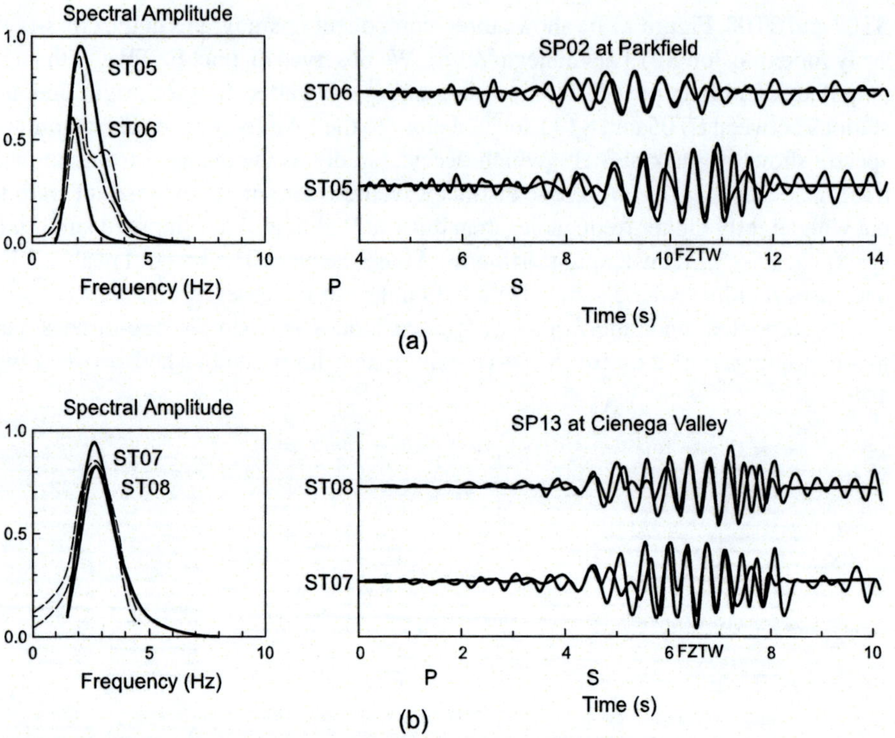

Fig. 11.7 **a** Synthetic SH-Love type FZTWs (solid lines), and < 4 Hz filtered seismograms (dotted lines) recorded at stations ST05 and ST06 within and out of the SAF for SP02 detonated within the SAF at Parkfield. Computed synthetic (solid line) and observed (dashed lines) amplitude spectra of FZTWs at ST05 and ST06 in a 5-s window starting from S arrivals. The observed amplitude spectra are normalized to the same unit of synthetics. **b** Same as in (**a**), but for stations ST07 and ST8 for SP13 at Cienega Valley. Two stations and explosion are located within the SAF

half-spaces with relatively high velocity. Figure 11.7a shows synthetic Love-type FZTWs at stations ST05 and ST06 for shot SP02 within the SAF at Parkfield. ST05 and ST06 were located on and 160 m northeast of the main fault trace, respectively. The best fit to the observations was obtained using model parameters: a low-velocity fault zone with the width of 160–180 m, S velocity of 0.8–0.9 km/s, and Q of 30–40 km/s; a southwest block with S velocity of 1.45 km/s and Q of 100 km/s; and a northeastern block with S velocity of 1.3 km/s and Q of 60 km/s. These parameters on the basis of observed S arrival times and the dispersive guided waveforms are not uniquely constrained by this modeling because there is a trade-off among them. Furthermore, these estimates are valid only for the upper few kilometers of the fault zone because the distances between the shot and arrays were short (~ 8 km at Parkfield, and ~ 4 km at Cienega Valley). However, comparing these model parameters obtained from explosion-excited FZTWs to those from the FZTWs generated by earthquakes (in Sect. 11.1) illuminates a low-velocity waveguide existing along the SAF from the surface to depth at Parkfield.

We also simulated FZTWs recorded at stations ST07 and ST08 located within the fault zone for explosion SP13 at Cienega Valley (Fig. 11.7b). The best-fit model parameters are: a ~ 120-m wide fault zone with S velocity of 0.85 km/s and Q of 35 km/s, a southeast block with S velocity of 1.4 km/s and Q of 100 km/s, and a northeast block with the S velocity of 1.2 km/s and Q of 60 km/s. Again, these model parameters are non-unique and only reflect the properties of the shallow part of the fault zone (Li et al., 1997).

11.3 2-D Finite-Difference Investgation of Trapping Efficiency of Low-Velocity Fault-Zones

We have used two-dimensional finite-difference code with a fourth-order in space and second-order in time (Vidale, 1989) to simulate SH-type fault-zone guided waves. For example, Fig. 11.8 shows that a low-velocity fault zone is capable to trap and focus seismic energy (Li et al., 1994a). In this instance, the most of seismic energy at frequencies 3–6 Hz is trapped within the fault zone and traveled slowly. The seismic energy at a frequency lower than 3 Hz, which penetrated a few tens to hundreds of meters into the surrounding rock and traveled more rapidly. Because of the dispersion

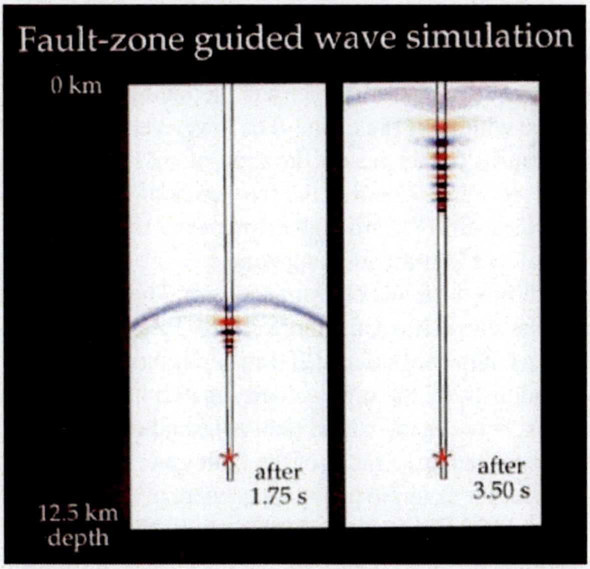

Fig. 11.8 Cross section of the fault zone and 2-D finite-difference simulation of fault-zone guided waves 1.75 s and 3.5 s after an earthquake (red star) at 9.75-km depth. Blue and red indicate the opposite polarities of displacement, and white indicates no displacement. The shear velocities are 2.1 km/s and 3.1 km/s within and outside the fault zone. The fault zone is 180 m wide at the surface and tapers to 120 m at a depth of 10 km. The simulation is two-dimensional finite-difference method and models the fault-parallel horizontal component of displacement

and concentration of guided waves within and near the fault zone, these waves can be used as a probe of the state of the low-velocity fault zone.

Li and Vidale (1996) then carried out a detailed investigation of various factors to influence trapping efficiency of a low-velocity fault-zone using 2-D finite-difference code: (1) source offsets from the center of the fault zone, (2) fault-zone width and velocity reduction, (3) thickness of the surface layer lying on the fault zone, (4) kink in a continuous fault zone, (5) non-planar fault zone, (6) fault branches, (7) fault step-over, (8) fault zone depth, and (9) source radiation.

Here, we review some examples to show the sensitivity of these factors to produce and record fault-zone trapped waves (FZTWs). In the simulations, our reference model has an 8-km source depth, the fault-zone width is 200 m, and the source is placed at the center of the low-velocity fault zone. The receiver array is centered on the surface fault trace. We set the wall rock shear-wave velocity to 3 km/s and the fault zone velocity to 2 km/s, so the velocity anomaly in the fault zone is about 40%. This structure most effectively traps 3- to 10-Hz shear waves. The maximum amplitude of fault zone waves generated in each case is listed in Table 11.1.

Figure 11.9a exhibits the effect of source location on trapping efficiency of a fault-zone low-velocity waveguide by placing a double-couple source at 0 and 300-m away from the center of fault zone. FZTWs are most efficiently excited by a seismic source that is located within the waveguide. A source location 200 m, or one fault-zone width, outside the fault zone still produces visible FZTWs, but their amplitude is about a factor of 3 weaker than for source within the fault zone. When a source 400 m or farther away from the fault zone, FZTWs are not visible clearly.

Figure 11.9b exhibits the effect of varying the fault-zone velocity contrast and width. The amount of reduction of velocity in the fault zone determines the amount of FZTW resonance within the fault zone. The lower-velocity material within a fixed width fault zone causes longer post-S duration of FZTWs. The width of the low-velocity fault zone, however, controls the frequency of the FZTWs. An 80-m-wide fault zone shows a clear shift toward higher-frequency trapped waves compared with the reference model of a 200-mwide fault zone.

Figure 11.9c exhibits the effect of a surface layer. The closer that the low-velocity fault zone approaches the surface, the more clearly FZTWs appear at the fault trace. The simulation for a fault zone buried at 200-m depth shows that FZTWs are distorted and attenuated obviously by the high-velocity materials atop the fault zone due to the velocity contrast between the buried fault zone and the surface layer. As the fault zone is buried deeper than three times of the fault-zone width, FZTWs are invisible at the ground surface, for example in our experiment at the blind Port Hills fault that ruptured in the 2011 *M*6.3 Christchurch earthquake in New Zealand (Li et al., 2014a). Downhole measurements may be desirable in the case of these complications. This tendency is also shown by the simulation with a 600-m-thick layer with 1.5 km/s shear-wave velocity on top of fault zone. 600 m of cover significantly distorts and attenuates the guided waves. This factor may make the characterization of fault zones difficult beneath sedimentary basins or other fault geometry that do not reach the surface.

Table 11.1 Perturbations in model parameters and maximum amplitudes

Source position: (offsets from the center of the fault zone)					
Source offset (m)	0	100	300	500	1100
Maximum amplitude	3.9829	2.5941	0.5719	0.4075	0.6215
Fault-zone width and velocity contrast					
Fault-zone velocity (km/s)	2.5	1.8	2.0		
Fault-zone thickness (m):	200	200	80		
Maximum amplitude	3.2891	3.5119	1.5625		
Thickness of the surface layer					
Surface-layer velocity (km/s)	1.5	1.5	1.5		
Layer thickness (m)	200	600	1000		
Maximum amplitude	3.6208	3.0158	2.2454		
Depth of the fault top (m)	100	200	600		
Maximum amplitude	3.5784	3.2174	2.3842		
Kink in the fault zone: (fault kink at a depth of 4 km)					
Fault bent from the vertical (°)	15	30	45		
Maximum amplitude	2.8609	3.1867	1.6935		
Fault branch: (two branches are 800-m apart at surface and connected at 4-km depth)					
Widths of two branches (m)	200/200	200/80			
Maximum amplitude	2.1008	2.9056			
Fault step-over: (fault step-over at a depth of 4 km)					
Step distance (m)	200	500	1000		
Maximum amplitude	1.5567	1.1959	1.1548		
Fault-zone depth					
Depth of the fault bottom (km)	7	4	2		
Maximum amplitude	2.1706	1.3171	0.8563		

Note Part of cases are exhibited in this section. Others refer to Li and Vidale (1996)

Figure 11.9d exhibits effect of the fault-zone depth. The depth to which a fault zone penetrates the crust is usually not well known. We examine the effect of the fault-zone depth on the excitation and propagation of guided waves. We did simulations with the fault zone depths of 7 and 4 km. The source is located at a depth of 8 km beneath the fault zone. The simulation with 7 km of fault zone depth preserves most of the characteristics of fault-zone guided waves, as shown in the reference model. But the amplitude of FZTWs is reduced by a factor of 2. The fault zone with a depth of 4 km, however, results in a weaker and shorter post-S duration of guided waves than the deeper fault zone. The wavetrain length and amplitude of FZTWs are reduced by a factor of 2 and 3 respectively. As the fault-zone depth reduced to 2 km, only a much shorter wavetrain of FZTWs is seen at receivers located within the fault zone. This numerical test helped us to evaluate the depth extension of rupture zones of the 1992 *M*7.4 Landers and the 2004 *M*6 Parfield earthquakes (Li et al., 2000).

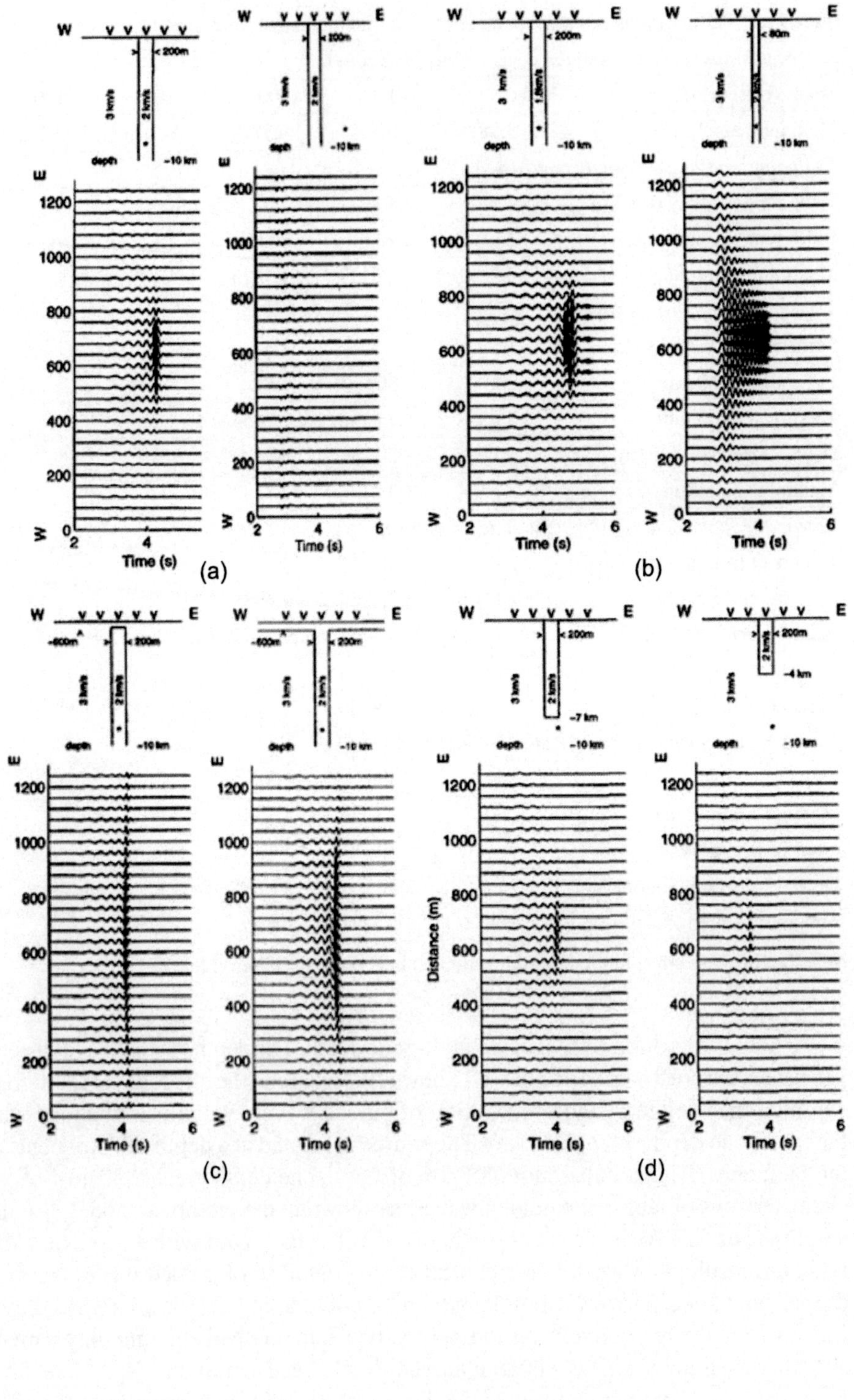
W
E
V V V V V
200m
80m
3 km/s
2 km/s
1.8km/s
depth
~10 km
~600m
~7 km
~4 km
Distance (m)
Time (s)
(a)
(b)
(c)
(d)

◂**Fig. 11.9** **a** The effect of source location relative to a low-velocity fault zone on a profile of seismometers run perpendicular to the 200-m-wide uniform fault zone with shear velocity 2 km/s in a simple half-space with shear velocity 3 km/s. The fault-parallel horizontal component is shown. The profiles show the motions for a source 0 and 300 m from the center of the fault zone in the left and right profiles. A schematic diagram at the top of each frame shows the geometry of source (*), receiver array (v), and fault zone. **b** The effect of varying the fault-zone velocity contrast and width. Left: waves for a fault zone of 1.8 km/s. Right: waves for a fault zone only 80-m wide to compare with waves for the reference model in (**a**). **c** left: the result from the fault zone buried at depths of 200 m. Right: the effect of capping the low-velocity fault zone with a 200-m-thick surface layer with velocity of 1.5 km/s. **d** The effect of the fault-zone depth extending from the surface to a depth of (left) 7 km and (right) 4 km. **e** The effect of a fault-zone step-over, showing the results of a (left) 200-m and (right) 500-m step-over in the fault zone at 4 km in depth. **f** Left: the effect of the fault kink at a depth of 4 km. The upper part of the fault remains vertical, and the lower part of the fault zone tilts 45° to the vertical. Right: the effect of fault branch structure, showing the result of two fault branches split at a depth of 4 km

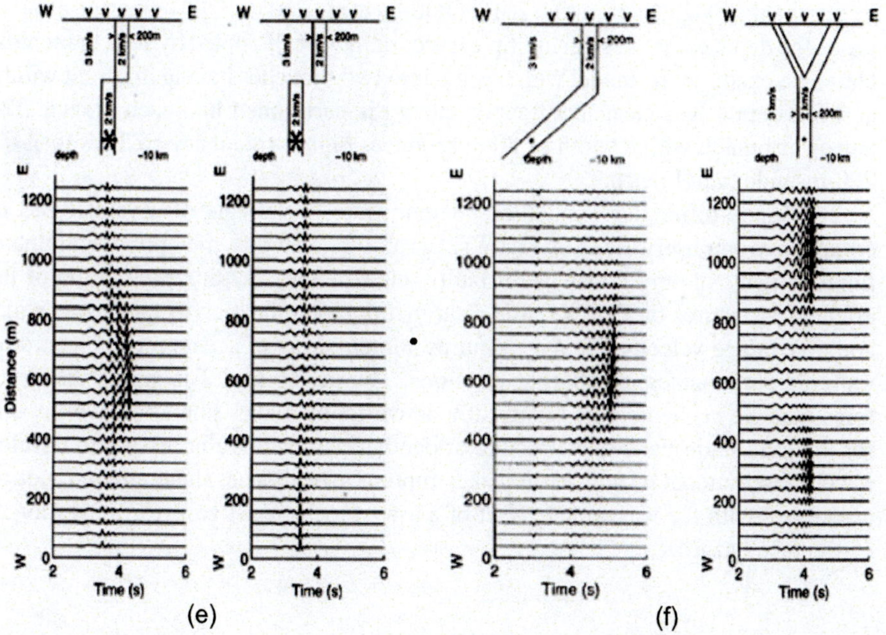

Fig. 11.9 (continued)

Figure 11.9e exhibits effect of a fault step-over. The estimation of the width of the step-overs may be crucial in processing dynamic rupture and predicting the extent of future earthquakes (Harris & Day, 1993). The lack of a continuous waveguide for the Landers fault zone, for example, has been linked to hesitation in the progression of rupture north along the fault during the Landers mainshock (Li et al., 1994a, 1994b). We show simulations of stepping the fault over by 200 and 500 m at a depth of 4 km. The 200-m step-over allows a lot of waveguided energy to proceed up the fault zone.

However the 500-m step-over effectively blocks the waveguide. So stepping over much more than twice fault zone width should result in the disappearance of the guided waves.

Finally, Fig. 11.9f exhibits effect of a kink in the fault zone. The effect of a kink is minor compared to the other structural complications we considered. As the lower part of the fault was bent even 45° from the vertical at the depth of 4 km, only a minor loss in the amplitude of the FZTWs is seen, for example in our experiment conducted at the Longmenshan fault that ruptured in the 2008 *M*8 Wenchuan earthquake in Sichuan, China (Li et al., 2014b). The more visible difference is that there is a first-arriving direct shear wave for the greater kinks that travels straight through the bedrock. Also, the bigger kinks produce longer travel paths along the fault, which result in later arrivals of the FZTWs. We also simulated effect of a fault branch. There are often multiple traces of faults at the surface, which merge at depth or may remain parallel to seismogenic depths as seen in our experiments at multiple rupture zones of the 1999 *M*7.1 Hector Mine earthquake and the 2014 *M*6 South Napa earthquake (e.g. Li et al., 2002, 2003a, 2004, 2018; Oglesby et al., 2003). FZTWs may be a tool to probe the depth extent of such multiple traces or may be disrupted by such branching. Here, we examine the model which splits into two branches having different widths at 4-km depth. It is clear that trapped energy is partitioned into each branch. The narrower branch with a width of 80 m produces higher-frequency FZTWs than the 200-m-wide fault branch.

2-D Finite-difference simulations demonstrate the effects of several types of complexity on observations of FZTWs. Careful field studies may provide dramatic improvements in our knowledge of fault-zone structure. By this exploration of the effect of structures on FZTWs, it is clear that the identification of fault-zone width and percentage velocity anomaly requires a linear array of seismometers recording multiple earthquakes in differing locations. The possibility of mapping the structures that are at the core of seismically active faults makes this effort worthwhile for imaging fault-plane fine structure at depth. Such data are necessary to develop a well-based model of how earthquakes rupture as well as evaluate ground motion associated with the waveguide effect of a low-velocity fault-zone for mitigation of earthquake hazard.

11.4 The Velocity Structure of the SAF at SAFOD Site Viewed by FZTWs Generated by Explosions and Earthquakes

11.4.1 Results from FZTWs Recorded at the Dense Seismic Array at the Surface

In the fall of 2003, as part of a site characterization program the San Andreas Fault Observatory at Depth (SAFOD), we deployed an array of 45 seismographs with

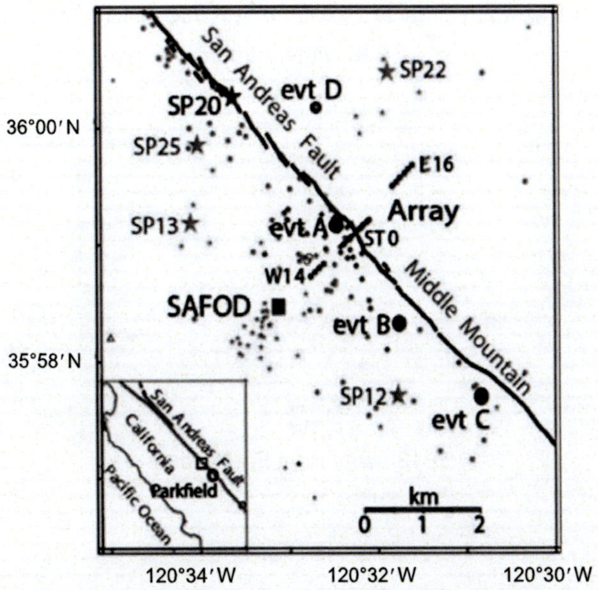

Fig. 11.10 Locations of the study area (box in inset map), 45 portable seismographs (small squares) across and along the SAF near the SAFOD site (square), five explosions (stars with number); and ~ 120 microearthquakes (dots) recorded in 2003. Array station ST0 located on the SAF surface trace; W14 and E16 on its west and east ends. Dots with crosses denote 33 events generating FZTWs. SP20 detonated within the fault zone generated FZTWs. The SAFOD drilling target event A (solid circle) occurred at ~ 3-km depth. Event B, a 7 km-deep aftershock of the 2004 *M*6 Parkfield earthquake was recorded by the SAFOD borehole seismograph, which waveforms are shown in Fig. 11.15a. Events C and D (solid and open circles) occurred at 9–11 km depths, and recorded at the surface array

2 Hz three-component geophones across and along the surface trace of the SAF near the SAFOD site approximately 14.5 km northwest of Parkfield (Fig. 11.10). Station spacing of the cross-fault array was 25 m for central part of the array near the main fault trace at the surface, and increased to 50 m, 100 m for stations farther away from the fault. This surface array recorded ~ 120 local earthquakes at depths between 2 km and 12 km over two months and 5 explosions detonated by the USGS. We observed clear FZTWs for one shot SP20 and 33 events that were located within the fault zone. The raypath incidence angles from these 33 events to ST0 are smaller than 30° from vertical. Figures in this section are adapted from Li et al., (2007b, 2008).

Figure 11.11 (top) shows seismograms of 3 shots detonated within and outside of the fault zone ~ 3 km NW of SAFOD and recorded at the surface array. We observed prominent FZTWs at stations within a 150–200-m wide zone near the main trace for shot SP20 detonated within the fault zone, but not for shots SP13 and SP22 detonated ~ 2 km away from the fault zone, indicating a distinct LVZ along the SAF between the shot and array at shallow depth. We note that the low-velocity damage zone is asymmetric with the main fault trace, with the LVZ extending farther on the southwest side. Figure 11.11 (bottom) shows three-component seismograms

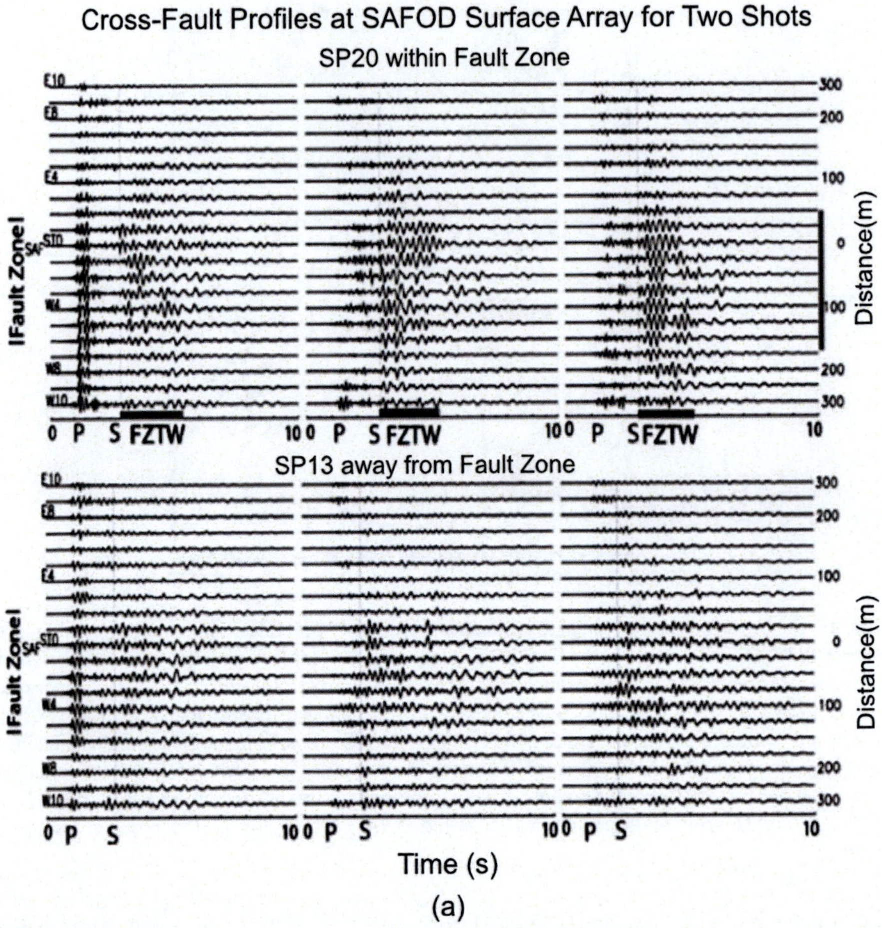

Fig. 11.11 **Top**: three-component seismograms recorded at stations between E10 and W10of the cross-fault array for shots SP20 and SP13 detonated within and away from the fault zone of Array station ST0 was located on the SAF main fault trace. Seismograms have been low-pass (< 3 Hz) filtered and are plotted using a fixed amplitude scale for each event. The recordings have been deconvolved by the sensor and instrument responses. FZTWs with large amplitudes and long post-S duration time (2 s marked by brackets) are recorded at stations within 150–200-m-wide fault zone (marked by a vertical bar) for shot SP20. In contrast, body waves with much shorter wavetrain dominate in seismograms for shots SP22 and SP13. **Bottom**: three-component seismograms for events C and D (see Fig. 11.10). Seismograms are < 6 Hz filtered, and aligned on the S-wave arrivals. FZTWs with large amplitudes and long post-S durations (2.2 s) are recorded at stations within fault zone for the on-fault event C but not for the off-fault event D

recorded at the cross-fault array for two earthquakes (events C and D in Fig. 11.10) are shown in. Event C whose source-and-receiver points are closer to the fault trace shows the more prominent, longer lasting, fault-guided waves than those for Event D away from the fault trace, implying a ~ 150–200-m wide LVZ existing along the SAF. Again, we note that the damage zone is not symmetric, but is broader on the southwest side of the main fault trace. We measured ~ 2.2 s of the post-S duration

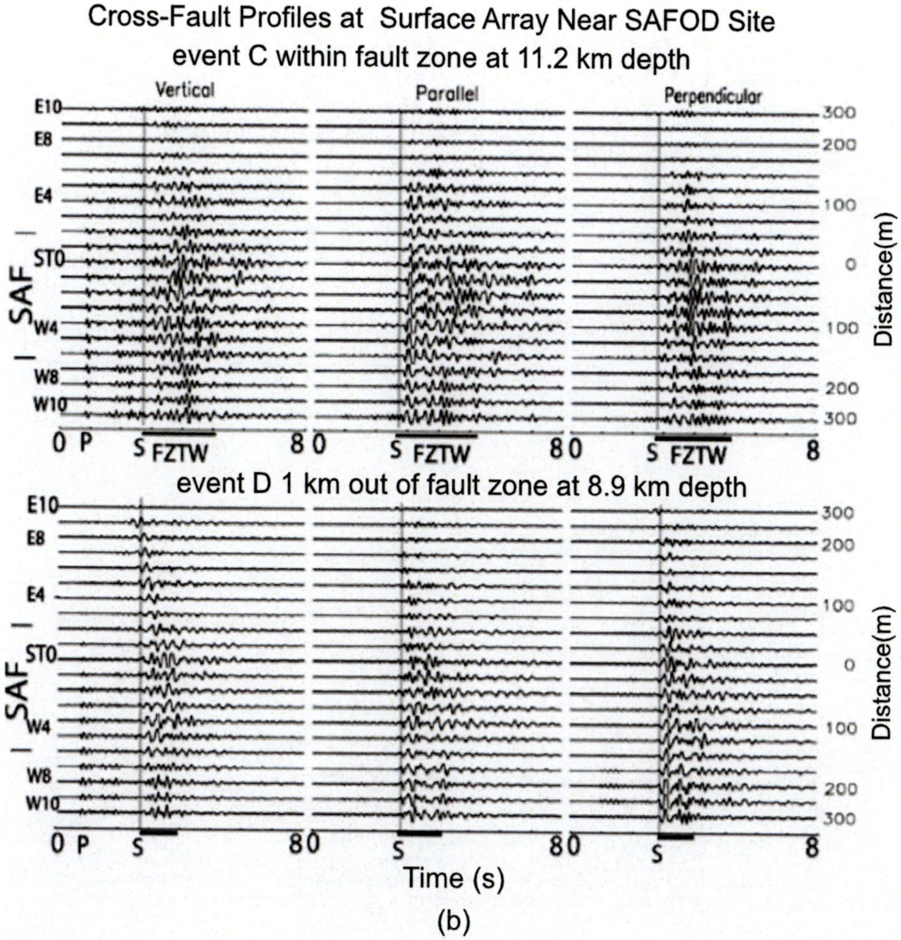

Fig. 11.11 (continued)

time of FZTWs, in which amplitude envelopes of FZTWs are above twice the level of the background noise coda.

In order to examine the depth extension of the LVZ on the SAF, we used the data from thirty-seven on-fault micro-earthquakes near the SAFOD site with the raypath incidence angles to the array smaller than 30° from vertical (Fig. 11.12a). Figure 11.12b shows seismograms and envelopes at station ST0 located on the SAF surface trace for eleven on-fault events at increasing depths. We observed that the post-S durations of FZTWs (i.e. post-S duration time of FZTWs) progressively increase from ~ 1.0 s to ~ 2.2 s as the event depths increase from 2.6 km to 11.7 km. In contrast, much shorter post-S durations in time are registered at the same station for eleven other events located away from the fault zone in a similar depth range. Figure 11.12c shows the measured post-S duration times of FZTWs registered at stations within the fault zone for thirty-seven on-fault events and thirteen off-fault events at depths between 2 km and 12 km. The durations of FZTWs for on-fault

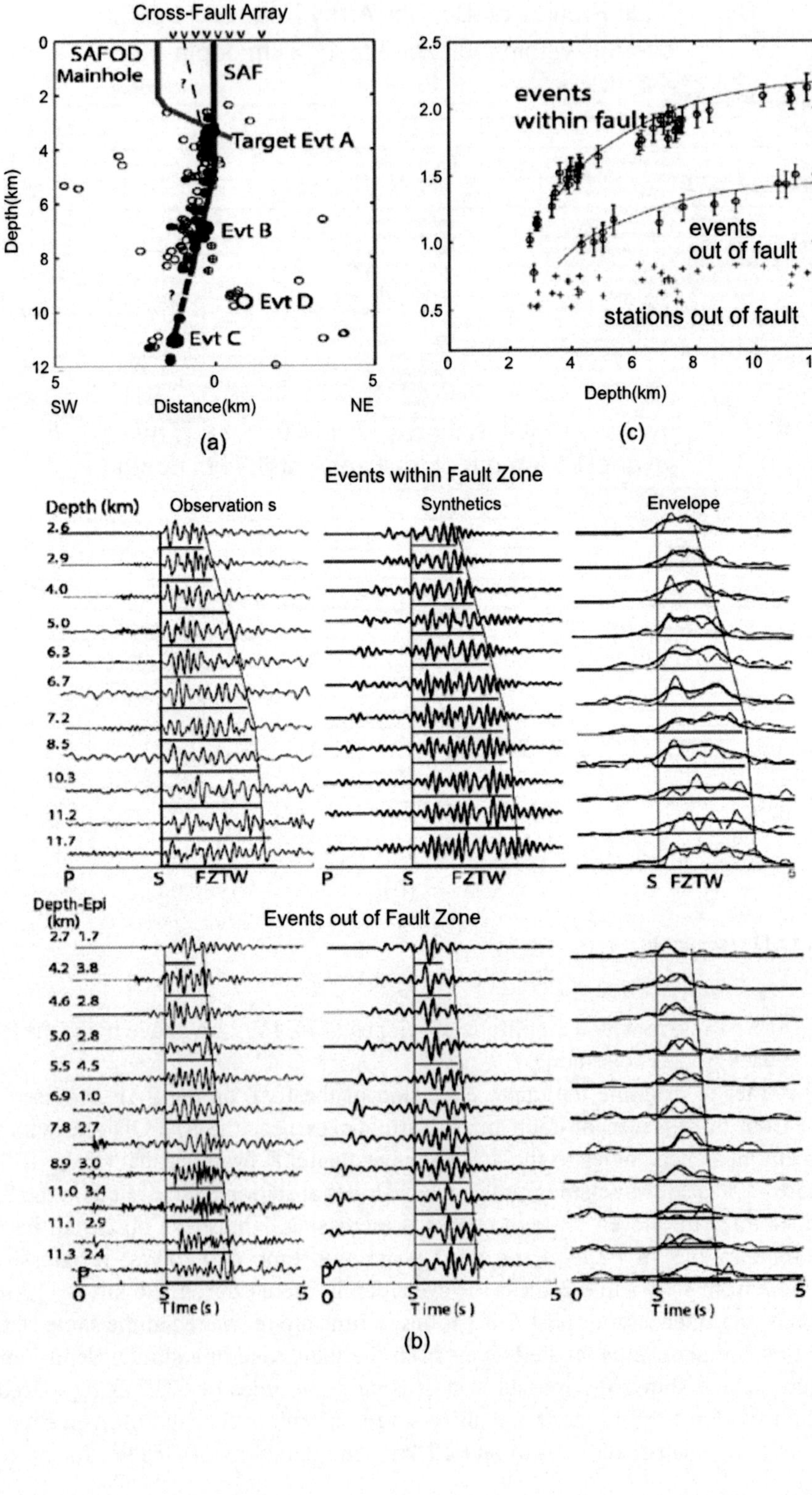
Cross-Fault Array
SAFOD Mainhole
SAF
Target Evt A
Evt B
Evt D
Evt C
Depth(km)
SW
Distance(km)
NE
(a)
events within fault
events out of fault
stations out of fault
Depth(km)
(c)
Events within Fault Zone
Depth (km)
Observation s
Synthetics
Envelope
P
S
FZTW
Depth-Epi (km)
Events out of Fault Zone
Time(s)
(b)

◀**Fig. 11.12** **a** Depth section across the SAF (gray vertical line) passed by the SAFOD borehole (gray curved line) shows locations of microearthquakes recorded at the surface seismic array in 2003. Solid and open circles denote events showing and without FZTWs. **b** Top: observed and synthetic vertical-component seismograms and envelopes at array station ST0 for eleven on-fault earthquakes show an increase in post-S duration time of FZTWs (denote by horizontal bars) as event depths increase. The time move-out is marked by the slop of dashed line. S-arrivals for these events are aligned at the same time. Bottom: the plot for eleven off-fault events in a similar depth range shows much shorter post-S durations and flat changes with event depths. The synthetic seismograms and envelopes are computed using the model in Fig. 11.16a. **c** The measured post-S durations versus focal depths recorded at stations within and out of the fault zone for on-fault (black circles) and off-fault (gray circles) events. Crosses denote the measurements at stations out of fault zone for these events. Each data point is averaged from measurements at four stations. Error bars are standard deviations. Curves are polynomial fits to the data. The post-S duration increases rapidly with focal depths increases, but this increasing trend becomes nearly flat as the event depth is larger than ~ 7–8 km

events increase from 1.0 s to 2.0 s (marked by an obvious time move-out) as the depth increases from ~ 2.5 km to ~ 8.0 km, but increase only ~ 0.2 s from 2.0 s to 2.2 s with nearly flat changes as events depths increase ~ 3 km from 8.4 km to 11.7 km in contrast, much shorter durations with flat depth-dependent changes are measured at station ST0 for the same events or at all stations for off-fault events. These observations show that the post-S duration of FZTWs is a strong function of event depth. The low-velocity waveguide on the SAF extends continuously from the surface to ~ 7–8 km depth, but diminishes below ~ 8 km probably due to the larger confined stress at greater depths.

11.4.2 Results from the FZTWs Recorded at the SAFOD Main-Hole Seismograph

In December of 2004 and afterwards a borehole seismograph with 4.5 Hz three-component geophones was installed in the Main-Hole (MH) of San Andreas Fault Observatory at Depth (SAFOD) at ~ 2.7 km depth below the sea level (~ 3 km beneath the wellhead) where the highly fractured and low velocity zone of the SAF was found (Hickman et al., 2005; Zoback, 2007, 2008). Figure 11.13a show the location of SAFOD site, ~ 14.5 km northwest of Parkfield and 1.7 km southwest of the SAF surface trace. At this location, the SAF separates Pacific Plate Salinian rocks from North American Plate Franciscan mélange. Vertical seismic profiling (VSP) in the vertical 2.1-km-deep SAFOD Pilot-Hole (PH) from nearby micro-earthquakes contain signals scattered by the local geologic structure, the strength of which suggests large contrasts in material properties, as in cracks and fluids of the SAF at depth (Fig. 11.13b, c) (Malin et al., 2006). Additionally, it is the transition between the creeping and locked segments of the SAF (Nadeau & McEvilly, 1997). The SAFOD's goal is to investigate the structure and mechanics of this unique section of fault.

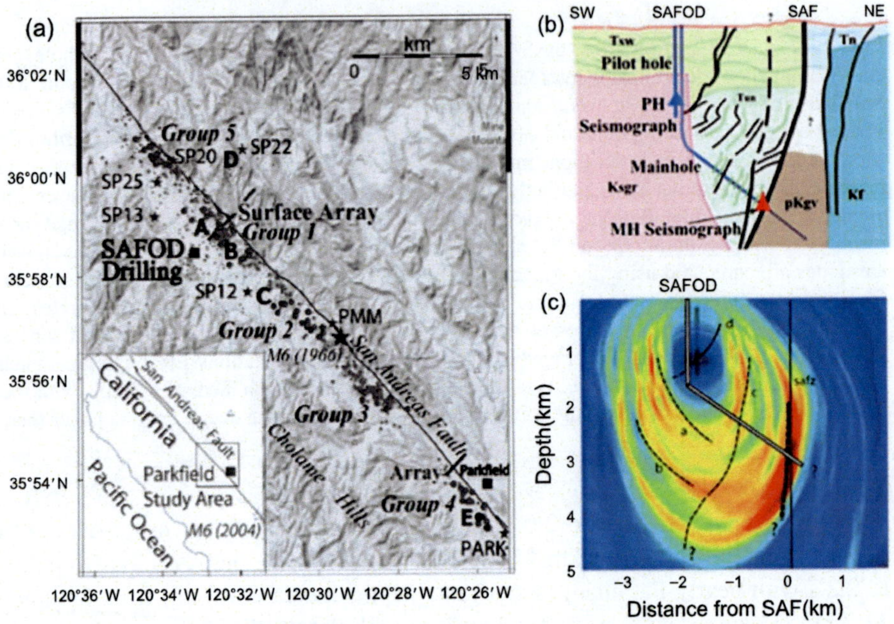

Fig. 11.13 **a** Location of the study area (box in inset map). ~ 120 micro-earthquakes (grey dots) and 5 explosions (stars) were recorded at the cross-fault surface array (black line) near the SAFOD site (square) in 2003. ~ 350 aftershocks (gray circles) of the 2004 *M*6 Parkfield earthquake in five groups were recorded at the SAFOD borehole seismographs. Signals from two explosions PMM and PARK (stars) were recorded at the cross-fault array deployed near Parkfield during our experiment in 2002. Event E is one of aftershocks recorded at Parkfield array redeployed in 2004. **b** Cross-section passing the SAFOD site towards the SAF. Red and blue triangles denote the SAFOD MH and PH seismographs. Ksgr = Gabilan Granite; pKgv = pre-Cretaceous Great Valley formation; Kf = Franciscan mélange; Tsw and Tn = Tertiary cover SW and NE of the SAF surface trace; Tun = unnamed tilted Tertiary encountered along MH drilling. **c** A cross section of migrated micro-earthquake VSP signals assuming P-to-S scattered waves recorded at the string array of 32-levels of seismographs (shown by squares) installed in the SAFOD PH (Malin et al., 2006). The solid line at $x = 0$ km is the extension of the surface trace of the SAF to depth. The colors indicate the amplitude of scattered energy. Black dash-line denotes the highest energy scattering cells interpreted as the down dip extension of the SAF

On September 28 2004, a *M*6 earthquake occurred ~ 15 km southeast of the town of Parkfield (Fig. 11.13a). ~ 350 aftershocks were recorded at the SAFOD MH and PH seismographs in forty days staring from December 17 of 2004. We divided these aftershocks into five groups, according to epicentral distances between the events and SAFOD site. Figure 11.14 shows seismograms recorded at MH and PH seismographs for four aftershocks in four Groups G1 to G4. FZTWs with large amplitudes and long post-S durations are seen in the MH seismograms while the PH seismograms are dominated by body waves with brief wavetrains for the same events. We measured post-S time durations of FZTWs in MH seismograms, in which wave amplitudes are above twice background level in later coda. Figure 11.14a, b exhibit seismograms for four on-fault aftershocks in groups G1 and G2, respectively,

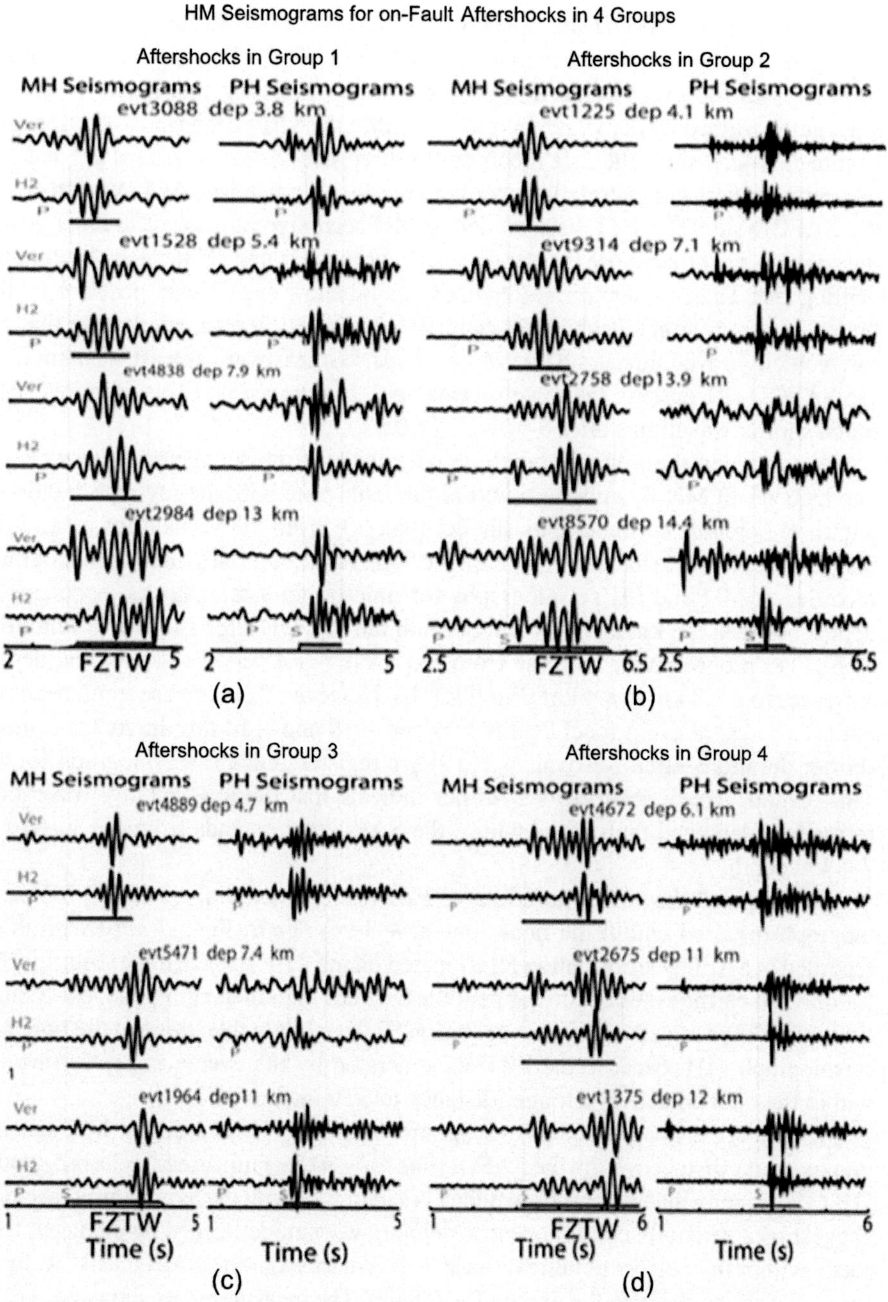

Fig. 11.14 **a–d** Vertical and horizontal seismograms recorded at the SAFOD borehole seismographs MH and PH for aftershocks occurring within the fault zone in Groups G1, G2, G3 and G4 located 1–2 km, 4–6 km, 8–10 km, and 14–16 km SE of the array. Earthquakes are ordered by increasing focal depths and aligned on the S-wave arrivals. Event index and focal depths are listed above seismograms that are < 6 Hz filtered and plotted in trace normalized. FZTWs with large amplitudes and long post-S duration (marked by horizontal bar) are recorded by MH seismograph while body waves with short post-S duration are dominant in PH seismograms

occurring near the SAFOD site. The post-S durations of FZTWs increase from ~ 0.5 s to ~ 1.4 s as the event depths increase from 3.8 km to ~ 7.9 km, and further increase to ~ 1.7 s as the event depth down to 14.3 km, suggesting that the low-velocity waveguide formed by damaged fault rocks within the SAF extends to 7–8 km depth. Figure 11.14c, d show the data recorded for three aftershocks in groups G3 and G4, respectively, with epicentral distances between 8–10 km and 14–16 km southeast of the SAFOD site. The FZTWs recorded at MH seismograph show 1.8–2.6 s post-S durations, longer than those for aftershocks in groups G1 and G2 located closer to the drilling site. The FZTWs from aftershocks in G4 show even longer post-S duration than those from aftershocks in G3 occurring at similar depths, suggesting that the low-velocity waveguide along the SAF extends southeastward at least ~ 20 km from the SAFOD site passing the town of Parkfield. In contrast, PH seismograms show much shorter wavetrains after S-arrivals (< 0.5 s).

We examined the SAFOD borehole data for aftershocks occurring at different depths beneath MH seismograph within the fault zone with the ray-path incidence angles to borehole seismographs smaller than 30° from the vertical for a carefully insight of the depth extension of LVZ on the SAF. Figure 11.15a exhibits seismograms recorded at MH and PH seismographs for nine on-fault aftershocks, occurring at depths between 3.8 km and 14.4 km beneath the SAFOD site. The post-S durations of FZTWs progressively increase from ~ 0.4 s to ~ 1.4 s as the event focal depths increase from 3.8 km to 8.4 km (Fig. 11.15b). However, this increase trend becomes flat (1.5–1.7 s) at event focal depths between ~ 10 and ~ 14 km. In contrast, much shorter durations after *S*-arrivals (<1.0 s) are registered at PH seismograph for the same events. These observations further indicate that the low-velocity waveguide formed by damaged fault rocks within the SAF likely extends from the surface to 7–8 km depth.

We compared the post-S durations of FZTWs recorded at the SAFOD MH seismograph installed within the fault zone at ~ 3-km depth (Fig. 11.15b) with those recorded at SAFOD array station ST0 located on the SAF surface trace (Fig. 11.12c) for on-fault earthquakes occurring near the SAFOD site. In general, the post-S durations of FZTWs registered at surface station ST0 are larger than those at the borehole seismograph MH, because the FZTWs generated by the events at depth travelled within the fault zone over a longer distance to ST0 than to MH.

Figure 11.15c is a plot of post-S durations of FZTWs recorded at MH seismographs versus distances from the SAFOD site for ~ 80 on-fault aftershocks in 5 groups G1–G5, showing that the duration of FZTWs increases with the hypocentral distance, suggesting a generally continuous low-velocity waveguide formed by damaged fault rocks within the SAF extending at least ~ 20 km through the SAFOD site (refer to Fig. 12.13a) and down to the depth of ~ 7–8 km. The measurements show the greater increase trend in post-S duration of FZTWs with distance for shallower aftershocks than that for deeper aftershocks, suggesting the greater seismic velocity reduction (greater rock damage magnitude) within the shallower portion of the fault zone.

We calculated coda-normalized spectral amplitudes of FZTWs recorded at MH seismograph versus hypocentral distances for aftershocks in five groups G1-G5 to measure Q values of fault rocks within the SAF at Parkfield, using the same method

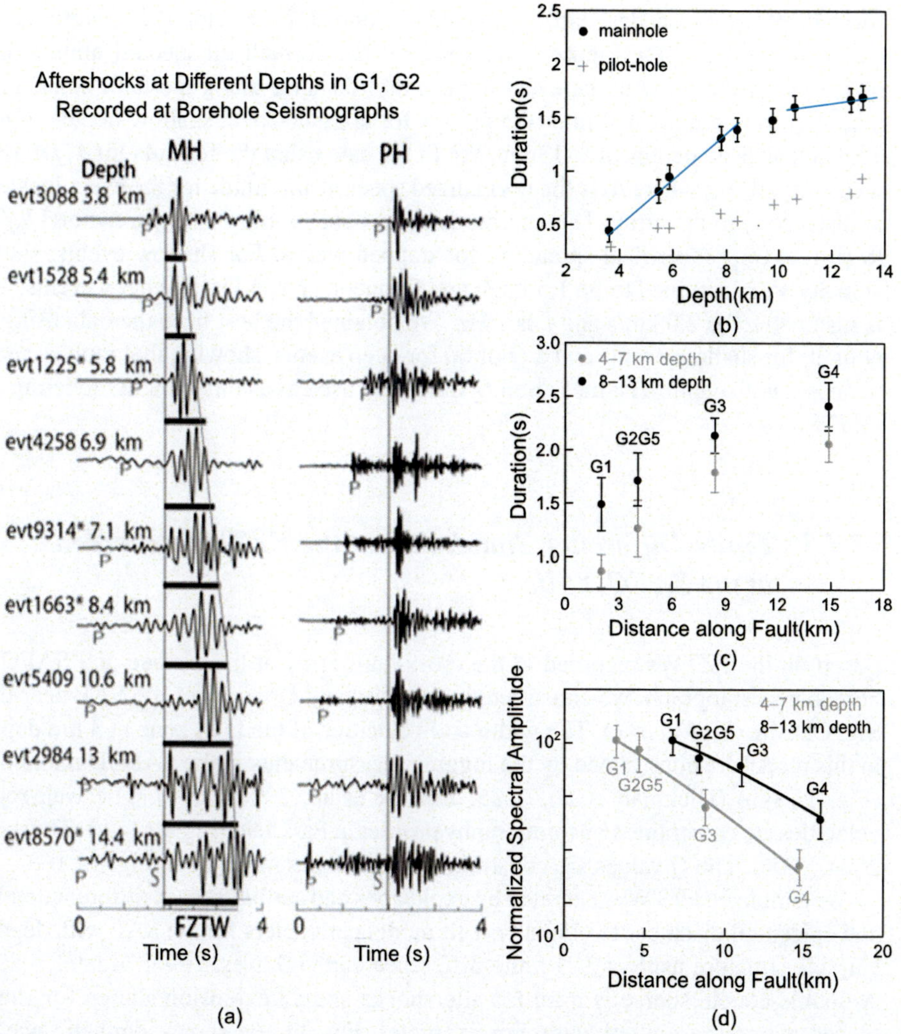

Fig. 11.15 **a** Seismograms recorded at SAFOD MH and PH seismographs for nine on-fault aftershocks at different depths in groups G1 and G2. Seismograms have been < 6 Hz filtered. The post-S duration times of FZTWs (denoted by horizontal bars) increase as event depths increase (the time move-out marked by the slop of gray lines). **b** Measured post-S durations for MH (dots) and PH (crosses) seismograms versus focal depths for these aftershocks. Each data point with error bar is the average of measurements from three component seismograms and standard deviation. **c** Measured post-S durations of FZTWs in MH seismograms for ~ 80 on-fault aftershocks in five groups G1-G5 versus hypocentral distances from SAFOD borehole seismographs. Each data point with error bar is the average for all on-fault aftershocks in each group and standard deviation. Gray and black dots denote the measurements for aftershocks at depths of 4–8 km and 8–13 km, respectively. **d** Normalized spectral amplitudes of FZTWs versus hypocentral distances for aftershocks in five groups G1-G5. Each data point with error bar denotes the mean of coda-normalized spectral amplitude peaks of FZTWs at 4–8 Hz in MH seismograms and standard deviation

for the 1992 $M7.4$ Landers earthquake and the 1999 $M7.2$ Hector Mince earthquake (Li et al., 1999, 2002). Figure 11.15d shows that normalized spectral amplitudes of FZTWs decrease with the event hypocentral distance along the fault zone. The measurements are divided into two groups for aftershocks at shallow depths of 4–7 km and at deep depths of 8–13 km. We fit the data using the formula $\ln(A_1/A_i) = \pi f(r_i - r_1)/QV_s$, where A_i is the normalized spectral amplitude for the event located at distance r_i to the array. The amplitude A_i has been multiplied by a factor $1/\sqrt{r_i}$ to correct for geometrical spreading for trapped waves. For shallow events, shear velocity V_s is assumed to be 1.5 km/s and frequency f is 5 Hz. For deep events, V_s is assumed to be 2.0 km/s and f is 7 Hz. We obtained the best fit to the data using a Q of 30 for shallow events and a Q of 60 for deep events, showing that fault zone Q increases with depth. The measured Q values are used as a constraint in modeling of FZTWs.

11.4.3 Finite-Difference Simulation of the FZTWs Recorded at the SAFOD Site

Based on the FZTWs recorded at the cross-fault array at the surface and SAFOD borehole seismograph, we constructed a velocity and Q model of the SAF near the SAFOD site (Fig. 11.16a). The width and velocities of the fault zone at 3 km depth in this model are constrained by the logging measurements in the SAFOD mainhole (Fig. 11.16b) (Hickman et al., 2005; Zoback et al., 2007, 2008). The wall-rock velocities are constrained by tomography profiles at Parkfield (Fig. 11.16c) (Thurber, 2004, 2006). The Q values are constrained by the measurements in Fig. 11.16d.

We simulated FZTWs generated by explosions and earthquakes at different depths and epicentral distances to obtain best-fit model parameters for the SAF with depth-variable structure using a 3-D finite-difference code (Graves, 1996; Vidale, 1989). A double-couple source is used for aftershocks and an explosion source for shots placed within the rupture zone. For examples, Fig. 11.16e shows synthetic waveforms, using the model in Fig. 11.16a, comparable with seismograms recorded at the cross-fault surface array for the SAFOD target event A at ~ 3-km depth beneath the drilling wellhead (refer to Fig. 11.16d). Figure 11.16f exhibits synthetic and recorded seismograms at MH seismograph for event B (035810) occurring within the fault zone at ~ 7-km depth and with its epicenter ~ 2 km SE of the SAFOD site. There is a good fit between the synthetics and observations. The FZTWs generated by the target event A occurring at ~ 3-km depth were received at stations located close to the SAF surface, showing that the low-velocity waveguide should exists on the SAF above 3-km depth. On the other hand, the FZTWs generated by event B occurring at ~ 7 km depth were recorded at the SAFOD MH seismograph located within the SAF core zone at ~ 3 km, showing that the low-velocity waveguide on the SAF should extend downward below 3 km, likely to ~ 7 km depth.

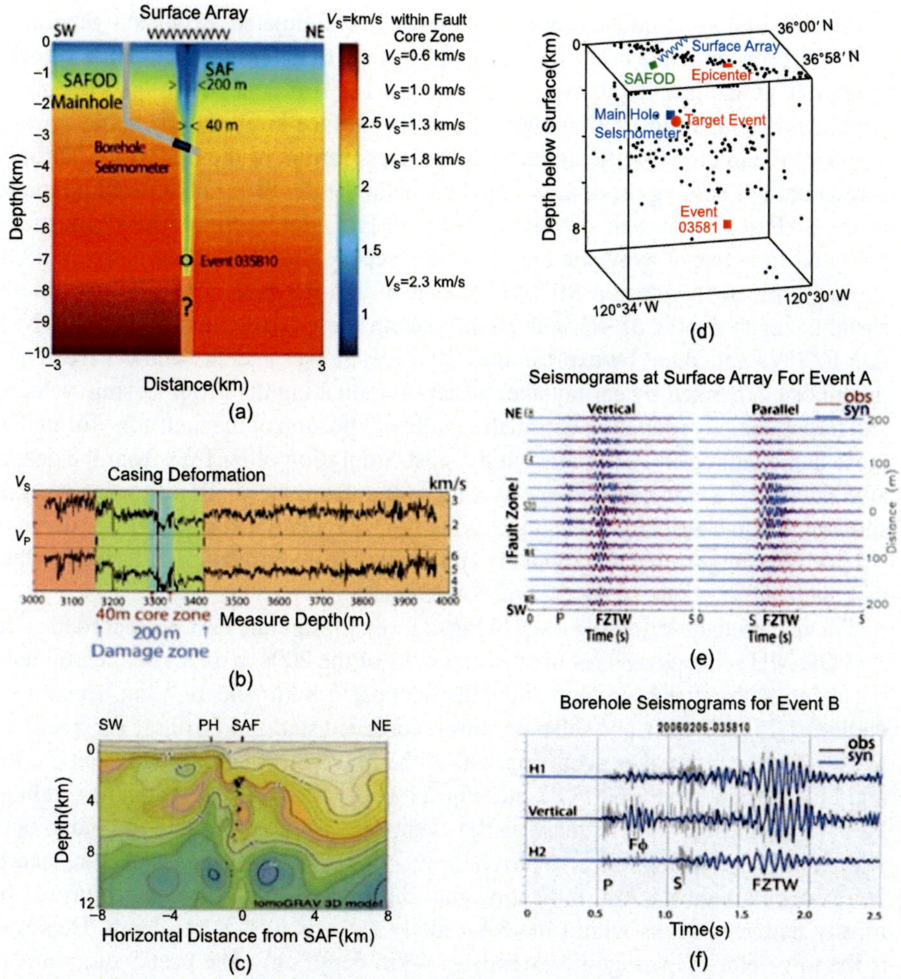

Fig. 11.16 **a** Cross-section of the S-wave velocity model across the SAF used to compute FZTWs. Model parameters (velocity, Q value and width of fault zone) are depth dependent. The LVZ along the SAF below ~ 7–8km depth are not well constrained in this model. Circle denotes the earthquake which waveforms are shown in (**f**). **b** SAFOD drilling log data show a 40-m-wide fault core (blue color) surrounded by a 200-m-wide LVZ (green color), in which velocities are reduced by 25–35%. The red line indicates the location where fault creep is deforming the borehole casing. **c** The cross-section through the SAFOD site from the tomography DD 3D velocity model. Filled circles denote earthquakes within 1 km-wide LVZ down to ~ 6 km along the SAF. **d** 3-D volume shows locations of aftershocks (black dots) of the 2004 *M*6 Parkfield earthquake recorded at the SAFOD surface array (green triangles) and MH seismograph (blue square). The red circle denotes the SAFOD drilling target event A located nearby the MH at ~ 3 km depth. The red square denotes the on-fault event B (035810) occurring at ~ 7 km depth with its epicenter ~ 3 km SE of the SAFOD site (black square). Locations of events A and B also shown in Fig. 11.13a. **e** Observed and synthetic seismograms at the surface array across the SAF for the target event A. Seismograms have been < 8-Hz filtered. **f** Observed and 3-D finite synthetic three-component seismograms at the SAFOD MH seismograph for the event B, contain S-type FZTWs and P-type fault-guided wave (marked by Fϕ) (Ellsworth and Malin, 2006)

In the forward modeling procedure, we first simulated explosion-generating FZTWs recorded at the cross-fault array at surface to determine fault zone velocity structures at shallow depth (approximately to 1–2 km depth). We then simulated earthquake-generating FZTWs recorded at the surface array and borehole seismograph to obtain the velocity structure of the deep portion of the fault zone. Through a systematic modeling process, we finally obtain the depth-variable structure model of the SAF at seismogenic depths as shown in Fig. 11.16a. Figure 11.17 illustrates the good fit between synthetic and observed seismograms recorded at the SAFOD surface array for explosion SP20 detonated within the fault zone and two micro-earthquakes at depths of ~ 3 and 7-km beneath the cross-fault surface array. Note that FZTWs generated by explosion SP20 travel slower and show lower frequency than those generated by earthquakes at depth, indicating the lower seismic velocity and larger velocity reduction within the shallower portion of the fault zone formed by more damaged rocks. Also note that the post-S duration of FZTWs from the deeper aftershock at 7 km is nearly twice as long as that from the shallower event at 3 km, showing the low-velocity waveguide along the SAF likely extending to the depth of ~ 7 km. We suggest that this model is applicable for interpretation of the subsurface damage zone structure of the SAF at SAFOD site.

We also used the velocity model in Fig. 11.16a to simulate FZTWs recorded at the SAFOD MH seismograph for nine aftershocks of the 2004 *M*6 Parkfield earthquake occurring within the fault zone at depths between 3.8 km and 14.4 km beneath the wellhead (Fig. 11.18a), and seismograms recorded at station ST0 of the surface array for 11 micro-earthquakes occurring within the fault zone at depths between 2.6 km and 11.7 km beneath the SAFOD site (Fig. 11.18b). The post-S durations of synthetic FZTWs for these events increase as the event depths increase (the time move-out), with greater increasing rate at depth shallower than 7–8 km, and then become nearly flat at depths below 8 km, indicating that the low-velocity waveguide formed by mostly damaged rocks within the SAF likely extends to 7–8 km depth. However, if the low-velocity waveguide extends to 4-km depth only, the post-S durations of synthetic FZTWs for events at depths below 4 km show shorter with flat change in duration than observations (Fig. 11.18c). These modeling results indicate that the low-velocity waveguide formed by damaged fault rocks within the SAF at the SAFOD site likely extends to the depth of 7–8 depth.

In summary, the SAFOD fault-zone trapped waves can be well-fit using a tapered, ~ 40 m wide fault core with velocities reduced by 40–50% inside a wider, 100–200 m zone with velocities reduced by ~ 30–40%. The relatively intact wall rocks surrounding this composite damage zone have different velocities on the east and west sides the SAF to match the geologic material velocities determined from tomographic modeling. Further, by matching the observed increase in FZTW duration with event depth, it would appear that the main low velocity zone, and probably its interior core, extends downward to at least ~ 7 km depth. This is the depth extension of the fault damage zone determined using the most clearly resolved and best fit FZTWs seen in the SAFOD borehole seismogram and surface array modeled in this section.

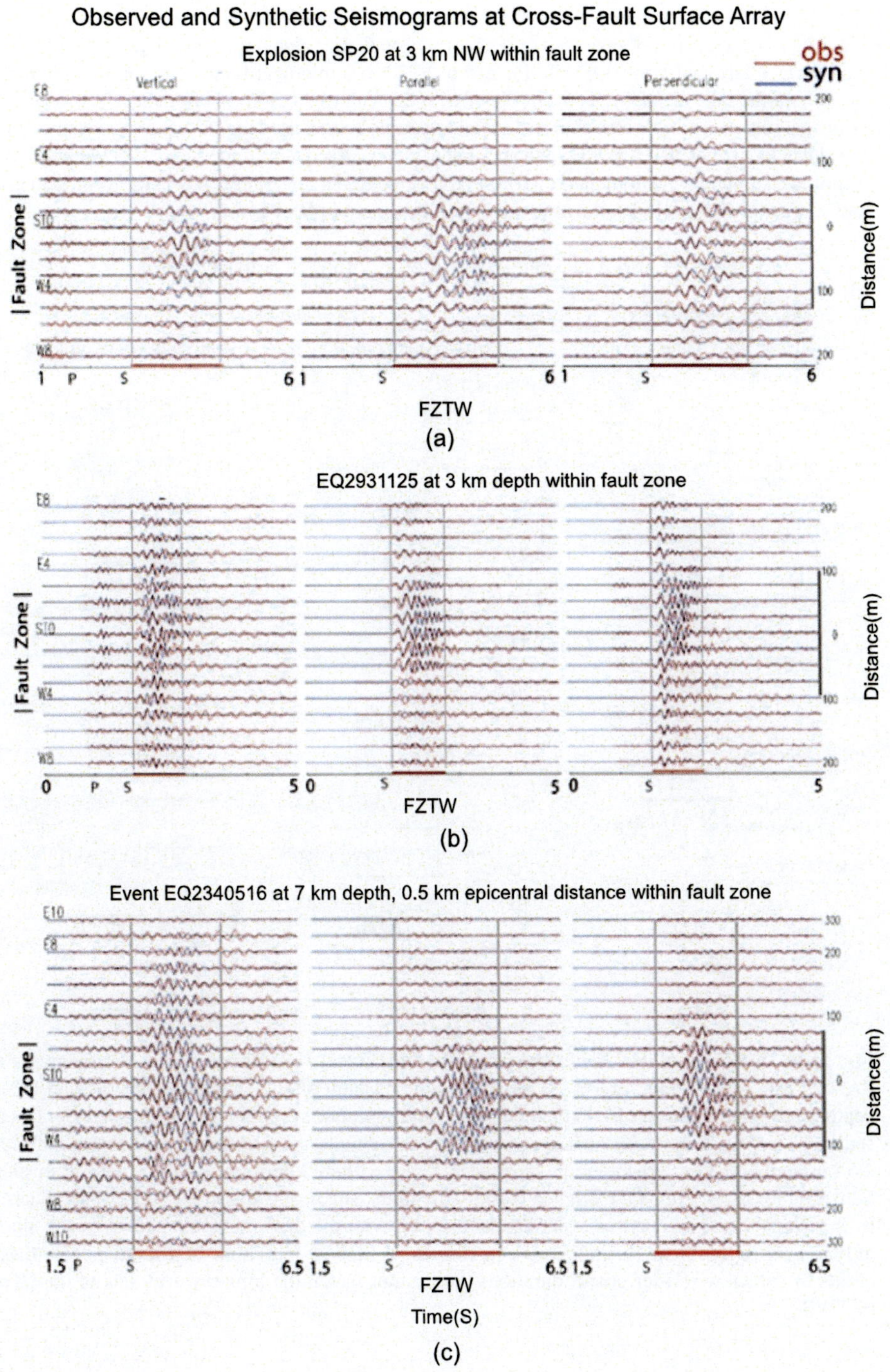
Observed and Synthetic Seismograms at Cross-Fault Surface Array
Explosion SP20 at 3 km NW within fault zone
obs
syn
Vertical
Parallel
Perpendicular
| Fault Zone |
Distance(m)
FZTW
(a)
EQ2931125 at 3 km depth within fault zone
FZTW
(b)
Event EQ2340516 at 7 km depth, 0.5 km epicentral distance within fault zone
FZTW
Time(S)
(c)

◂**Fig. 11.17** Comparison of observed (red lines) and synthetic (blue lines) seismograms at the cross-fault surface array near SAFOD site for **a** shot SP20, **b** a microearthquake at 3 km depth, and **c** a micro-earthquake occurring at 7 km depth with epicenters ~ 0.5 km from the surface array. An explosive source is used for the shot and a double-couple source used for earthquakes. Seismograms have been low-pass filtered at 6 Hz and are plotted using a fixed amplitude scale. FZTWs with large amplitude and long wavetrain marked by horizontal red bar in time range between two vertical gray lines appear at stations located within the ~ 150–200-m wide fault zone

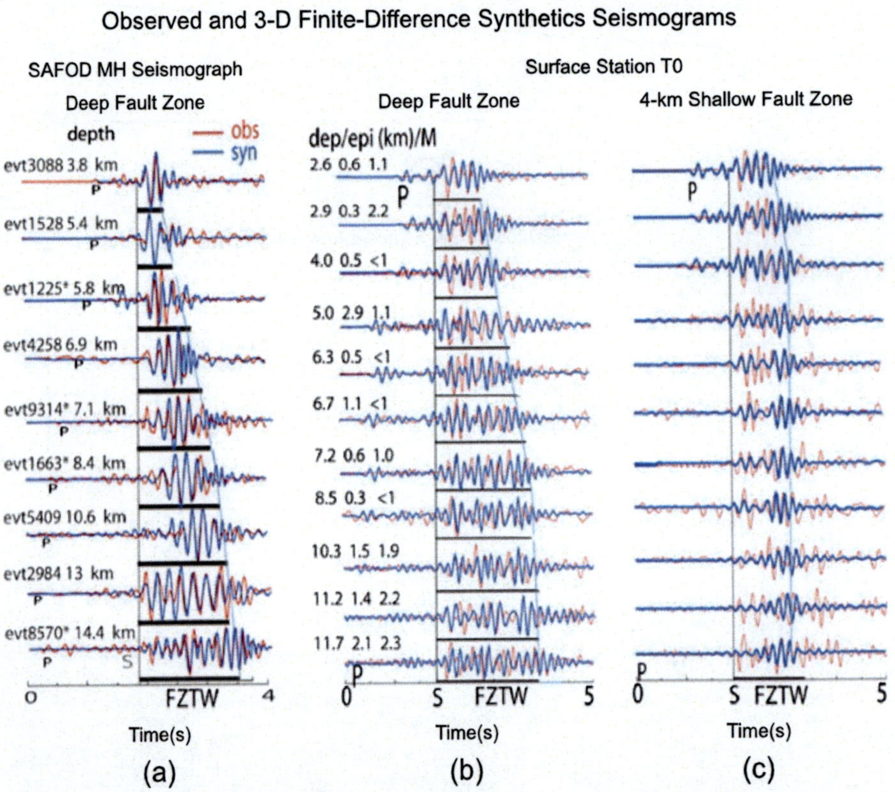

Fig. 11.18 **a** Observed (red lines) and synthetic (blue lines) vertical-component seismograms at SAFOD MH seismograph at ~ 3 km depth for nine on-fault aftershocks in groups G1 and G2 at depths between 3.8 km and 14.4 km. **b** Observed and synthetic vertical-component seismograms at station ST0 of the cross-fault surface array for eleven on-fault microearthquakes at depths between 2.6 km and 11.7 km. Seismograms have been < 8 Hz filtered and are trace-normalized in plot. S-arrivals for these events are aligned at the same time. The post-S durations of FZTWs (marked by horizontal bars) increase as event depths increase with the time move-out shown by the slope of blue lines. **c** Synthetic seismograms at ST0 using a shallow fault zone to 4-km depth for these events in **(b)** show shorter post-S durations that cannot match the longer wavetrains of observed FZTWS

11.5 Velocity Structures of the SAF at Parkfield Viewed by FZTWs Generated by Explosions and Earthquakes

11.5.1 Results from FZTWs Recorded at the Dense Array Near Parkfield in 2002

In 2002, a team of USC and UCLA researchers (Li et al., 2004) deployed dense linear seismic arrays across and along the SAF near the town of Parkfield (Fig. 11.19) for several days to record FZTWs generated by explosions detonated around the SAFOD drilling site in the PASO experiment (Thurber et al., 2004). Array A is 850-m-long across the SAF zone. Arrays B and C are 400-m along the main and north strands. Array A consists of 35 stations with 2-Hz sensors and station spacing of 25 m. Station ST0 of array A was located on the SAF main trace. Arrays B and C consist of 5 stations spaced at 50 m, respectively. Three components of the sensor were aligned vertical, parallel, and perpendicular to the fault trace. Figures in this section udapted from Li et al. (2004).

Prominent FZTWs with large amplitudes and long post-S durations excited by explosions PMM and PARK detonated within the fault zone were recorded at stations located close to the SAF main and north surface traces (Fig. 11.20a, b). During the experiment, the arrays also recorded FZTWs generated by a couple of micro−earthquakes that occurred within the fault zone (Fig. 11.20c). The amplitude of FZTWs decreases rapidly away from the ~150–200-m wide fault zone. In contrast, FZTWs are not obvious at any station for shots detonated out of the fault zone. In order to eliminate near−surface site effects on FZTWs, we computed amplitude ratios of trapped waves to P waves for the events recorded at the array (Fig. 11.20d). The amplitudes are computed in 2 s time windows that include P and dominant trapped waves, respectively, for low−pass (< 5 Hz) filtered seismograms at all array stations. The maximum amplitude ratios appear at stations within the SAF main fault zone and decreases rapidly away from the main fault for events within the fault zone. A second peak is seen at the north strand for the $M1.5$ earthquake and explosion PMM from which the FZTWs propagate southeastward and partition some energy into the north strand of the SAF, suggesting that the north strand may connect or approach the main fault of the SAF at depth. In contrast, amplitude ratios for events far from the SAF are low and flat across the fault.

Base on observations of these FZTWs and constrained by results from other experiments conducted at Parkfield (e.g. Catchings et al., 2002; Li et al., 1997; Thurber et al., 2004; Unsworth et al., 1997), we constructed a depth-variable velocity and Q section across the fault to a depth of 5 km for the structure of the SAF segment at Parkfield (Fig. 11.20e). The observed FZTWs were simulated using a 3-D finite−difference code (Graves, 1996; Vidale, 1989). As we did for simulation of FZTWs recorded at the SAFOD site, we first synthesized the FZTWs excited by the explosions to determine the shallowest 1 or 2 km fault zone structure and then stripped the shallow effects to resolve deeper structure of the fault zone. The

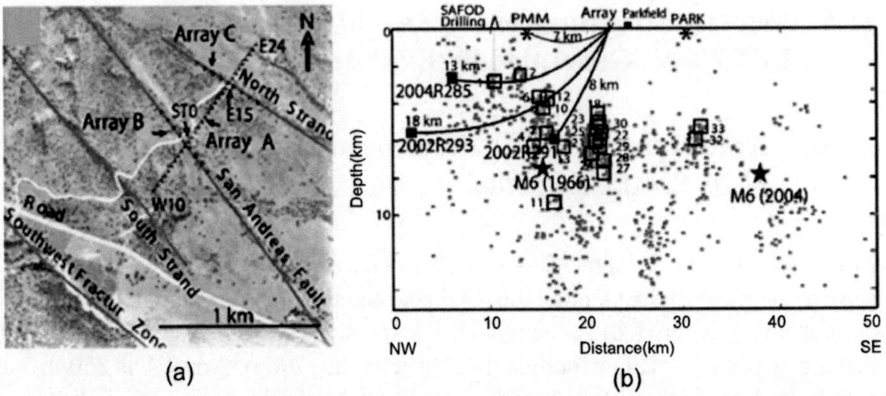

Fig. 11.19 **a** Map shows locations of seismic arrays A, B and C deployed across and along the SAF main and north traces at the surface near the town of Parkfield in 2002 and 2004. **b** The vertical section along the SAF strike at Parkfield area shows locations of the SAFOD site, array A near the town of Parkfield, shots PMM and PARK, the 1966 and 2004 $M6$ mainshocks, aftershocks (gray circles) of the 2004 $M6$ earthquake. Open squares labeled by numbers denote 21 clusters of repeated aftershocks; each cluster includes at least 5 similar aftershocks located within 200 m. Three black squares denote repeated earthquake pairs with events 2002R291, 2002R293 in 2002 and 2004R285 in 2004 with ray-paths to Parkfield array

depth-variable structure includes two low-velocity waveguides on the main and north fault strands. The synthetic FZTWs generated by explosions and the $M1.5$ micro-earthquake appearing within the ~150–200-m wide SAF main fault zone, are agreeable with the recordings at along- and across-fault arrays. FZTWs within the north strand of the SAF show smaller amplitudes and shorter post-S durations than those within the main fault zone. We tested the trapping effect of fault zone depth extension in modeling procedure. For example, synthetic seismograms generated by the $M1.5$ event for a shallow fault zone to 3-km depth cannot match observed FZTWs with long post-S durations.

Consequently, we interpret that the distinct low-velocity core zone of the matured SAF was formed by repeated damage during recurrent $M6$ events and other large earthquakes on the principal slip plane between the Pacific and North America plates at Parkfield. A less-developed narrower low-velocity zone on the SAF north strand might experience minor surface breaks in the 1966 $M6$ event, most likely due to secondary slip and strong shaking from dynamic rupture on the main fault. The width (~150–200 m) of the low-velocity waveguide inferred by FZTWs probably represents the macroscopic damage extent in dynamic rupture and microscopic fault process zone that accumulate mechanical, chemical, thermal, and other kinematical processes. The variation in velocity reduction along the fault zone and with depth may be caused by changes in overburden pressure, rock type, stress and slip rate, fault geometry, fluid content, and dynamic rupture during past earthquakes.

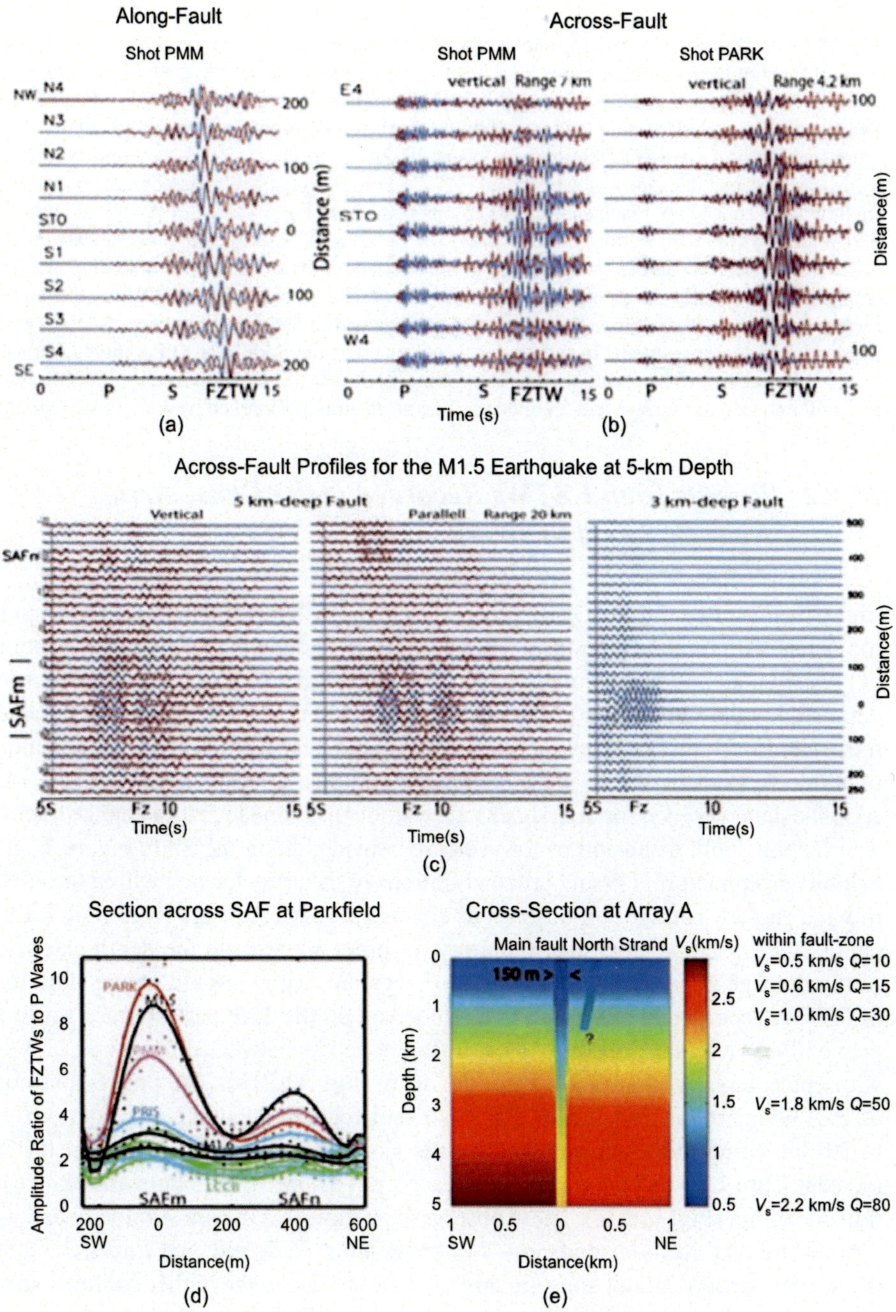
Along-Fault
Shot PMM
NW
N4
N3
N2
N1
STO
S1
S2
S3
S4
SE
200
100
0
100
200
0
P
S
FZTW
15
(a)
Across-Fault
Shot PMM
vertical
Range 7 km
E4
STO
W4
Distance (m)
Shot PARK
vertical
Range 4.2 km
100
0
100
Distance(m)
Time (s)
(b)
Across-Fault Profiles for the M1.5 Earthquake at 5-km Depth
5 km-deep Fault
Vertical
Parallell
Range 20 km
3 km-deep Fault
SAFn
SAFm
5S
Fz
10
15
Time(s)
Distance(m)
(c)
Section across SAF at Parkfield
Amplitude Ratio of FZTWs to P Waves
PARK
SAFm
SAFn
200
SW
0
200
400
600
NE
Distance(m)
(d)
Cross-Section at Array A
Main fault North Strand
V_s(km/s)
within fault-zone
V_s=0.5 km/s Q=10
V_s=0.6 km/s Q=15
V_s=1.0 km/s Q=30
V_s=1.8 km/s Q=50
V_s=2.2 km/s Q=80
150 m > <
Depth (km)
1
SW
0.5
0
0.5
1
NE
Distance(km)
(e)

◀**Fig. 11.20** **a** Observed (red lines) and synthetic (blue lines) vertical-component seismograms at array B for shot PMM. Seismograms have been < 3 Hz filtered and are trace-normalized in plot. **b** Same as in (**a**), but at 9 stations of array A across to the main fault for two shots PMM and PARK. Seismograms are plotted using a fixed amplitude scale in each profile. **c** Same as in (**b**), but at the cross-fault array A for a *M*1.5 earthquake (event 2002R293 in Fig. 11.19b). Synthetic waveforms using a 5-km deep fault zone match observed FZTWs. However, synthetic seismograms from s 3-km shallow fault zone show much shorter wavetrains than those observations. **d** Amplitude ratios of FZTWs to P waves at all stations of three arrays for 12 shots and 3 local earthquakes located within and out of the SAF versus distance from the main fault trace. Dots are the data points computed from amplitude ratios at all stations of 3 arrays for each event. Curves are a polynomial fit to the data for each event. The peak amplitude ratio is seen at stations within the main fault zone (SAFm). **e** The depth-variable velocity model in the vertical section across the SAF trace at Parkfield. The main fault is marked by a 150-m-wide waveguide within which velocities are reduced by 30–40% and Q is 10–50 between the surface and 5 km depth. The north strand is marked by a minor waveguide

11.5.2 Results from FZTWs Recorded at the Dense Array Near Parkfield in 2004

Immediately after the *M*6 Parkfield earthquake on September 28 2004, 45 seismometers with three−component 2-Hz sensors were deployed at the same locations of those in our experiment conducted in 2002 to record aftershocks and an explosion detonated at the same place PMM in 2002 (refer to Fig. 11.19b). The data recorded the re-deployed seismic array for aftershocks and repeated shots provide us a unique opportunity to study the co-seismic damage and post-mainshock heal of the SAF. Also the data recorded for aftershocks at seismogenic zone give us a change to characterize the depth extension of low-velocity waveguide on the SAF. Figure 11.21a exhibits prominent FZTWs recorded at stations of the array located within the ~200-m wide rupture zone for an on-fault aftershock (event E in Fig. 11.13a) at 7.5 km depth and ~ 3 km southeast of the array. In order to evaluate the depth extension of the damage zone on the Parkfield SAF segment, we measured post-S durations of FZTWs recorded at array station ST0 located on the SAF surface trace for fourteen on0fault aftershocks occurring at different depths between 3.6 km and 11.3 km with epicenters less than 3 km from the array (Fig. 11.21b). The post-S durations of FZTWs increase from 1.5 to 2.5 s as event focal depths increase from 3.6 km to 8.0 km (time move−out marked by the slope of a blue line), and then further increase from 2.7 s to 3.0 s as depth increase from 8.4 km to 11.3 km with nearly flat time move-out (Fig. 11.21c). These observations show that the low−velocity damage zone of the SAF likely extends to ~ 7–8 km depth at Parkfield, and fractures within the deeper portion of fault zone are might be closed due to the higher confined stress (Angevine & Turcotte, 1983). We have simulated FZTWs from these aftershocks using the velocity model in Fig. 11.20e, but in which the low−velocity waveguide extends to 8-km depth, to fit observations (Fig. 11.21b).

We note that FZTWs recorded the Parkfield array show the longer post-S durations (by ~ 15%) than those recorded at the SAFOD surface array for the events at similar focal depths and epicentral distances (refer to Fig. 11.12), indicating that the

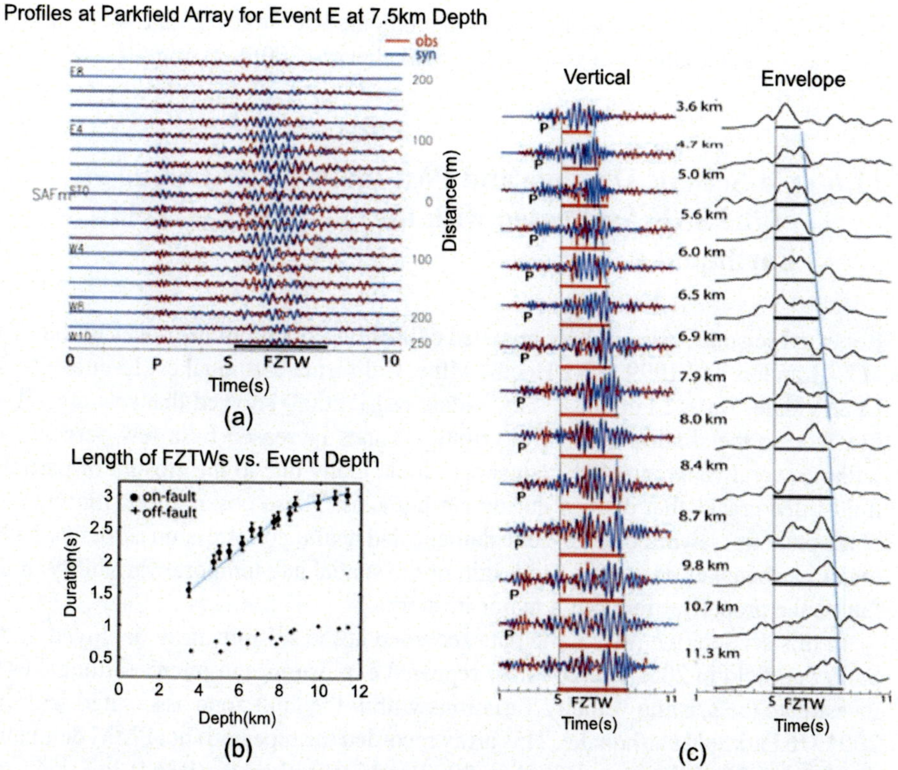

Fig. 11.21 **a** Observed and synthetic vertical-component seismograms recorded at the array near Parkfield for an on-fault aftershock (event E in Fig. 11.13a), showing FZTWs with large amplitudes and long post-S durations (marked by solid horizontal bars) at stations close to the SAF main fault trace. Seismograms have been < 6 Hz filtered. **b** Observed and synthetic seismograms and envelopes at the on−fault station ST0 for 14 aftershocks occurring within the fault zone nearby the array at different depths show an increase in post-S duration of FZTWs as event depths increase. The slop of blue line denotes the time move-out. S-arrivals for these events are aligned at the same time. **c** The measured post-S durations of FZTWs versus focal depths for the 14 aftershocks recorded at on-fault (black circles) and off-fault (grey crosses) stations. Each data point is averaged from measurements at 4 stations within or away from the fault zone for each aftershock. Error bars are standard deviations. Curves are polynomial fits to the data

velocity structure and rock damage magnitude of the SAF vary laterally along the fault strike, with lower values within the SAF Parkfield segment than the SAFOD segment in Middle Mountain. We also note that the differential post-S duration of FZTWs recorded at Parkfield is ~ 1.0 s (refer to Fig. 11.21b) slightly longer than ~ 0.95 s (refer to Fig. 11.15b) registered at the SAFOD site for aftershocks at the same depth range between ~ 4 km and ~ 8 km, suggesting that fault rocks within the SAF Parkfield segment experienced more damage than those within the SAFOD segment because larger slip and the stronger shaking happened on the SAF Parkfield segment that closer to the 2004 *M*6 epicenter than the SAFOD site. Indeed, the large slip and anomalous high peak acceleration were documented at Turkey Trail road ~ 2 km SE of the town of Parkfield (Li et al., 2004; Shakal et al., 2005).

11.6 Co-Seismic Damage and Post-mainshock Healing of the SAF Associated with the 2004 *M*6 Parkfield Earthquake

Repeated seismic surveys using repeated explosions along rupture zones of the 1992 *M*7.4 Landers and 1999 *M*7.1 Hector Mine, California earthquakes (Li et al., 1998; Li & Vidale, 2001; Li et al., 2003b; Vidale & Li, 2003) showed that seismic velocities within the shallow part of the rupture zones increased by a few percentages within several years after the mainshock, most likely due to the closure of partially fluid-filled cracks that opened during earthquakes. These observations illuminate a conceptual framework of fault-zone damage and healing progression associated with major earthquakes. However, the origin of the spatial and temporal variability in the fault zone properties remains a major mystery.

In this section, we review the data recorded at the seismic array deployed at the SAF, Parkfield in 2002 and 2004 for repeated explosions and micro-earthquakes to investigate the seismic velocity variations within the fault zone associated with the 2004 *M*6 Parkfield earthquake. This array recorded the repeated shot PMM detonated on October 16 (Julan date R289) in 2002 and December 30 (R364) in 2004, and two microearthquakes occurring within the fault zone on October 18 and 20 (Julian date R291 and R293) in 2002 (Fig. 11.19b). In the experiment carried out in 2004, this array operated in a continuous mode for 3.5 months, starting 1 week after the mainshock. We culled 21 clusters of repeated events from ~ 800 aftershocks in the recorded aftershock data (refer to Fig. 11.19b). Each cluster includes at least five repeated aftershocks occurring at the same place, with the location difference among them being smaller than 100 m and the difference in magnitude being smaller than 0.5. Repeated aftershocks in each cluster show similar waveforms with a correlation coefficient higher than 0.8. We used the moving-window waveform cross-correlation method to measure the changes in travel times of waves for the repeated events recorded in two experiments in 2002 and 2004 to evaluate temporal variations in seismic velocity within the SAF. Figures in this section are updated from Li et al. (2006, 2007a).

Figure 11.22a exhibits seismograms recorded at cross-fault Array A for a pair of repeated events at a depth of 6 and ~5 km northwest of the array on R291 in 2002 and on October 14 (Julian date R288) in 2004, 16 days after the *M*6 Parkfield earthquake occurred on September 28 (Julian date R272), showing similar waveforms (cross−correlation coefficient ~ 0.95) but longer traveltime for the aftershock in 2004 than the event in 2002. S-waves arrive at ~ 4.5 s while FZTW are dominant between 5.5 s and 8.5 s at stations within the ~200-m-wide SAF main fault zone. In order to avoid errors in event origin times provided by the seismic network catalog, the first

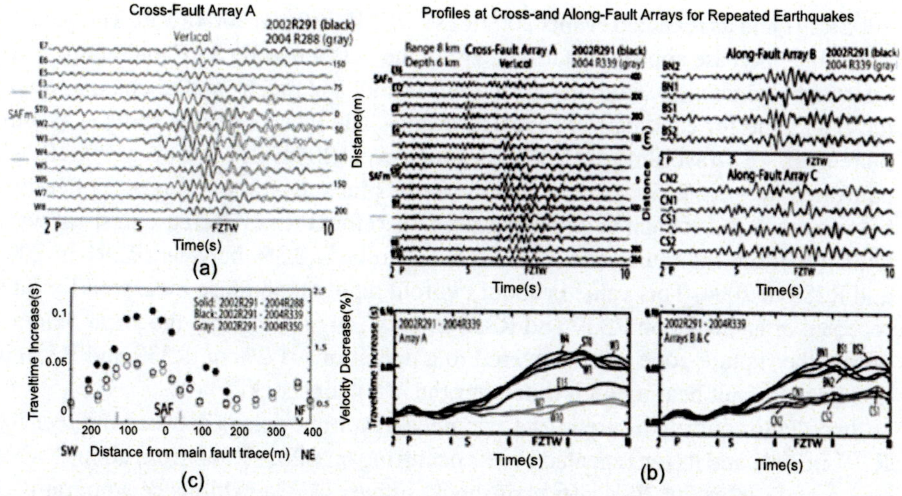

Fig. 11.22 **a** Overlapped seismograms at stations in the central part of cross-fault array A for the event on R291 in 2002 and its repeated event on R288 in 2004. Seismograms from repeated events have been < 5 Hz filtered with first P-arrivals aligned at 2.2 s. Stations ST0 was located on the SAF main fault trace. **b** Seismograms recorded at cross- and along-fault arrays A, B and along C for event 2002R291 and it repeated event 2004R339. Stations ST0 and E15 were located on the main fault (SAFm) and north strand (SAFn). Cross-correlations of seismograms for repeated events 2002R291 and on 2004R339 show larger traveltime increases (due to seismic velocities decrease) registered at stations within the SAF main fault zone. **c** The maximum traveltime increases of seismic waves measured by waveform cross-correlations between event 2002R291 and three repeated events versus distances from the main fault trace

P-arrivals for repeated events are aligned in the profile of seismograms. Obviously, seismic waves traveled slower after the *M*6 mainshock than before.

Figure 11.22b exhibits seismograms recorded at cross- and along-fault arrays A, B and C for the event on R291 in 2002 and its repeated events on R339 in 2004, showing similar waveforms but waves traveling slower after the 2004 *M*6 quake. We applied waveform cross-correlation for repeated events using a 1-s window with the window center moving from the first P-arrival to the later S-coda. Results show ~ 35–65 ms retarded in traveltime between 5.5 and 8.5 s at stations within the main fault zone, ~ 25–35 ms retarded in traveltime at stations on the north fault strand, and less than 20 ms retarded in traveltime at stations away from the fault zone for the aftershock on R339 in 2004 than the event at the same place in 2002. Assuming that velocity changes were uniform in the crust sampled by these waves, the increases in traveltime are straightforward to interpret. FZTWs arrived between 5.5 and 8.5 s, with 35–65 ms delay for the aftershock, so the shear wave velocity within the fault zone decreased by ~ 0.7% plus changes for the P-wave between 2002R291 and 2004R339. In contrast, the shear wave velocity in surrounding rocks decreased by < 0.25% during the same time period.

Figure 11.22c shows measured maximum increases in traveltime by moving window cross-correlation for seismograms recorded at all array stations for event

2002R291 and its repeated events 2004R288, 2004R339 and 2004R350. The greatest traveltime increase appear at stations within the ~200-m-wide SAF main fault zone, intermediate increase within the north fault strand, and much smaller changes out of the fault zone for each pair of repeated events. The measurements show ~ 110 ms increase in ~ 8 s traveltime of FZTWs between event 2004R288 and event 2002R291, equivalent to ~ 1.4% decrease in shear velocity plus traveltime changes for P-waves between repeated events. In a Poisson solid, therefore the net decrease in shear wave velocity within the fault zone was estimated to be ~ 2.3% between R291 in 2002 and R288 in 2004. This value is about twofold higher than those measured for later repeated aftershocks on R339 and R350 in 2004, suggesting that the shear velocity within the rupture zone was recovered to a deficit of ~ 1.2% on R339 and R350 in 2004 due to fault healing with time after the *M*6 mainshock.

In order to confirm this trend, we examined traveltime delays between the event on R291 in 2002 and its ten repeated events occurring on R281–R360 from about 1 week to 3 months after the 2004 *M*6 mainshock. Figure 11.23a exhibits seismograms of these repeated aftershocks recorded at stations W1 and E7 located within and 175 m northeast of the SAF, respectively. Prominent FZTWs appear between 6.5 s and 8.5 s in seismograms at station W1, but they are weak at station E7. Waveforms for these repeated events are similar (correlation coefficient > 0.85), but seismic waves traveled slower after the 2004 *M*6 earthquake on September 28 (R272) 2004 than before this quake. Moving-window cross-correlations of waveforms show the largest delay of ~ 125 ms in ~ 8 s traveltime for FZTWs arriving at station W1 between the earliest aftershock 2004R281 and event 2002R291 plus changes in the P-wave (Fig. 11.23b). The traveltime delays gradually decreased to ~ 50 ms for event 2004R350, approximately three months after the *M*6 mainshock. Based on the assumption for a Poisson solid, we estimate that the net shear-velocity within the fault zone decreased by ~ 2.5% between October 18 (Julian date R291) in 2002 and October 7 (Julian date R281) nine days after the *M*6 mainshock in 2004, and then increased by ~ 1.2% in the following ~ 3 months. In contrast, the changes in shear-velocity registered at station E7 was smaller than 0.5% in the same time period. The data recorded at our arrays starting from about a week after the 2004 Parkfield *M*6 earthquake do indicate that shear-velocities within the fault zone were reduced by at least 2.5%, most likely due to the damage of fault-zone rocks during dynamic rupture in the mainshock. On the other hand, seismic velocity increased with time after the mainshock, indicating that the damaged rock had been healing due to its rigidity recovery with crack closure in the post-seismic interval. The healing rate was not constant, but decreased logarithmically with time, similar to our observations at rupture zones of the 1992 Landers and 1999 Hector Mine earthquakes (Li & Vidale, 2001; Vidale & Li, 2003).

The trend of healing rate decreasing with time is solidly seen in Fig. 11.24a. Seismograms recorded at station W4 of array A within the fault zone for four on-fault repeated aftershocks in cluster 29 at 6.5 km depth and ~ 2 km northwest of the array (refer to Fig. 11.19b), showing very similar waveforms, with the correlation coefficient ~ 0.95. Travel-time advances measured by moving window cross-correlation for the first pair of repeated events on 8 October (Julian date R282) and 7 November

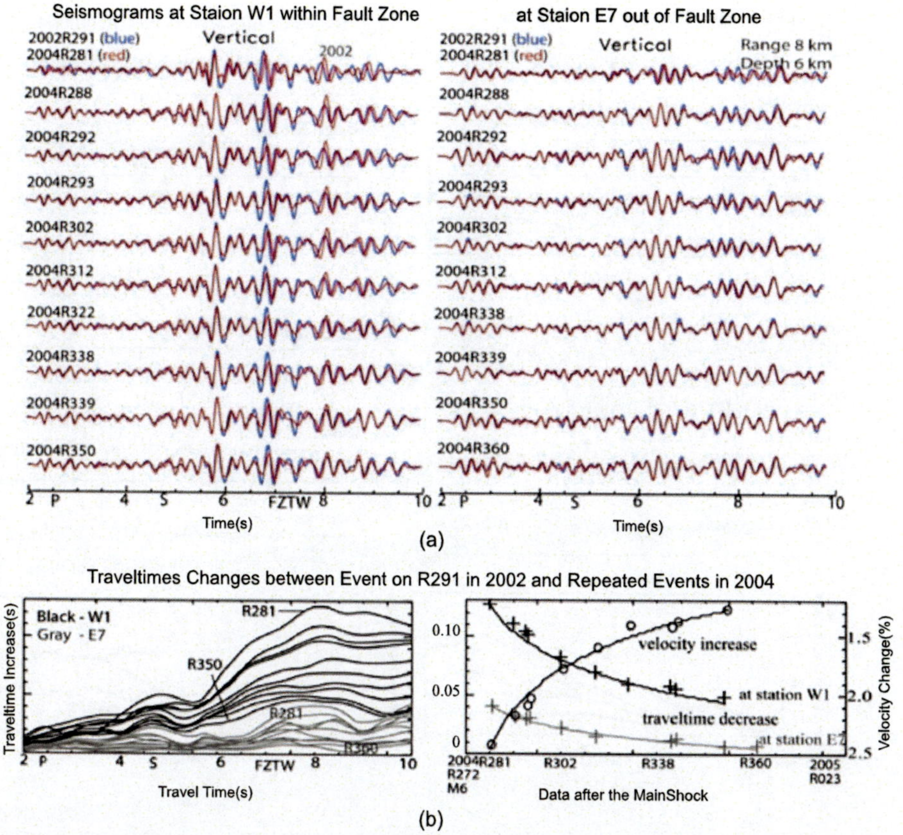

Fig. 11.23 **a** Vertical-component seismograms recorded at stations W1 and E7 of array A for event R291 in 2002 (blue line) and its ten repeated events from R281 to R360 in 2004 (red lines), showing larger traveltime increases registered at station W1 within the fault zone than at station E7 175 m off the fault. Seismograms have been filtered (< 5 Hz). First P-arrivals from repeated events are aligned at ~ 2.2 s. **b** Left: moving window waveforms cross-correlation measurements show traveltime increases between event 2002R291 and ten repeated events in 2004, with the largest increase between 2002R291 and the earliest repeated event 2004R281, and then the increase rate becoming gradually smaller with dates after the mainshock. Right: the maximum traveltime delays and shear velocity changes in percentages measured at stations W1 and E7 between event 2002R291 and its repeated events in 2004. The curve is the logarithmic fit to measurements, with a velocity change 0.012 in logarithm

(R312) are ~ 30–50 ms in traveltimes of S and FZTWs plus the change in P traveltime. These values are nearly twice as large as those for the second pair of repeated events on 29 November (R334) and 27 December (R362), indicating that the healing rate was larger in the earlier stage after the *M*6 mainshock. Figure 11.24b exhibits seismograms recorded at station W4 for four repeated aftershocks in cluster 28, occurring on 10 October (R284), 4 November (R309), and 3 and 17 December (R339 and R353) in 2004. Moving window cross-correlations of waveforms show the decreases in travel time to be ~ 30–50 ms in the first month, ~ 20–30 ms in the second month, and less

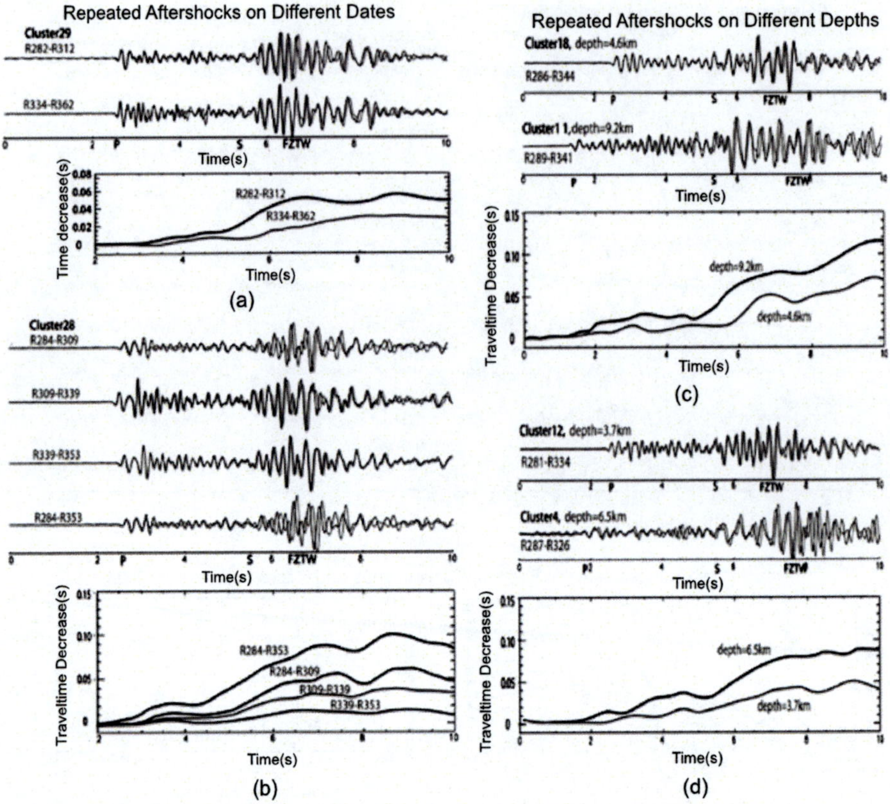

Fig. 11.24 **a** Vertical-component seismograms recorded at station W4 of cross-fault array A for four repeated aftershocks in cluster 29, occurring on Julian date R282, R312, R334, and R362 in 2004. Seismograms have been low-pass filtered and are plotted for the earlier (gray lines) and later (black) events in each pair. Moving-window waveform cross-correlations show the larger traveltime advance for the pair of earlier repeated aftershocks. **b** Same results from four repeated aftershocks in cluster 28, occurring on R284, R309, R339, and R353, show greater traveltime advance in the earlier post-seismic stage. **c** Vertical-component seismograms recorded at station W4 for two repeated aftershocks in cluster 18 at 4.6 km depth and other two repeated events in cluster 11 at 9.2 km depth with similar epicentral distances. Moving-window waveform cross-correlations show that the larger traveltime decreases for the pair of deeper aftershocks. **d** Same as in **(c)** but for two repeated aftershocks in cluster 12 at 3.7 km depth and another two repeated events in cluster 4 at 6.5 km depth, showing the larger travel-time decreases for the pair of deeper aftershocks

than ~ 15 ms in the third month after the mainshock. The healing rate in the third month after the mainshock reduced to a third of that in the first month indicates that the post-seismic fault healing rate is not constant but decreases with time.

We further examined the data for variations in travel-time with depth. Figure 11.24c exhibits the seismograms recorded at station W4 for a pair of repeated aftershocks in cluster 18 (Fig. 11.19b) at 4.6 km depth on 12 October (R286) and 9 December (R344), and another pair of repeated events in cluster 11 at 9.2 km depth on 15 October (R289) and 6 December (R341). The time span between repeated events in two pairs are nearly the same. Moving-window waveform cross-correlations show that traveltimes of S and guided waves were decreased by ~ 30–50 ms for the pair of shallower events and by ~ 50–80 ms for the deeper pair of deeper. Figure 11.24d shows seismograms recorded at station W4 for two repeated aftershocks in cluster 12 at 3.7 km depth on 7 October (R281) and 29 November (R334), and another two repeated aftershocks in cluster 4 at 6.5 km on 12 October (R286) and 20 November (R334). Again, the larger travel-time decreases is seen for the pair of deeper repeated events, suggesting that the post-mainshock healing occurred on the ruptured fault that likely extends to the depth of at least ~ 7 km, but with the smaller healing magnitude within the deeper portion of fault zone because the increasing pressure at depth reduces the crack density and probably influences the mineralogy of rocks and the development of fault gouge (Marone, 1998; Scholz, 1990) (Angevine & Turcotte, 1982).

Finally, we used the data from repeated explosion PMM detonated within the main fault zone of the SAF in Middle Mountain at 7 km northwest of the seismic array site near Parkfield before and after the 2004 *M*6 earthquake to confirm the observation of co-seismic velocity change caused by this earthquake. Figure 11.25 shows similar waveforms (cross-correlation coefficient ~ 0.95) recorded at arrays A, B and C for the repeated shot PMM detonated on October 16 (Julan date R289) in 2002 and December 30 (R364) in 2004. Seismic waves traveled slower after the 2004 *M*6 earthquake, with the largest traveltime increases occurring within the ~ 150–200-m-wide main fault zone, moderate increases on the north fault strand, and much smaller changes in surrounding rocks. Moving-window cross-correlations of waveforms for the repeated shots show the maximum decrease in shear-velocity by ~ 1.3% within the SAF main fault zone; this value is twice as large as that within the SAF north branch strand.

The percentage increases in traveltimes of P, S, and FZTWs at all stations on arrays A, B and C for repeated explosion PMM are plotted in Fig. 11.25f, showing the largest changes in traveltimes at stations within the ~ 150–200-m wide rupture zone along the SAF. Assuming velocity changes were uniform in the crust sampled by these waves, the P wave arrived with a 45-ms delay in ~ 3 s traveltime for the repeated shot PMM in 2004, so the P wave velocity decreased by ~ 1.5% within the fault zone between the repeated shots on October 16, 2002 and December 29, 2004. The S wave arrived with 85-ms delay in ~ 6.8 s traveltime in 2004, so the S wave velocity decreased by ~ 1.25%. FZTWs in window 3 with longer travel times had larger time advances than P and S waves, again resulting in ~ 1.25% increase in shear velocity within the rupture zone. In contrast, the S-wave velocity decreased by < 0.5% in the surrounding rocks.

After accounting for the shear velocity increase of ~ 1.2% measured from repeated aftershocks between October 7 (R281) and December 25 (R360) due to post-mainshock healing (refer to Fig. 11.23), and assuming no significant changes in seismic velocity at the Parkfield area during the time period between the repeated shots on October 16 2002 and the *M*6 Parkfield earthquake on September 28 2004 (Rubinstein & Beroza, 2005), we estimate a net decrease in shear-velocity of at least ~ 2.5% owing to co-seismic damage of fault zone rocks during dynamic rupture in

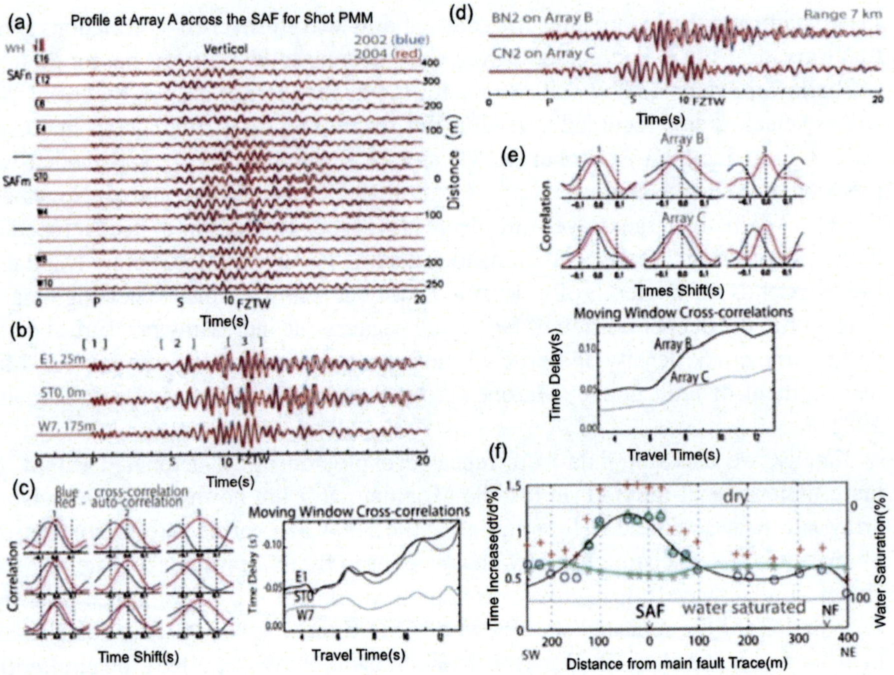

Fig. 11.25 **a** Seismograms recorded at array A across the SAF for repeated shot PMM in 2002 and 2004. Seismograms have been < 3 Hz filtered. The shot origin time is at 0 s aligned by the signals recorded at a sensor installed at the wellhead (WH) of shot-hole. FZTWs are dominant at stations within the fault zone marked by two bars. **b** Seismograms recorded at stations E1, ST0 and W7 located within and out of fault zone for the repeated shot PMM. **c** Autocorrelations of seismograms in 2004 and cross-correlations of seismograms between 2002 and 2004 at stations E1, ST0 and W7 in three time windows 1–3 including P, S, and FZTWs, respectively. The peak of the autocorrelation curve is at zero lag time taken as the reference. The negative time shift indicates traveltime advance. Moving-window cross-correlations of waveforms show the greater traveltime delays at stations E1 and ST0 than that at station W7 in 2004. **d** and **e** Seismograms recorded at station BN2 of array B and station CN2 of array C for the repeated shot PMM. Waveforms cross-correlations show the larger traveltime delay in 2004 within the SAF main fault zone than the north fault strand. **f** Traveltime increases in percentage for P (red crosses), S (blue circles) and FZTWs (green circles) measured by waveform cross-correlations of recordings at all array stations for the repeated shot PMM in 2002 and 2004, showing seismic velocities decrease after the 2014 *M*6 earthquake. The solid curve is a polynomial fit to traveltime increases of S waves in 2004. Ratio of traveltime changes for P waves to S waves (green stars with a curve) indicates the degree of water saturation in cracks. Two horizontal light lines indicate the ratios predicted for a range of water percentage for Poisson solid

the *M*6 earthquake although we have no data recorded in the first week after the mainshock. We calculated the ratio of traveltime changes for P- to S-waves to be 0.57 within the rupture zone and ~ 0.65 in the surrounding rocks according to equations for the elastic moduli of the medium with cracks (Garbin & Knopoff, 1975), indicating that cracks within the fault zone are more wet than those outside of the zone (Fig. 11.25f). Water may be coming up from depth through the highly fractured fault-zone acting as a fluid channel.

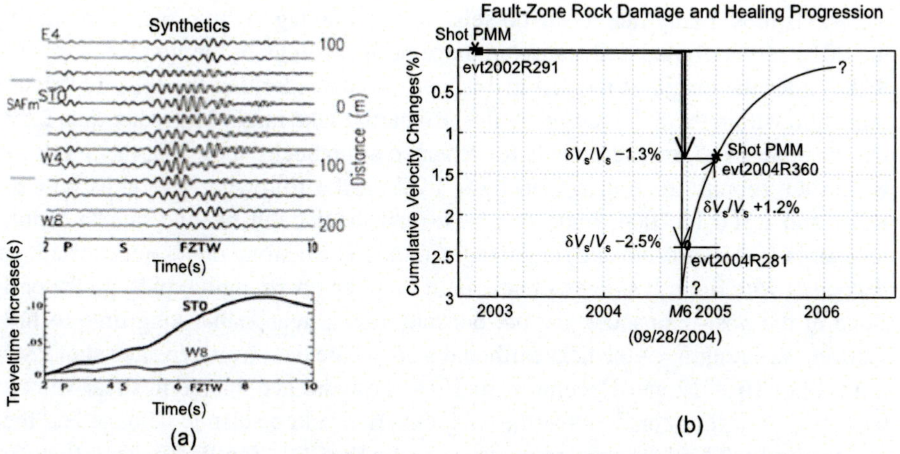

Fig. 11.26 **a** Vertical-component of synthetic seismograms at central part of cross-fault array A for the repeated events on R291 in 2002 (black lines) and on R281 in 2004 (gray lines). Seismograms have been < 5 Hz filtered. First P-arrivals from repeated events are aligned. Moving-window cross-correlations of seismograms for the repeated events at stations STO and W8 within and out of the fault zone. **b** S-wave velocity changes within the rupture zone associated with the *M*6 Parkfield earthquake on September 28 (Julian date R272) 2004. The S velocity decreased by ~ 2.5% due to the co-seismic fault rock damage during the *M*6 earthquake, and then increased by ~ 1.2% owing to post-mainshock fault rock healing in three months between R281 and R360 in 2004. The shear-velocity net increased by ~ 1.3% in the time period between the two repeated shots on October 16 (R289) in 2002 and December 30 (R364) in 2004

In addition, we have simulated seismograms for the event on R291 in 2002 using a 3-D finite-difference code in terms of the depth-variable structure model shown in Fig. 11.20e and for its repeated event on R281 in 2004, with the velocities reduced by 2.5% within the fault zone and by 0.5% in wall-rocks from original model parameters. Figure 11.26a shows synthetic seismograms at the cross-fault array A and traveltime delays measured by moving-window cross-correlation of waveforms at stations STO and W8 located within and outside of the fault zone using the model parameters before and after the *M*6 Parkfield earthquake. The measured maximum traveltime delay (~ 110 ms) of synthetic FZTWs for the model after the *M*6 quake is in accordance with the observation at station STO (refer to Fig. 11.25c).

In summary, Fig. 11.26b shows seismic velocity changes measured from the data recorded at our seismic arrays deployed at Parkfield before and after this earthquake. Waveform cross-correlation measurements for repeated events between the microearthquake on R291 in 2002 and the earliest aftershock recorded on R281, about one week after the 2004 *M*6 mainshock on September 28 (R272) show an ~ 2.5% decrease in S velocity. As two experiments were carried out in the fall of 2002 and 2004, there would be not significant variations in water table in the same season at Parkfield. It is reasonable to assume that there was no significant seasonal change in seismic velocity at Parkfield area in the time period between the shot in 2002 and the 2004 *M*6 earthquake. Therefore, our observation of seismic velocity decrease by

~ 2.5% is most likely due to the co-seismic rock damage within fault zone during the 2014 *M*6 earthquake. Because we have no recordings in the first week (R273–R280) after the origin of the *M6* mainshock, we consider that the seismic velocity reduction within the SAF caused by this *M*6 earthquake might be greater than 2.5%. On the other hand, measurements for repeated aftershocks show that the S velocity within the rupture zone increased by ~ 1.2% in the following 3 months between R281 and R360 in 2004, indicating post-mainshock fault healing due to rigidity recovery (with cracks closure) of damaged rocks. The healing rate is not constant but decreases with time. It suggests that a fault may regain strength rapidly in the early stage in the post-mainshock period, but may take a reasonable long time to fully recover the strength for the next earthquake on it. The data from repeated shot PMM in October 16 2012 and December 30 2014 show the net change in shear-velocity by ~ 1.36% further confirm that the SAF experienced co-seismic damage and then post-mainshock healing process associated with the 2004 *M*6 Parkfield earthquake.

11.7 Conclusions

In this article, we have reviewed the results from a series of seismic experiments that we carried out at the San Andreas Fault near Parkfield, Central California, a scientific seismic experiment site in the past decades to investigate the subsurface structure and physical properties of the fault zone in 4-D, using fault zone trapped (guided) waves generated by near-surface explosions and earthquakes at depth, and recorded at network stations and dense linear arrays deployed across and along the fault at surface as well as the SAFOD borehole seismograph installed within the fault core zone at ~ 3-km depth.

Through identification, collection, analysis and numerical simulations of the FZTW data in a systematic procedure, we quantitatively characterized the dimensions and magnitude of the highly damaged zone on the mature SAF near Parkfield (Li et al., 2004, 2007a, 2008). The zone along the main fault is marked by a low-velocity waveguide ~150–200-m wide, extending to the depth of ~ 7–8 km, within which Q is 15–50 and S velocities are reduced by 30–40% from wall-rock velocities, varying with depth and along the fault. The seismic velocity reduction within the SAF Parkfield segment in the lock-to-creeping transition is larger by ~ 10–15% than the fault segments at SAFOD site and Cienega Valley with creeping mechanism. We interpret that the distinct low-velocity core zone at Parkfied was formed by repeated damage during recurrent *M*6 earthquakes and other large events on the principal slip plane. Although the damage zone along the SAF has accumulated over geological time, it was in part weakened by the latest *M*6 earthquake in 2004 due to inelastic deformation around the propagating crack tip in the mainshock, as predicted by existing fault-zone rupture models (e.g., Andrews, 2005; Harris & Day, 1993; Scholz, 1990). The width of the low-velocity waveguide inferred by FZTWs likely represents the macroscopic damage extent in dynamic rupture and microscopic fault process zone accumulating mechanical, chemical, thermal, and other kinematical processes. The coincidence of low seismic velocity, high conductivity and Poisson's ratio within the SAF (e.g. Michelini & McEvilly, 1991; Thurber et al., 2006; Unsworth et al., 1997) suggests that a zone of fluid saturated fractured fault rock associated may extend to the depth of ~ 5 km. Results from the SAFOD drilling project and borehole logs at Parkfield show a high porosity and multiple slip planes in an ~ 200-m-wide low-velocity damage zone and 40-m-wide fault core with a velocity reduction of ~ 25–35% on the SAF at a depth of approximately 3.2 km (Hickman et al., 2005; Zoback et al., 2007, 2008; Lockner et al., 2011). While, the FZTWS generated by micro-earthquakes occurring at seismogenic depths depict a depth-variable low-velocity waveguide on the SAF that likely extends to the depth of ~ 7–8 km. The variation in velocity reduction along the fault zone and with depth may be caused by changes in overburden pressure, rock type, stress and slip rate, fault geometry, fluid content, and dynamic rupture during past earthquakes.

Our study using repeated events shows that the SAF at Parkfield co-seismically weakens and consequently heals when it undergoes strong dynamic stresses and pervasive cracking during the 2004 *M*6 earthquake (Li et al., 2006; 2007a). The data recorded for repeated explosions detonated within the SAF in 2002 and after the 2004 *M*6 Parkfield earthquake as well as the micro-earthquakes before and after the 2004 *M*6 mainshock show decreases of ~ 2.5% in seismic wave velocity within an ~ 200-m-wide rupture zone along the SAF strike, most likely due to the co-seismic damage of rocks during dynamic rupture in this *M*6 earthquake. The ~ 200-m width of the damage zone showing larger velocity changes is consistent with the low-velocity waveguide model on the SAF near Parkfield derived from FZTWs (Li et al., 2004). The data from clustered aftershocks at depths down to ~ 7 km show velocity recovery by ~ 1.2% within the fault zone in a 3-month period after the mainshock, indicating that the damaged rock has been post-mainshock healing.

The spatial extent of fault weakness, and the loss and recouping of strength across the earthquake cycle are critical ingredients in our understanding of fault mechanics and earthquake physics. In order to relate present day crustal stresses and fault motions to the geological structures formed in their past earthquake histories, we must understand the evolution of fault systems on many spatial and time scales. In Addition, characterization of ground-motion path effects due to low-velocity fault zones, is highly dependent on their subsurface velocity structure. A better knowledge of detailed damage and rheology structure of mature faults at seismogenic depth would further help with estimation of shaking near the faults and seismic hazard from future earthquakes.

The results from our previous study using fault-zone trapped (guided) waves for the matured San Andreas Fault at the earthquake scientific experimental site in Central California will be valuable for on-going experiments carried out by the China Seismic Experimental Site (CSES) to develop the community velocity and fault models as well as earthquake hazard and forecasting models for mitigation of earthquakes in China.

Acknowledgements The scientific experiments carried out at the San Andreas Fault at Parkfield, California and reviewed in this article were supported by multiple grants of the U.S. National Science

Foundation and Geological Survey, and the Southern California Earthquake Center (SCEC) in the past decades. SCEC is funded by NSF and USGS Cooperative Agreements. The author thanks many researchers and graduate students from multiple institutions for their collaborations and joining in this long-tern research project and field experiments, including the late Professor Keiiti. Aki, P. Malin, W. Ellsworth, C. Thurber, M. Zoback, S. Hickman, J. Vidale, C. Marone, W.H.K. Lee, E. Cochran, P. Chen, D. Adams, F. Xu, T. Burdette, K. Gross, M. Alvarez, S. Roecker, M. Rymer, R. Catchings, A. Snyder, R. Russell, L. Powell, B. Nadeau, N. Boness, and D. McPhee. Special thanks to SAFOD PIs for their coordination of our experiments to seismically characterize the SAFOD drill site, and to the IRIS for the use of PASSCAL instruments.

References

Aki, K. (1984). Asperities, barriers, characteristic earthquakes, and strong motion prediction. *Journal of Geophysical Research, 89*, 5867–5872.

Aki K., Richards P.G., 1980. Quantitative Seismology: Theory and methods, W.H. Freeman and Co.

Anderson, J. G., & Brune, J. N. (1999). Probabilistic seismic hazard analysis without the ergodic assumption. *Seismological Research Letters, 70*(1), 19–28. https://doi.org/10.1785/gssrl.70.1.19

Andrews, D. J. (2005). Rupture dynamics with energy loss outside the slip zone. *Journal of Geophysical Research, 110*, B01307. https://doi.org/10.1029/2004jb003191

Angevine, C. L., & Turcotte, D. L. (1983). Porosity reduction by pressure solution: A theoretical model for quartz arenites. *Geological Society of America Bulletin, 94*, 1129–1134.

Ben-Zion, Y. (1998). Properties of seismic fault zone waves and their utility for imaging low-velocity structure. *Journal of Geophysical Research, 103*, 12567–12585.

Blanpied, M. L., Marone, C. J., Lockner, D. A., Byerlee, J. D., & King, D. P. (1998). Quantitative measure of the variation in fault rheology due to fluid-rock interactions. *Journal of Geophysical Research, 103*, 9691–9712.

Byerlee, J. (1990). Friction, overpressure and fault-normal compression. *Geophysical Research Letters, 17*, 2109–2112.

Catchings, R. D., Rymer, M. J., Goldman, M. R., Hole, J. A., Huggings, R., & Lippus, C. (2002). High-resolution seismic velocities and shallow structure of the San Andreas fault zone at Middle Mountain, Parkfield, California. *Bulletin of the Seismological Society of America, 92*, 2493–2503.

Catchings, R. D., Goldman, M. R., Li, Y. G., & Chan, J. H. (2016). Continuity of the West Napa-Franklin fault zone inferred from guided waves generated by earthquakes following the 24 August 2014 Mw 6.0 South Napa earthquake. *Bulletin Seismological Society America, 106*(6). https://doi.org/10.1785/0120160154

Chester, F. M., Evans, J. P., & Biegel, R. L. (1993). Internal structure and weakening mechanisms of the San Andreas fault. *Journal of Geophysical Research, 98*, 771–786.

Cochran, E. S., Li, Y. G., Shearer, P. M., Barbot, S., Fialko, Y., & Vidale, J. E. (2009). Seismic and geodetic evidence for extensive, long-lived fault damage zones. *Geology, V9.37*(4), 315–318. https://doi.org/10.1130/G25306A.1; Data Repository item 2009082.

Das, S., & Aki, K. (1997). Fault plane with barriers: A versatile earthquake model. *Journal of Geophysical Research, 82*, 5658–5670.

Dieterich, J. H. (1978). Time-dependent friction and the mechanics of strike-slip. *Pure and Applied Geophysics, 116*, 790–806.

Eberhart-Phillips, D., & Michael, A. J. (1993). Three-dimensional velocity structure, seismicity, and fault structure in Parkfield region, central California. *Journal of Geophysical Research, 98*, 15737–15758.

Ellsworth, W. L., & Malin, P. E. (2011). Deep rock damage in the San Andreas fault revealed by P- and S-type fault zone guided waves, Sibson's volume. *Geological Society Special Publication, 359*, 39–53.

Garbin, H. D., & Knopoff, L. (1975). Elastic moduli of a medium with liquid-filled cracks. *Quarterly of Applied Mathematcis, 32*, 301–303.

Graves, R. W. (1996). Simulating seismic wave propagation in 3D elastic media using staggered-grid finite differences. *Bulletin of the Seismological Society of America, 86*, 1091–1106.

Harris, R. A., & Day, S. M. (1993). Dynamics of fault interaction: Parallel strike-slip faults. *Journal of Geophysical Research, 98*, 4461–4472.

Heaton, T. H. (1990). Evidence for and implications of self-healing pulses of slip in earthquake rupture. *Physics of the Earth and Planetary Interiors, 64*, 1–20.

Hickman, S. H., & Evans, B. (1992). *Growth of grain contacts in halite by solution-transfer: Implications for diagenesis, lithification, and strength recovery, in fault mechanics and transport properties of rocks* (pp. 253–280). Academic.

Hickman, S., Zoback, M., Younker, L., & Ellsworth, W. (1994). Deep scientific drilling in the San Andreas fault zone. *Eos, 75*, 137.

Hickman, S., Sibson, R., & Bruhn, R. (1995). Introduction to special section: mechanical involvement of fluids in faulting. *Journal of Geophysics Research, 100*, 12831–12840.

Hickman, S. H., Zoback, M. D., & Ellsworth, W. L. (2005). Structure and Composition of the San Andreas fault zone at Parkfield: Initial results from SAFOD Phase 1 and 2. EOS, *Transactions of American Geophysics Union, 83*(47), 237.

Korneev, V. A., Nadeau, R. M., & McEvilly, T. V. (2003). Seismological studies at Parkfield IX: Fault-zone imaging using guided wave attenuation. *Bulletin Seismological Society America, 80*, 1245–1271.

Laehenbruch, A. H. (1980). Frictional heating, fluid pressure and the resistance to fault motion. *Journal of Geophysical Research, 85*, 6097–6112.

Lambert, V., Lapusta, N., & Perry, S. (2021). Propagation of large earthquakes as self-healing pulses or mild cracks, Nature. *Nature, 591*(7849):252–258. https://doi.org/10.1038/s41586-021-03248-1

Leary, P. C., Li, Y.-G., & Aki, K. (1987). Observation and modelling of fault zone fracture seismic anisotropy—I. P, SV, SH travel times. *Geophysics Journal of Royal Astronomical Society, 91*, 461–484.

Lees, J. M., & Malin, P. E. (1990). Tomographic images of P wave velocity variation at Parkfield, California. *Journal of Geophysical Research, 95*, 21793–21804.

Li, Y. G., Leary, P. C., & Aki, K. (1987). Observation and modeling of fault zone fracture seismic anisotropy—II. P-wave polarization anomalies. *Geophysics Journal of Royal Astronomy Society, 91*, 485–492.

Li, Y. G., Leary, P. C., Aki, K., & Malin, P. E. (1990). Seismic trapped modes in the Oroville and San Andreas fault zones. *Science, 249*, 763–766.

Li, Y. G., & Leary, P. C. (1990). Fault-zone trapped seismic waves. *Bulletin of the Seismological Society of America, 80*, 1245–1271.

Li, Y. G., Vidale, J. E., Aki, K., Marone, C. J., & Lee, W. H. K. (1994a). Fine structure of the Landers fault zone: Segmentation and the rupture process. *Science, 256*, 367–370.

Li, Y. G., Aki, K., Adams, D., Hasemi, A., & Lee, W. H. K. (1994b). Seismic guided waves trapped in the fault zone of the Landers, California, earthquake of 1992. *Journal of Geophysical Research, 99*, 11705–11722.

Li, Y. G., & Vidale, J. E. (1996). Low-velocity fault zone guided waves; numerical investigations of trapping efficiency. *Bulletin Seismological Society America, 86*, 371–378.

Li, Y. G., Ellsworth, W. L., Thurber, C. H., Malin, P. E., & Aki, K. (1997). Observations of fault-zone trapped waves excited by explosions at the San Andreas fault, Central California. *Bulletin Seismological Society America, 87*, 210–221.

Li, Y. G., Vidale, J. E., Aki, K., Xu, F., & Burdette, T. (1998). Evidence of shallow fault zone strengthening after the 1992 M7.5 Landers. *California, Earthquake, Science, 279*, 217–219.

Li, Y. G., Aki, K., Vidale, J. E., & Xu, F. (1999). Shallow structure of the Landers fault zone from explosion-generated trapped waves. *Journal of Geophysical Research, 104*, 20257–20275.

Li, Y. G., Vidale, J. E., & Aki, K. (2000). Depth-dependent structure of the Landers fault zone using fault zone trapped waves generated by aftershocks. *Journal of Geophysical Research, 105*, 6237–6254.

Li, Y. G., & Vidale, J. E. (2001). Healing of the shallow fault zone from 1994–1998 after the 1992 M7.5 Landers, California, earthquake. *Geophys Res. Letter, 28*, 2999–3002.

Li, Y. G., Vidale, J. E., Day, S. M., & Oglesby, D. (2002). Study of the M7.1 Hector Mine, California, earthquake fault plan by fault-zone trapped waves, Hector Mine earthquake special issue. *Bulletin Seismological Society America, 92*, 1318–1332.

Li, Y. G., Vidale, J. E., Oglesby, D., Day, S. M., & Cochran, E. (2003a). Multiple-fault rupture of the M7.1 Hector Mine, California, earthquake from fault-zone trapped waves. *Journal of Geophysics Res, 108*, 1–25.

Li, Y. G., Vidale, J. E., Day, S. M., Oglesby, D. D., & Cochran, E. (2003b). Post-seismic fault healing on the 1999 M7.1 Hector Mine, California earthquake. *Bulletin Seismological Society America, 93*, 854–869.

Li, Y. G., Vidale, J. E., & Cochran, S. E. (2004). Low-velocity damaged structure of the San Andreas fault at Parkfield from fault-zone trapped waves. *Geophyics Research Letters, 31*, L12S06.

Li, Y. G., Chen, P., Cochran, E. S., Vidale, J. E., & Burdette, T. (2006). Seismic evidence for rock damage and healing on the San Andreas fault associated with the 2004 M6 Parkfield earthquake, special issue for Parkfield M6 earthquake. *Bulletin Seismological Society America, 96*(4), S1–15.https://doi.org/10.1785/0120050803

Li, Y. G., Chen, P., Cochran, E. S., & Vidale, J. E. (2007a). Seismic velocity variations on the San Andreas Fault caused by the 2004 M6 Parkfield earthquake and their implications. *Earth, Planets and Space, 59*, 21–31.

Li, Y. G., Malin, P. E., Vidale, J. E. (2007b). Low-velocity damage zone on the San Andreas fault at depth near SAFOD site at Parkfield by fault-zone trapped waves. *Scientific Drilling, 1*https://doi.org/10.2204/iodp.sd.s01.09,73-77

Li, Y. G., & Malin, P. E. (2008). San Andreas fault damage at SAFOD viewed with fault-guided waves. *Geophysical Research Letters, 35*, L08304. https://doi.org/10.1029/2007GL032924

Li, Y. G., De Pascale, G., Quigley, M., & Gravely, D. (2014a). Fault damage zones of the M7.1 Darfield and M6.3 Christchurch earthquakes characterized by fault-zone trapped waves. Tectonophysics. https://doi.org/10.1016/j.tecto.2014.01.029

Li, Y. G., Xu, Z. Q., & Li, H. B. (2014b). Rock damage structure of the south Longmen-Shan fault in the 2008 M8 Wenchuan earthquake viewed with fault-zone trapped waves and scientific drilling. *Acta Geological Sinica, V88*(2), 444–467.

Li, Y. G., Catchings, R. D., & Goldman, M. R. (2016). Subsurface fault damage zone of the 2014 Mw 6.0 South Napa, California, earthquake viewed from fault-zone trapped waves. *Bulletin Seismological Society America, 106*. https://doi.org/10.1785/0120160039

Lockner, D. A., Morrow, C., Moore, D., & Hickman, S. (2011). Low strength of deep San Andreas fault gouge from SAFOD core. *Nature, 472*, 82–85.

Malin, P., Shalev, E., Balven, H., & Lewis-Kenedi, C. (2006). Structure of the San Andreas fault at SAFOD from P-wave tomography and fault-guided wave mapping. *Geophysical Research Letters, 33*, L13314. https://doi.org/10.1029/2006GL025973

Marone, C. (1998). The effect of loading rate on static friction and the rate of fault healing during the earthquake cycle. *Nature, 391*, 69–72.

McBride, J. H., & Brown, L. D. (1986). Reanalysis of the COCORP deep seismic imaging of steeply dipping structure near the San Andreas fault, Parkfield, California, Bull Seism. *Society America, 76*, 1668–1686.

Michelini, A., & McEvilly, T. V. (1991). Seismological studies at Parkfield, I, Simultaneous inversion for velocity structure and hypocenters using cubic B-splines parameterization. *Bulletin of the Seismological Society of America, 81*, 524–552.

Mooney, W. D., & Ginzburg, A. (1986). Seismic measurements of the internal properties of fault zones, Pure Appl. *Geophysics, 124*, 141–157.

Nadeau, R. M., & McEvilly, T. V. (1997). Seismological Studies at Parkfield V: Characteristic microearthquake sequences as fault-zone drilling targets. *Bulletin Seismological Society of America, 87*, 1463–1472.

Noda, H., Dunham, E. M., & Rice, J. R. (2009). Earthquake ruptures with thermal weakening and the operation of major faults at low overall stress levels. *Journal of Geophysics Research Solid Earth, 114*, B07302.

Nur, A. (1972). Dilatancy, pore fluid, and premonitory variations of ts/tp travel times. *Bulletin Seismological Society of America, 62*, 1217–1222.

Oglesby, D. D., Day, S. M., Li, Y. G., & Vidale, J. E. (2003). The 1999 Hector Mine earthquake: The dynamics of a branched fault system. *Bulletin Seismological Society of America, 93*, 2459–2476.

Papageorgiou, A. S., & Aki, K. (1983). A specific barrier model for the quantitative description of inhomogeneous faulting and the prediction of strong motion, I, description of the model. *Bulletin of the Seismological Society of America, 73*, 693–722.

Perry, S. M., Lambert, V., & Lapusta, N. (2020). Nearly magnitude-invariant stress drops in simulated crack-like earthquake sequences on rate-and-state faults with thermal pressurization of pore fluids. *Journal of Geophysics Research Solid Earth, 125*, e2019JB018597.

Rader, D., Schott, W., Dresen, L., & Ruter, H. (1985). Calculation of dispersion curves and amplitude-depth distributions of Love channel waves in horizontally-layered media. *Geophysical Prospecting, 33*, 800–816.

Raleigh, C. B. (1977). Frictional heating, dehydration and earthquake stress drops. In *Proceedings of conference II: Experimental studies of rock friction with application to earthquake prediction* (pp. 291–304). USGS, Menlo Park, California.

Scholz, C. H. (1990). *The mechanics of earthquakes and faulting*. Cambridge University Press.

Shakal, A. F., Graizer, V. G., Huang, M., Borcherdt, R., Haddadi, H. R., Lin, K. W., Stephens, C., & Roffers, P. (2005). Preliminary analysis of strong-motion recordings from the 28 September 2004 Parkfield, California earthquake. *Seismological Research Letters, V76*(1), 27–40.

Shaw, B. E. (2006). Initiation propagation and termination of elastodynamic ruptures associated with segmentation of faults and shaking hazard. *Journal of Geophysical Research, 111*(B08302), 1–14. https://doi.org/10.1029/2005JB004093

Sibson, R. (1977). Fault rocks and fault mechanisms. *Journal of the Geological Society of London, 133*, 191–213.

Sleep, N. H., Richardson, E., & Marone, C. (2000). Physics of friction and strain rate localization in synthetic fault gouge. *Journal of Geophysical Research, 105*, 25875–25890.

Spudich, P., & Olsen, K. B. (2001). Fault zone amplified waves as a possible seismic hazard along the Calaveras fault in central California. *Geophysical Research Letters, 28*, 2533–2536.

Thurber, C. H., Roecker, S., Lutter, W., & Ellsworth, W. (1996). Imaging the San Andreas fault with explosion and earthquake sources. *Eos, 77*, 45.

Thurber, C., Roecker, S., Zhang, H., Baher, S., & Ellsworth, W. (2004). Fine-scale structure of the San Andreas fault zone and location of the SAFOD target earthquakes. *Geophysics Research Letter, 31*, L12S02. https://doi.org/10.1029/2003GL019398

Thurber, C., Zhang, H., Waldhauser, F., Hardbeck, J., Michael, A., & Eberhart-Phillips, D. (2006). Three-dimensional compressional wavespeed model, earthquake relocations, and focal mechanisms for the Parkfield, California, Region. *Bulletin Seismological Society, 96*, S38–S49. https://doi.org/10.1785/0120050825

Unsworth, M., Malin, P., Egbert, G., & Booker, J. (1997). Internal structure 260 of the San Andreas fault at Parkfield, CA. *Geology*, 356– 362.

Vidale, J. E. (1989). Finite-difference seismograms for SH waves. *Bulletin Seismological Society of America, 78*, 1765–1782.

Vidale, J. E., & Li, Y. G. (2003). Damage to the shallow Landers fault from the nearby Hector Mine earthquake. *Nature, 421*, 524–526.

Zoback, M. D., Hickman, S., & Ellsworth, W. L. (2007). The role of fault zone drilling. In H. Kanamori (Ed.), *Earthquake seismology. Treatise on geophysics* (Vol. 4, pp. 649–679). Elsevier.
Zoback, M., Hickman, S., & Ellsworth, W. (2008). *In-situ fault zone observations*. EarthScope National Office, Oregon State University.

Chapter 12
A Brief Introduction to the Major Science and Technology Products and the 2019 Highlights Scientific and Technological Progress of China Seismic Experimental Site (CSES)

Chaozhong Hu, Long Wang, Ke Sun, and Lingxue Tai

Abstract The China Seismic Experimental Site (CSES) is a "field laboratory for the research of mainland earthquake science". In 2019, CSES brought together 28 scientific research teams to participate in the construction and scientific research tasks. At the same time, CSES organized scientific research teams from five research institutes of the China Earthquake Administration to participate in the work. The experimental site also cooperated with the Sichuan Earthquake Administration, the Yunnan Earthquake Administration and other units produced the first batch of products, laying a solid foundation for the experimental field to promote numerical earthquake prediction, strong ground motion simulation and earthquake rupture process simulation in the next step. CSES has formed impressive scientific and technological progress from 90 projects supported by CSES in 2019.

Keywords China seismic experimental site (CSES) · Science and technology products · Highlights scientific and technological progress

12.1 Scientific and Technology Products

The first-generation scientific and technology products includes four public models: Community velocity model version 1.0, Active faults model, Community deformation model, Rheological model; three basic data sets: relocation earthquake catalog, focal mechanism solution catalog, and "de-cluster" earthquake catalog; ground motion prediction model: a 30-year-scale strong ground motion probability prediction model (Table 12.1).

Every product contains complete data, research and development reports, model establishment and usage instructions, calculation procedures, etc. Every product has undergone strict review by peer experts, DOI coding is carried out, and it is publicly released to scholars around the world on CSES website (http://www.cses.ac.cn/).

C. Hu (✉) · L. Wang · K. Sun · L. Tai
Institute of Earthquake Forecasting, China Earthquake Administration, Beijing 100036, China

© Higher Education Press 2022

Y.-G. Li et al. (eds.), *China Seismic Experimental Site*,
https://doi.org/10.1007/978-981-16-8607-8_12

12.2 Highlights Scientific and Technological Progress

The CSES has carried out basic research in an open and cooperative manner. From 2015 to 2019, CSES funded 90 projects to carry out research in the Sichuan-Yunnan region. According to the process of expert nomination, voting, joint office discussion and determination, and publicity, the 2019 CSES has formed outstanding scientific and technological progress (Table 12.1).

The most important progress is the Scientific Design of CSES. The 2019 top highlights of scientific and technological progress in CSES include the scientific design of CSES, which has pointed out the direction for the development of CSES; in addition, including five excellent technology products of CSES: CSES community velocity models version 1.0, kinematics model of CSES and dynamic model of key structural parts, three-dimensional community active faults model of CSES, 3D density structure model of CSES, and fine crustal structure of CSES based on geophysical section; also including three scientific issues; two new technology experiments in the experimental site: field experiment of AI technology in earthquake monitoring system, field experiment of new geochemical sensor; popular science product in the experimental site.

12.3 Prospects of Scientific Products in CSES

The scientific goal of the CSES is to study the environment of continental strong earthquakes, the process of earthquake occurrence, and the mechanism of disasters, etc. The business goal of the CSES is to build an upgraded prototype of the existing seismic business system. The technical goal of the CSES is to explore the development and construction of new seismic theories and technologies. The ultimate proposition of the CSES is to promote "earthquake control" as an opportunity to monitor and analyze the mechanism of human activities. The CSES will be built into an innovation-driven earthquake science research plateau. The CSES will systematically promote the update of existing scientific products and the output of a series of new technological products.

Table 12.1 The 2019 highlights scientific and technological progress in CSES

S. No.	Type	Name of highlight scientific and technological progress	Author unit
1	Top-level design of CSES	Scientific design of CSES	CSES Scientific Design Group from
2	Excellent technology products of CSES	CSES community velocity models version 1.0	University of Science and Technology of China
3		Kinematics model of CSES and dynamic model of key structural parts	Institute of Earthquake Forcasting, China Earthquake Administration
4		Three dimensional community active faults model of CSES	Institute of Geology, China Earthquake Administration
5		3D density structure model of CSES	Institute of Geology, China Earthquake Administration
6		Fine crustal structure of CSES based on geophysical section	Geophysical Exploration Center of China Earthquake Administration
7	Important scientific issues of CSES	Study on the influence and mechanism of water injection in changing earthquake window, Sichuan	China Earthquake Network Center; Earthquake Administration of Sichuan Province; Institute of Geology, China Earthquake Administration
8		The locked state of the Xianshuihe-Xiaojiang fault zone	China University of Geosciences (Wuhan)
9		The deep tectonic activity behavior of the Xiaojiang fault zone and the earthquake risk estimation	Peking Univeristy
10	New technology experiment of CSES	Field experiment of AI technology in earthquake monitoring system	University of Science and Technology of China; Institute of Earthquake Forcasting, China Earthquake Administration
11		Field experiment of new geochemical sensor	Institute of Earthquake Forcasting, China Earthquake Administration; Sichuan Earthquake Administration
12	Popular science products of CSES	Remote sensing image and main structure video of CSES	Management Department of CSES

Acknowledgements This work was completed under the leadership of Zhongliang Wu, Xiaodong Zhang, Zhigang Shao, Ying Li. Many experts participated in the review of Science and Technology Products and progress. Qian Li and many colleagues jointly completed the work. The comments of Yongxian Zhang and Yonggang Li have improved the quality of the article. Thank you.

Correction to: CSES Community Deformation Models in Southwest China

Hua Wang, Jean-Philippe Avouac, Zhigang Shao, Xiaoxia Liu, Wenxin Wei, Songhui Zhan, Xiaolei Dai, and Yidong Lou

Correction to:
Chapter 6 in: Y.-G. Li et al. (eds.),
***China Seismic Experimental Site*,**
https://doi.org/10.1007/978-981-16-8607-8_6

The original version of the book was inadvertently published with an incorrect figure in Chapter 6, which has now been corrected as below. The book and the chapter have been updated with the change.

The updated version of this chapter can be found at
https://doi.org/10.1007/978-981-16-8607-8_6

© Higher Education Press 2022
Y.-G. Li et al. (eds.), *China Seismic Experimental Site*,
https://doi.org/10.1007/978-981-16-8607-8_13

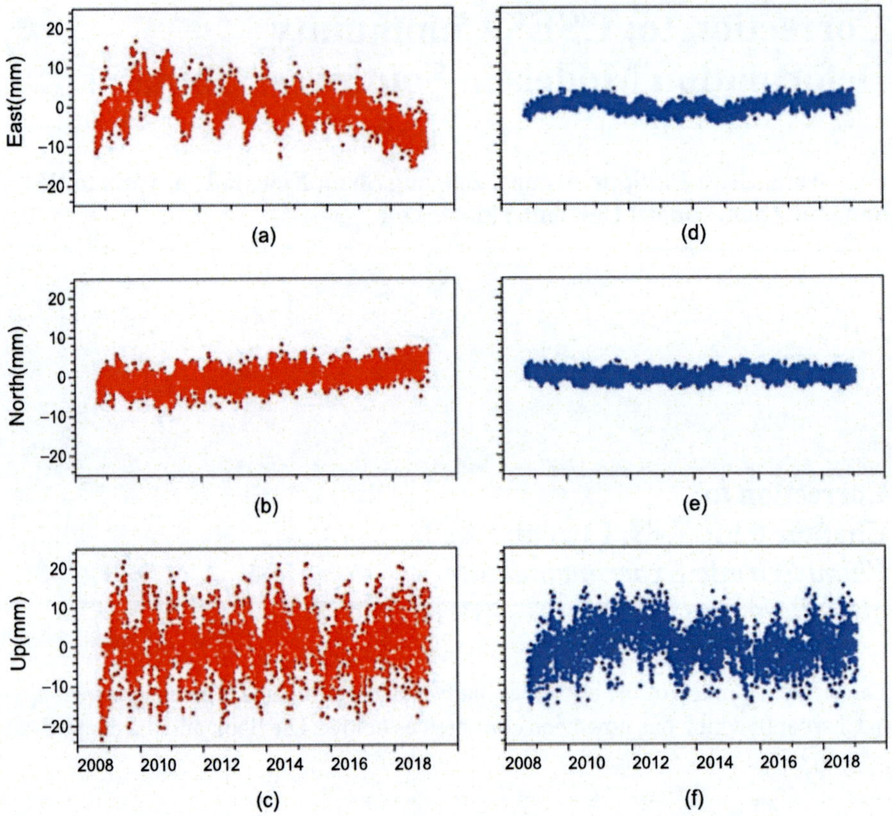

Fig. 6.5 Time series of LS03 detrended by a linear velocity **a–c** without and **d–f** with corrections. Note that we didn't correct postseismic deformation for vertical components as we only used horizontal GNSS velocities in this study